Communication System Design Using DSP Algorithms

with Laboratory Experiments for the TMS320C6713™ DSK

Information Technology: Transmission, Processing, and Storage

Series Editors: Robert Gallager
                  Electrical Engineering & Computer Science
                  Massachusetts Institute of Technology
                  Cambridge, Massachusetts

                  Jack Keil Wolf
                  Electrical & Computer Engineering
                  University of California at San Diego
                  La Jolla, California

Communication System Design Using DSP Algorithms

with Laboratory Experiments for the TMS320C6713™ DSK

Steven A. Tretter
University of Maryland
College Park, MD

 Springer

Steven A. Tretter
Department of Electrical Engineering
University of Maryland
College Park, MD, 20742, USA

Series Editors

Robert Gallager
Electrical Engineering & Computer Science
Massachusetts Institute of Technology
Cambridge, Massachusetts

Jack Keil Wolf
Electrical and Computer Engineering
University of California at San Diego
La Jolla, California

ISBN 978-0-387-74885-6   ISBN 978-0-387-74886-3 (eBook)
DOI 10.1007/978-0-387-74886-3

Library of Congress Control Number: 2007940172

The following are trademarks of Texas Instruments: Code Composer Studio, TMS320C30, C6000, C6x,
TMS320C5000, TMS320C3x, TMS320C40, C67x, C2x, C24x, C5x, C8x, C54x, C55x, and TMS320C67x.

MATLAB is a registered trademark of MathWorks.

Printed on acid-free paper

9 8 7 6 5 4 3 2 1

springer.com

To Teresa, Anne, Jeffrey, Max, Norah, and in memory of David

Preface

The first edition of this book began in January 1993 when I was "elected" by Dr. William Destler, who was then Chairman of the Electrical Engineering Department at the University of Maryland and is now President of the Rochester Institute of Technology, to set up and write experiments for a new senior elective laboratory course, ENEE418c Communications Laboratory, scheduled to be given for the first time in the Fall 1993 semester. At that time, we chose to use the state-of-the-art Texas Instruments TMS320C30 EVM (evaluation module) DSP board. In January 2001 we upgraded the lab to use the new state-of-the-art Texas Instruments TMS320C6701 EVM DSP board which is now no longer supported by TI. Starting with the Fall 2007 semester, we will use TI's TMS320C6713 DSK which is relatively inexpensive and connects conveniently to a USB port of a modern PC. The lab's PC's are all connected to the campus network. Each lab group is given a private workspace on a departmental server. Students are given only read/execute privileges for the standard utility and development software on the PC's so they do not inadvertently alter these files.

In 1993, books for hardware based laboratory courses with standard digital signal processing and filter design experiments existed, but no book focusing on analog and digital communications techniques was available. This is still largely true.

Laboratories in the Department of Electrical and Computer Engineering at the University of Maryland are separate courses. Each week they have a one hour lecture given by a regular faculty member to introduce the theory and explain the experiments followed by a three hour laboratory period run by a graduate teaching assistant. Students in this lab work in pairs. We have found that this works well because both group members actively participate. With groups of three or more, some members just sit and watch. Students have card key access to the laboratory from 8:00 AM to 11:00 PM seven days a week and so they can work outside of regular class hours if they wish.

One section of the lab was first offered in the Fall 1993 semester and two in the Spring 1994 semester. Then five sections a week were offered for several years until a couple of years ago when a new communications capstone design course was offered in addition. The students have been highly enthusiastic and often spend extra hours working on the experiments because they find them to be very interesting and challenging. They also have realized that this course will help them get jobs and provide them with the skills required to perform well in their future jobs. The lab was designed for seniors, but 1/4 to 1/3 of the class is now graduate students who want to learn some real-world practical skills in addition to the purely theoretical concepts presented in the typical graduate communications and signal processing courses. When asked why they are taking this senior class, the graduate students often say they think it will help them get jobs.

The goal of this set of experiments is to explore the digital signal processing and communication systems theoretical concepts presented in typical senior elective courses on these subjects by implementing them with actual hardware and in real time. In the process, students will gain experience using equipment commonly used in industry, such as, oscilloscopes, spectrum analyzers, signal generators, error rate test sets, digital signal processors, and analog-to-digital and digital-to-analog converters. They will also learn about typical software development tools. In addition, they will learn that there is a big step in going from an equation on paper to a real working system.

This book differs from any others on the market in that its primary focus is on communication systems. Fundamental digital signal processing concepts like digital filters and FFT's are included because they are required in communication systems. Approaches that are particularly useful for DSP implementations are presented. While the experiments, particularly the earlier ones, are described for the TMS320C6713 DSK, they can be easily modified for any DSP board with an A/D and D/A converter.

There are several books on digital signal processing experiments for stable software packages like MATHCAD and MATLAB. In my view, one of the purposes of a laboratory course is to help prepare students for industrial jobs. Off-line software simulation is no substitute for making actual hardware work in real-time. It does not present students with the strange unexpected and often frustrating things that occur when using actual hardware in real-time which can not be explained by nice equations, nor does it teach them how to use standard lab equipment.

The prerequisites for this course are an understanding of linear systems and transform methods at a level that is often presented in a junior required course on Signals and Systems and a working knowledge of PC's and C programming. Students who have programmed in other languages like BASIC, PASCAL, or FORTRAN can quickly learn enough C to do the experiments if they are willing to make the effort. Corequisite are a senior level elective course in Digital Signal Processing and/or Communication Systems. Ideally, both courses should be taken before the Communications Laboratory. However, this is not usually possible for our seniors. We wanted our students to have the opportunity to take this lab, so we made just one of them a corequisite. With the engineering background of a senior, the presentation of the necessary theory in the text, and the one hour lab lecture to explain the theory, students have quickly learned the signal processing and communication system concepts required for the experiments. In addition, it can be argued that a lab course should help prepare students for the work world where they will have to figure out new things for themselves so the experiments should have some uncertainty and require students to fill in some of the details.

There is a large initial hurdle for the students to get over while learning the details required to use the lab's hardware and software tools. Chapters 1 and 2 gradually introduce them to these tools and the architecture of the TMS320C6713 floating-point DSP. An attempt has been made to reduce this hurdle by including some basic programs on the program disk for initializing the DSK that can be used as a starting point for the experiments.

FIR and IIR filter design and implementation are explored in Chapter 3. Filters are required in many communication system signal processing algorithms. Experiments comparing the relative merits of C and assembly language implementations are performed. In particular, TI's *linear assembly* is briefly discussed. Modern DSP applications in industry are often written primarily in C with only numerically intensive critical functions written in assembly to reduce development time and improve portability to new platforms. TI's optimizing C compiler generates relatively efficient software pipelined executable code that is adequate for the experiments in this course. Therefore, assembly programming is not emphasized.

Chapter 4 investigates the FFT and power spectrum estimation. A simple spectrum analyzer is made.

Chapters 5 through 8 explore the classical analog communication methods of amplitude modulation, double-sideband suppressed-carrier amplitude modulation, single-sideband modulation, and frequency modulation. Transmitters and receivers are built using DSP techniques. Noncoherent receivers using envelope detectors and coherent receivers using phase-locked loops are implemented. The use of Hilbert transforms and complex signal representations in modulation systems are explored.

Chapters 9 through 16 introduce some digital communication techniques. These experiments focus on methods used in high-speed wire-line data modems where DSP's have been extensively used. Topics covered include linear shift register scramblers, the RS232C interface, pulse amplitude modulation (PAM), variable phase interpolation, and quadrature amplitude modulation (QAM). The experiments lead up to building almost a complete V.22bis transmitter and receiver. Symbol clock recovery and tracking, carrier tracking, and adaptive equalizer receiver functions are implemented. The echo canceling technique used in V.32, V.34, V.90, and V.92 modems is studied in Chapter 16. Enough details are included so that this set of experiments could form a good practical guide to engineers in industry interested in wire-line modem design. I learned many of these techniques while consulting since 1970 for companies that build high-speed wire-line modems and have seen them employed in hundreds of thousands of modems.

Multi-carrier modulation has become popular in a variety of systems. It is employed in several types of Digital Subscriber Line (DSL) systems which use copper telephone lines where it is called Discrete Multi-Tone (DMT) modulation. It is a popular choice for wireless systems transmitting over fading channels where it is called Orthogonal Frequency Division Multiplexing (OFDM). These include existing HF radio and Wi-Fi systems as well as soon to be deployed WiMax systems. The European cellular 3GPP committees are working to finalize a multi-carrier system called LTE. Multi-carrier modulation is explored in Chapter 17.

Chapter 18 briefly presents some ideas for additional projects related to high-speed wire-line modem design, error-control coding, and speech codecs. These ideas can be expanded to satisfy the capstone design project requirements of the ABET accrediting committee.

Appendix C contains a complete list of the equipment used for this laboratory at the University of Maryland. It has been included as a guide to others setting up a similar lab and is not intended to be an endorsement of any specific manufacturer. Clearly, any equipment with equivalent capabilities can be substituted for items in the list.

There are many more experiments in this book than can be performed in one semester. Based on our experience, an ambitious goal is to have all students do Chapters 1, 2, and 3 followed by a choice of any three additional experiments. In some semesters, we have limited the choice to three of the classical analog modulation chapters and in others to three of the digital communication chapters. It would be nice if students could continue in the lab for a second semester for additional credit and build on their earlier experiments.

Utility programs, software updates, text corrections, lab lecture slides, and supplementary material can be found on my web site `www.ece.umd.edu/~tretter`.

Steven A. Tretter

Acknowledgements

I would like to thank Mark Kohler and Sonjai Gupta for helping to unpack equipment and install and debug hardware and software in the PC's during the Spring 1993 semester when the lab was initially being set up. The students in the original Communications Laboratory section offered in the Fall 1993 semester as well as their Teaching Assistant, Yifeng Cui, deserve a great deal of thanks for being extremely patient and enthusiastic Guinea pigs. They helped correct and significantly improve the first few experiments. In particular, I want to thank Mike Barr who, with his partner Brian Silverman, forged well ahead of the rest of the students and was so enthusiastic by the end of the semester that he asked about helping with the lab in the Spring 1994 semester. The Electrical Engineering Department was able to make an exception and assign Mike as a senior to be the TA for one section and he did an outstanding job. Mrs. Tahereh Fazel was also an excellent Graduate TA for a second lab section in the Spring 1994 semester. I also want to thank those who have been TA's for the lab since that time and have all done excellent jobs, as well as Dr. Adrian Papamarcou and Dr. Jerome Gansman who have shared in teaching the lab. Jay Renner of the University of Maryland ECE staff also deserves thanks for laying out and getting manufactured the RS232/TTL daughter cards, and for securing the TMS320C6713 DSK's inside the PC cases. Dr. Brian Evans of the University of Texas at Austin also deserves thanks for reviewing preliminary versions of the previous edition of this book and making good suggestions for improvements.

I would also like to thank my friends from RIXON (which was in Silver Spring, MD, but has long since dissipated in a chain of corporate take-overs), from Penril Datability Networks (which was in Gaithersburg, MD, and was bought by Bay Networks, which was in turn bought by Nortel Networks, became the Signal Processing Group of Nortel in Germantown, MD, and is now gone), and at Texas Instruments in Germantown, MD (which was formerly Telogy Networks) for helping me stay at the state-of-the-art in DSP applications to wire-line and wireless modems by using me as a consultant.

I also want to thank Texas Instruments for designating our lab as one of ten Elite Digital Signal Processing Labs in the country. TI has been very generous in supporting our lab with TI hardware and software tools. In particular, Torrence Robinson, formerly in charge of University Programs at TI in Houston, and now Cathy Wicks deserve many thanks for their constant and cheerful support.

Contents

Chapter 1

Overview of the Hardware and Software Tools

The purpose of this initial chapter is to introduce you to the main features of the hardware and software tools that will be used in this course. A variety of signal processing and communication system components will be implemented by writing C and/or assembly language programs for the TMS320C6713 floating-point DSP in our lab. The TMS320C6713 resides on a board Texas Instruments (TI) calls the TMS320C6713 DSK (DSP Starter Kit) which is a small external board that connects to the PC through a USB port. The TMS320C6713 communicates with the analog world through a TI AIC23 stereo codec on the DSK board. The experiments in this book can be easily modified for other DSP boards. It is important to have a general picture of the hardware platform so you will understand how to write programs to accomplish the desired tasks.

First, a very brief history of DSP chips is presented and some typical applications are discussed. Then, the important features of the TMS320C6713 DSP are described. As a short cut, this DSP will sometimes be referred to as the 'C6713. It is a part of a larger family of floating-point DSP's including the TMS320C6701, TMS320C6711, TMS320C6722, TMS320C6726, and TMS320C6727. This family will be referred to as the 'C6x or 'C6000 family. Next, a block diagram for the DSK is discussed followed by an introduction to some of the software tools. For the first experiment, you are asked to work through the Code Composer Tutorial. If you work through all parts of the tutorial, it will more than fill up the three hour lab period.

Details of the 'C6713 architecture and instruction set and of the DSK board will be gradually introduced in the first few experiments as they are required. Complete details for the DSP's, DSK, and TI software can be found in the Texas Instruments manuals when required. You can find them on the PC hard drive, TI web site, and lab bookshelf. (The ability to read manuals is an important skill to acquire!) Extensive documentation for the DSK board, 'C6713 DSP hardware, and software development tools is included with the special version of TI's Code Composer supplied with the DSK. Look at Code Composer's "Help" menu for this extensive documentation.

There are a large number of things to be learned initially and you may feel overwhelmed. Please be assured that the majority of your classmates feel the same way. Very shortly you will get over this initial hurdle and find the experiments interesting and challenging. As the semester progresses, you will build up a bag-of-tricks that can be used in following experiments so they will almost seem to get easier.

1.1 Some DSP Chip History and Typical Applications

Digital signal processing chips (DSP's) were introduced in the early 1980's and have caused a revolution in product design. Current major DSP manufacturers include Texas Instruments (TI), Motorola, Lucent/Agere, Analog Devices, NEC, SGS-Thomson, and Conexant (formerly Rockwell Semiconductor). DSP's differ from ordinary microprocessors in that they are specifically designed to rapidly perform the sum of products operation required in many discrete-time signal processing algorithms. They contain hardware parallel multipliers, and functions implemented by microcode in ordinary microprocessors are implemented by high speed hardware in DSP's. Since they do not have to perform some of the functions of a high end microprocessor like an Intel Pentium, a DSP can be streamlined to have a smaller size, use less power, and have a lower cost. Low cost DSP's have made it more economical to implement functions by digital signal processing techniques rather than by hard-wired analog circuits, particularly for audio band applications like speech compression and telephone line modems. The speeds of the latest generations of DSP's have increased to the point where they are being used in high speed applications like DSL and wireless base stations and hand sets. Some of the advantage results from the fact that integrated digital circuits are very reliable and can be automatically inserted in boards easily. In addition, programmable DSP's can implement complicated linear and nonlinear algorithms and easily switch functions by jumping to different sections of program code. The complexity of the algorithms is only limited by the imagination of the programmer and the processing speed of the DSP. Once the program is perfected, the chip function does not change with age unless a very rare failure occurs. On the other hand, analog components require more board space, sometimes must be trimmed to the correct values after insertion, and change with temperature and age. Analog circuits are designed to perform specific functions and lack the flexibility of the programmable DSP approach. Another advantage is that small changes in the DSP function or bug fixes can be made by changing a few lines of code in a ROM or EPROM while similar changes may be very difficult with a hard-wired analog circuit.

Digital signal processing algorithms were used long before the advent of DSP chips. They were implemented on large main-frame computers and, later, on expensive "high-speed" mini-computers. Depending on the signal bandwidths, these implementations were real or non-real-time. As semiconductor technology evolved, custom processors were built with many TTL MSI chips including cascadable ALU sections and stand alone multiplier chips. A typical system contained over 100 MSI chips. These systems were big and expensive to manufacture because of the large chip count, and consumed many watts of power requiring cooling fans. The most popular first generation DSP chips, the NEC μPD7720 and Texas Instruments TMS32010, became commercially available in late 1982. These chips performed 16-bit integer arithmetic at the rate of 5 million instructions per second (MIPS) and had limited internal RAM, ROM, and I/O capabilities. They initially cost about $600 not including the software and hardware development tools. Many current DSP's that are orders of magnitude more advanced cost less than $20. They reduced the chip count by a very large percentage resulting in smaller circuit boards using significantly less power, more reliable systems, and reduced manufacturing complexity and cost to the point where DSP's are present in many consumer products. Cell phones and telephone line modems are probably the largest DSP markets.

DSP's have continually evolved since they were first introduced as VLSI technology improved, as users requested additional functionality, and as competition arose. More internal RAM and ROM has been added and the total address space has been increased. Additional functions have been added like hardware bit-reversed addressing for FFT's, hardware circular buffer addressing, serial ports, timers, DMA controllers, and sophisticated interrupt systems including shadow registers for low overhead context switching. Analog Devices has included switched capacitor filters and sigma-delta A/D and D/A converters on some DSP chips. Instruction rates have increased dramatically. State-of-the-art integer DSP's like the TMS320C5000 series are available with members that can operate at clock rates of 200 MHz and cost around $20 in quantity. The even newer TMS320C6000 family, which has a very long instruction word architecture (VLIW), has members with clock rates up to 1 GHz and cost about $150. The speed increase is largely a result of reduced geometries and improved CMOS technology. In the last couple of years, DSP manufactures have been developing chips with multiple DSP cores and shared memory for use in high-end commercial applications like network access servers handling many voice and data channels. DSP chips with special purpose accelerators like Viterbi decoders, turbo code decoders, multimedia functions, and encryption/decryption functions are appearing. The rapid emergence of broadband wireless applications is pushing DSP manufacturers to rapidly increase DSP speeds and capabilities so they do not lose out to FPGA's.

The introduction of CMOS technology after the first couple of generations of DSP's significantly reduced power consumption. Also, lower supply voltages are now being used. Applications like telephone line modems that required at least two DSP's and an additional ordinary microprocessor acting as a controller fifteen years ago can now be implemented using a single DSP and at lower cost! Power consumption has been significantly reduced for some DSP's to conserve battery life in cell phones. Custom TI chips currently in production with six DSP cores can do around 90 full duplex V.90 modems concurrently. With concurrent process running in DSP's, software issues involving real-time multi-tasking operating systems have become as important as hardware issues.

A milestone in DSP history occurred around 1986 when AT&T introduced the first commercial floating-point DSP, the DSP32. In 1988, TI shipped initial samples of the TMS320C30 at a price of $1,300 to begin its first generation TMS320C3x floating-point DSP family. Both processors have a 32-bit word length. The TMS320C30 family has a member that can run at 25 million instructions per second (MIPS) and costs about $200. A stripped down version, the TMS320C31, followed and has a member that can perform 40 MIPs with a price of about $35. TI started its second generation floating-point DSP family with the TMS320C40 which contains extensive support for parallel processing. At least eight or nine years ago, TI introduced the TMS320C67x series of floating point DSP's. The first of these was the TMS320C6701 which was followed by the TMS320C6711, TMS320C6713, and TMS320C672x family of floating point DSP's. The 'C67x DSP's have a VLIW (Very Long

Instruction Word) architecture with eight execution units in their CPU's and clock speeds currently range up to 350 MHz.

Floating-point DSP's are used in some applications because of their ease in programming. However, integer DSP's are used most frequently. Some reasons are because they are smaller, cheaper, faster, and use less power. Cost is a major concern in the competitive commercial market particularly for mass produced products like modems, cell phones, and hard disk drives. Power consumption is of particular concern for cell phones. Care must be taken with integer DSP's to scale signals to avoid overflow and underflow. However, this is not much of a problem in most digital communication system applications where signal power levels remain relatively constant. Floating-point DSP's automatically perform this scaling.

DSP's are used in a wide variety of offline and real-time applications. Some typical areas and specific applications are:

- **Telecommunications**: telephone line modems, FAX, cellular telephones, speaker phones, ADPCM transcoders, digital speech interpolation, broadband wireless systems, and answering machines

- **Voice/Speech**: speech digitization and compression, voice mail, speaker verification, and speech synthesis

- **Automotive**: engine control, antilock brakes, active suspension, airbag control, and system diagnosis

- **Control Systems**: head positioning servo systems in disk drives, laser printer control, robot control, engine and motor control, and numerical control of automatic machine tools

- **Military**: radar and sonar signal processing, navigation systems, missile guidance, HF radio frequency modems, secure spread spectrum radios, and secure voice

- **Medical**: hearing aids, MRI imaging, ultrasound imaging, and patient monitoring

- **Instrumentation**: spectrum analysis, transient analysis, signal generators

- **Image Processing**: HDTV, pattern recognition, image enhancement, image compression and transmission, 3-D rotation, and animation

Texas Instruments DSP's can be grouped into the following three categories:

- Low Cost, Fixed-Point, 16-Bit Word Length

  Motor control, disk head positioning, control
  TMS320C1s, 'C2x, 'C24x

- Power Efficient, Fixed-Point, 16Bit Words

  Wireless phones, modems, VoIP
  'C5x, 'C54x, 'C55x

- High Performance DSP's

  Communications infrastructure, xDSL, imaging, video
  'C62x (16-bit fixed-point)
  'C3x, 'C4x, 'C64x, 'C67x (32-bit floating-point)

Clearly, real-time DSP applications are limited to cases where the required signal sampling rate is sufficiently less than the DSP instruction rate so a reasonable number of instructions can be performed between samples. For example, a wide-band radio frequency (RF) signal with a high carrier frequency can not be directly sampled and demodulated with a DSP. However, when the bandwidth of the RF signal is sufficiently less than the instruction rate, analog front-end circuits can be used to demodulate it to baseband inphase and quadrature components which can then be sampled at a rate equal to the bandwidth and processed by a DSP. Alternatively, an analog filter can be used to form the Hilbert transform of the RF signal and then the original signal and its Hilbert transform can be sampled at a rate equal to the bandwidth and processed with a DSP. DSP's have been extensively used in audio frequency applications where many instructions can be performed between samples. However, they are being used to process increasingly wide-band signals as the instruction rates of new generations increase. Special purpose VLSI chips and FPGA's have been used to implement limited DSP functions at very high rates.

A very important attribute of the DSP approach is the flexibility of a programmable device. An example illustrating an ideal application for this flexibility will now be presented. In June 1994, the ITU approved the final draft of the V.34 telephone line modem recommendation. This modem can transmit data at rates varying from 2400 to 33600 bits/sec in multiples of 2400 bits/sec. A variety of modulation schemes are used during handshaking and normal data transmission. During the initial handshaking phase, binary, continuous phase, frequency shift keying is used to exchange information about modem capabilities using the V.8 protocol. Later, the called modem, usually referred to as the answer modem, transmits a signal called the answer tone to the calling modem. In the V.34 case, a small sinusoidal amplitude modulation is impressed on the answer tone to identify it as a V.34 modem, so the receiver in the calling modem needs to perform the function of an envelope detector. In another phase of the handshake, a channel probing sequence consisting of a sum of sine waves is transmitted first by one modem and then by the other. The calling and answer modem transmitters must synthesize this signal and the receivers must process it to estimate channel characteristics such as frequency response, noise level, and nonlinearities. Another special sequence is transmitted to rapidly adjust adaptive equalizers in the receivers. During still another portion of the handshake, binary differential phase shift keying (DBPSK) is used to exchange configuration information between the modems. Quadrature amplitude modulation (QAM) is used during normal data transmission. Transmit and receive signal separation is achieved by using adaptive echo cancellers. Additional tasks that must be performed during normal data transmission are scrambling and descrambling the input bit stream, mapping the scrambled data bits to transmitted signal points by a technique called shell mapping, trellis encoding, nonlinear precoding, and soft decision Viterbi decoding. All these tasks, and more, can now be performed by a single integer DSP. In fact, a state-of-the-art DSP can run several full duplex V.34 modems concurrently. In addition to the V.34 recommendation, these chips often contain code to act as at least six older modem types, have FAX modes, and do V.90 and V.92 modems. All that is required to include new functions is to "add a few more lines of code" and, possibly, some more memory. Consumer modem costs have plummeted to less than $100. Programmable DSP's have made it possible to manufacture products with tremendous complexity and sophistication at prices that the ordinary consumer can easily afford.

Finally, to maintain a balanced viewpoint, it should be pointed out that the DSP approach is not always the best solution to a problem even if a DSP can accomplish the task. For example, a commercial AM radio signal, which has a carrier frequency in the order of 1 MHz, can trivially be demodulated by a simple envelope detector consisting of little more than a diode, resistor, and capacitor. As engineers, you should always look for the most reasonable and economical method of solving a design problem.

1.2   The TMS320C6713 Floating-Point DSP

The experiments in this book are explained for the Texas Instruments TMS320C6713 DSK floating-point DSP boards. Floating-point DSP's with 32-bit words were chosen for this lab to simplify the programming tasks, especially in a high-level language like C. Most integer DSP's have 16-bit words so underflow, overflow, and dynamic range must be taken into account when programming them. In addition, emulating 32-bit floating-point arithmetic on a 16-bit integer DSP generates inefficient machine code. In practice, integer arithmetic can be used for most applications without very much difficulty. Therefore, high volume commercial products almost always use 16-bit integer DSP's because they are faster, use less power, and are cheaper. Floating-point DSP's are used for high-end low volume applications. A functional block diagram of the TMS320C6713 is shown in Figure 1.1, respectively. Their major components and features are briefly described in the following subsections. For complete details, see the *TMS320C6000 CPU and Instruction Set Reference Guide* [I.7], *TMS320C6000 Peripherals Reference Guide* [I.10], and the *TMS320C6713B Floating-Point Digital Signal Processor* [I.12].

1.2.1   The 'C6000 Central Processing Unit (CPU)

Members of the TMS320C67x family of DSP's all have essentially the same central processing unit (CPU) which is also called the DSP *core*. The CPU has a *very long instruction word* (VLIW) architecture that TI calls VelociTi. The CPU always fetches eight 32-bit instructions at once and there is a 256-bit bus to the internal program memory. Each group of eight instructions is called a *fetch packet*. The CPU has eight functional units that can operate in parallel and are equally split into two halves called the A or 1 and B or 2 sides. All eight units do not have to be given instruction words if they are not ready. Therefore, instructions are dispatched to the functional units as *execute packets* with a variable number of 32-bit instruction words. This variable length feature distinguishes the 'C6000 CPU from other VLIW architectures. The eight functional units include:

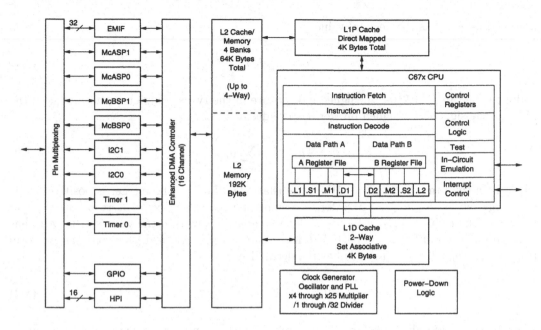

Figure 1.1: Functional Block Diagram of the TMS320C6713 DSP

- Four arithmetic and logic units (ALU's) that can perform fixed and floating-point operations (.L1, .L2, .S1, .S2)

- Two ALU's that perform only fixed-point operations (.D1, .D2)

- Two multipliers that can perform fixed or floating-point multiplications (.M1, .M2)

The General Purpose Register File

The CPU has thirty-two 32-bit general purpose registers split equally between the A and B sides. The CPU has a load/store architecture in which all instructions operate on registers. The data-addressing units .D1 and .D2 are in charge of all data transfers between the register files and memory. The four functional units on a side can freely share the 16 registers on that side. Each side has a single data bus connected to all the registers on the other side, so the functional units on one side can access data in the registers on the other side. Access to a register on the same side uses one clock cycle while access to a register on the other side requires a read and write cycle.

Integer and Floating-Point Word Formats

An *integer* in the 'C6000 family is a 32-bit word. Suppose the word has the form $\mathbf{d} = (d_{31}, d_{30}, \ldots, d_1, d_0)$. The left-most bit, d_{31}, is the most significant bit (MSB) and the right-most bit, d_0, is the least significant bit (LSB). Integers can be unsigned or signed. An unsigned integer has the decimal value

$$v = \sum_{n=0}^{31} d_n 2^n \tag{1.1}$$

and is in the range $[0, 2^{32} - 1]$. Signed integers are in the two's complement format with the MSB as the sign bit. A signed integer has the decimal value

$$v = -b_{31} 2^{31} + \sum_{n=0}^{30} d_n 2^n \tag{1.2}$$

which is in the range $[-2^{31}, 2^{31} - 1]$. Notice that when $b_{31} = 0$ the integer is positive, while it is negative for $b_{31} = 1$. These forms assume the binary radix point is on the right of the LSB. The radix point can be assumed to be anywhere within the word so the word has both an integer and fractional part. If L fractional bits are assumed, the word has the form $\mathbf{b} = (b_{31}, \ldots, b_L . b_{L-1}, \ldots, b_0)$ and has the decimal value

$$v = 2^{-L}(-b_{31} 2^{31} + \sum_{n=0}^{30} d_n 2^n) \tag{1.3}$$

The integer is said to have Q_L format. It is up to the programmer to keep track of the location of the radix point when integer arithmetic is performed. Integers occupy one register in the register file.

The 'C6000 DSP's also support 40-bit *long integers*. Long integers are stored in two successive registers in the register file. The least significant 32 bits are located in an even numbered register and the most significant 8 bits are located in the lower 8 bits of the next higher register which is odd numbered. The floating-point DSP's ('C67x) can also form double precision 64-bit integers that occupy all bits in a pair of consecutive registers. Some instructions also operate on 16-bit halves of registers.

The 'C67x DSP's support IEEE single and double-precision floating-point numbers . IEEE floating-point numbers can be normal numbers, denormalized (or subnormal) numbers, NaNs (not a number), and infinity numbers. Denormalized numbers are nonzero numbers that are smaller in magnitude than the smallest nonzero positive normal number. Single-precision numbers are 32 bits long and double-precision numbers are 64 bits long. Normal single-precision numbers are accurate to at least six decimal places and normal double-precision numbers to at least 15 places.

The single-precision word format is shown in the Figure 1.2.

31	30	23	22	0
s	$e_7, e_6, \ldots, e_0$		$f_1, f_2, \ldots, f_{22}, f_{23}$	

Figure 1.2: IEEE Single-Precision Floating-Point Number Format

The fields in this word will now be described.

- The one-bit field s is the *sign bit*. Numbers are positive for $s = 0$ and negative for $s = 1$.

- The eight-bit field $(e_7, \ldots, e_0)$ is the *biased exponent* which has the decimal value

$$e = \sum_{k=0}^{7} e_k 2^k$$

so $0 \leq e \leq 255$. The biased exponent for a normal number can not be 0 or 255. The value $e = 0$ is reserved for numbers that are identically 0. The value $e = 255$ is reserved for infinity and NaN.

- The 23-bit field $(f_1, \ldots, f_{23})$ is the *mantissa* or fractional part and has the decimal value

$$f = \sum_{k=1}^{23} f_k 2^{-k}$$

corresponding to the binary word $0 . f_1 f_2 \ldots f_{23}$. The mantissa is in the range $[0, 1 - 2^{-23}]$

The numerical value, x, of the single-precision floating-point word is determined by the following rules:

1. If $e = 255$ and $f \neq 0$, then x is NaN independent of s.

2. If $e = 255$ and $f = 0$, then $x = (-1)^s \infty$.

3. If $0 < e < 255$, then $x = (-1)^s 2^{e-127}(1 + f)$. These are the normal numbers. Notice that $1 + f$ corresponds to the binary word $1 . f_1 \ldots f_{23}$. For normal numbers, the exponent is chosen to move the leading 1 to the 1's position just to the left of the binary point. Since the 1 is know to be present, it is not included in the f field allowing for one extra bit of accuracy for the mantissa.

4. If $e = 0$ and $f \neq 0$, the $x = (-1)^s 2^{-126} f$. These are the denormalized numbers.

5. If $e = 0$ and $f = 0$, then $x = (-1)^s \times 0$. Notice that a positive and negative 0 are defined.

The double-precision word format is shown in Figure 1.3. The fields are similar to the

63	62 52	51 0
s	$e_{10}, e_9, \ldots, e_0$	$f_1, f_2, \ldots, f_{22}, f_{52}$

Figure 1.3: IEEE Double-Precision Floating-Point Number Format

single-precision fields and are:

- The one-bit field s is the *sign bit*. Numbers are positive for $s = 0$ and negative for $s = 1$.

- The eleven-bit field $(e_{10}, \ldots, e_0)$ is the *biased exponent* which has the decimal value

$$e = \sum_{k=0}^{10} e_k 2^k$$

so $0 \leq e \leq 2047$. The biased exponent for a normal number can not be 0 or 2047. The value $e = 0$ is reserved for numbers that are identically 0. The value $e = 2047$ is reserved for infinity and NaN.

- The 52-bit field $(f_1, \ldots, f_{52})$ is the mantissa or fractional part and has the decimal value

$$f = \sum_{k=1}^{52} f_k 2^{-k}$$

corresponding to the binary word $0.f_1 f_2 \ldots f_{52}$. The mantissa is in the range $[0, 1 - 2^{-52}]$

The numerical value, x, of the double-precision floating-point word is determined by the following rules:

1. If $e = 2047$ and $f \neq 0$, then x is NaN independent of s.

2. If $e = 2047$ and $f = 0$, then $x = (-1)^s \infty$.

3. If $0 < e < 2047$, then $x = (-1)^s 2^{e-1023}(1 + f)$. These are the normal numbers. Notice that $1 + f$ corresponds to the binary word $1.f_1 \ldots f_{52}$. For normal numbers, the exponent is chosen to move the leading 1 to the 1's position just to the left of the binary point. Since the 1 is know to be present, it is not included in the f field allowing for one extra bit of accuracy for the mantissa.

4. If $e = 0$ and $f \neq 0$, the $x = (-1)^s 2^{1022} f$. These are the denormalized numbers.

5. If $e = 0$ and $f = 0$, then $x = (-1)^s \times 0$. Notice that a positive and negative 0 are defined.

In the 'C67x DSP's, single-precision floating-point numbers are stored in a single register. Double-precision floating-point numbers are stored in a pair of adjacent registers. The least significant 32 bits are stored in an even numbered register and the most significant 32 bits are stored in the next higher register. Therefore, the even numbered register stores the lower 32 bits of the mantissa while the odd numbered register contains, starting from its LSB, the upper 20 bits of the mantissa, the eleven biased exponent bits, and the sign bit.

The Multiplier

The TMS320C67x DSP's have two multipliers, .M1 and .M2. Each multiplier can perform a variety of 16-bit integer × 16-bit integer products resulting in a 32-bit integer product that is stored in one register. They can perform the integer product of the lower 16 bits of one register and the lower 16 bits of another, the product of the lower 16 bits of one register and the upper 16 bits of another register, etc. They can also perform the product of two 32-bit integers resulting in a 32-bit product which is stored in one register and is the lower 32 bits of the product, or a 64-bit product which is stored in two adjacent registers with the lower 32 bits in the even register and upper 32 bits in the odd register.

The multipliers can also perform the product of two single-precision floating-point numbers resulting in a single-precision floating-point product. In addition, they can perform the product of two double-precision floating-point numbers resulting in a double-precision product.

Interrupts

The 'C6000 CPU's contain a vectored priority interrupt controller. The highest priority interrupt is RESET which is connected to the hardware reset pin and cannot be masked. The next priority interrupt is the non-maskable interrupt (NMI) which is generally used to alert the CPU of a serious hardware problem like a power failure. Then, there are twelve lower priority maskable interrupts INT4–INT15 with INT4 having the highest and INT15 the lowest priority. These maskable interrupts can be selected from up to 32 sources for the 'C6000 family. The sources vary between family members. For the TMS320C6713, they include external interrupt pins selected by the GPIO unit, and interrupts from internal peripherals such as timers, McBSP serial ports, McASP serial ports, EDMA channels, and the host port interface. The CPU's have a multiplexer called the *interrupt selector* that allows the user to select and connect interrupt sources to INT4 through INT15. The interrupt system is discussed in detail in Chapter 2.

1.2.2   Memory Organization for the TMS320C6713 DSK

The 'C6713 DSP has an L1/L2 memory architecture consisting of a 4K-byte L1P Program Cache (direct-mapped), a 4K-byte L1D Data Cache (2-way set associative), and an L2 memory with 256K-bytes total. The L2 memory is partitioned into a 64K-byte L2 unified cache/mapped RAM which is up to 4-way set associative, and 192K-bytes of additional L2 mapped RAM. The L1P cache has a 256-bit wide bus to the CPU so the CPU can read a fetch packet (eight 32-bit instructions) each cycle.

The 'C6713 DSP has a 32-bit External Memory Interface (EMIF) unit that provides a glueless interface to SDRAM, Flash, SBSRAM, SRAM, and EPROM. The DSP has a 512 M-byte total addressable external memory space. Data is byte (8-bit), half-word (16-bit), or word (32-bit) addressable. Table 1.1 shows the default memory map for the TMS320C6713 DSK.

1.2.3   Enhanced Direct Memory Access Controller (EDMA)

The TMS320C6713 has an *enhanced direct memory access controller* (EDMA) that can transfer data between any locations in the DSP's 32-bit address space independently of the CPU. See the *TMS320C6000 Peripherals Reference Guide* [I.10] for complete details. The EDMA handles all data transfers between the L2 cache/memory controller and the peripherals. These include cache servicing, non-cacheable memory access, user-programmed

Table 1.1: Memory Map for the TMS320C6713

Address	C67x Family Memory Type	C6713DSK
0x00000000	Internal Memory	Internal Memory
0x00030000	Reserved Space or Peripheral Regs	Reserved or Peripheral
0x80000000	EMIF CE0	SDRAM
0x90000000	EMIF CE1	Flash
0x90080000		CPLD
0xA0000000	EMIF CE2	Daughter Card
0xB0000000	EMIF CE3	

data transfers, and host access. It can move data to and from any addressable memory spaces including internal memory (L2 SRAM), peripherals, and external memory. The EDMA includes event and interrupt processing registers, an event encoder, a parameter RAM, and address generation hardware. It has 16 independent channels and they can be assigned priorities. Data transfers can be initiated by the CPU or *events* from the peripherals and some external pins. The user can select how events are mapped to the channels. The EDMA can transfer elements that are 8-bit bytes, 16-bit halfwords, or 32-bit words. Very sophisticated block transfers can be programmed including transfers of 1-dimensional and 2-dimensional data blocks consisting of multiple frames. The EDMA is described in more detail in Chapter 2 where you will use it to repetitively transfer an array to the codec.

1.2.4   Serial Ports

The TMS320C6713 contains two bidirectional multichannel buffered serial ports (McBSP0 and McBSP1). The serial ports operate independently and have identical structures. They can be set to transfer 8, 12, 16, 20, 24, or 32 bit words. The bit clocks and frame synchs can be internal or external and the McBSP includes programmable hardware for generating shift clocks and frame synchs. The McBSP's allow direct interface to high-speed data links like T1 and E1 channels, codecs, and Motorola Serial-Peripheral-Interface (SPI) devices. They can multiplex up to 128 channels. The McBSP can also perform μ-Law and A-Law companding and de-companding. For complete details see the *TMS320C6000 Peripherals Reference Guide* [I.10].

The 'C6713 DSP also includes two multi-channel audio serial ports (McASP0 and McASP1). The McASP is a serial port optimized for the needs of multi-channel audio applications. The two McASP's can support two completely independent audio zones simultaneously. Each McASP includes a pool of 16 shift registers that may be configured to operate as either transmit data, receive data, or general-purpose I/O (GPIO). The McASP's can use a time-division multiplexed (TDM) synchronous serial format or a digital audio interface (DIT) format. Both the transmit and receive sections of the McASP also support burst mode which is useful for non-audio data, for example, transferring data between two DSPs. See the *TMS320C6713B Floating-Point Digital Signal Processor* [I.12] for details.

1.2.5   Other Internal Peripherals

The TMS320C6713 DSP contains two 32-bit general-purpose timers that can be used to time events, count events, generate pulses, interrupt the CPU, and send synchronization events to the EDMA. They can be clocked by an internal or external source. Each has an input and output pin. A periodic clock signal can be generated on the output pin and the input can be used to count events.

The TMS320C6713 also has a Host Port Interface (HPI). The HPI provides a 16-bit interface to a host. The host functions as a master and can access the entire memory map of the DSP. Accesses are accomplished by using the EDMA.

Two I2C serial ports are included for control purposes. Each I2C port is compatible with Philips I^{2}C Specification Revision 2.1. They can operate up to 400 Kbps, have noise filters, seven and ten-bit device addressing modes, master and slave functionality, can generate events, and include slew-rate limited open-drain output buffers.

The TMS320C6713B has a PLL and a flexible PLL controller consisting of a prescaler and four dividers. The controller can generate different clocks for different parts of the DSP. A wide range of frequencies can be achieved with the dividers and prescaler.

There is also a general purpose input/output (GPIO) module with 16 external pins that can be individually programmed to be inputs or outputs. Some of the pins can be mapped to interrupts and events. See the *TMS320C6713B Floating-Point Digital Signal Processor* [I.12] for more details.

The DSP's are also IEEE-1149.1 (JTAG) boundary scan compatible.

1.2.6   Brief Description of the TMS320C6000 Instruction Set

The TMS320C6000 DSP's have an extensive instruction set which is tailored to the algorithms used in digital signal processing. There are instructions for fixed and floating-point addition, subtraction, and multiplication as well as for logical operations, circular buffering, and data loading and storing. All instructions can be conditionally executed. Tables 1.2 and 1.3 show the mnemonics for these instructions and how they are mapped to functional units. For complete details see the *TMS320C6000 CPU and Instruction Set Reference Guide* [I.7]. The function of an instruction is often somewhat obvious from the mnemonic. For example, instructions with the prefix ADD do addition, MPY is a multiplication, LDW loads a word from memory into a register, STB stores a byte from a register to memory, and MPYSP does a single-precision floating-point multiplication.

The addressing modes for the 'C67x are linear, circular using BK0, and circular using BK1. Linear addressing can be used with all register, but circular addressing can only be used with registers A4–A7 and B4–B7. Addresses can be formed in four basic ways: register indirect, register relative, register relative with 15-bit constant offset, and base + index. In addition, the addresses can be left unmodified, preincremented or predecremented, or

Table 1.2: Fixed-Point Instructions Common to the 'C62x and 'C67x

.L unit	.M Unit	.S Unit		.D Unit	
ABS	MPY	ADD	SET	ADD	STB (15-bit offset)
ADD	MPYU	ADDK	SHL	ADDAB	STH (15-bit offset)
ADDU	MPYUS	ADD2	SHR	ADDAH	STW (15-bit offset)
AND	MPYSU	AND	SHRU	ADDAW	SUB
CMPEQ	MPYH	B disp	SSHL	LDB	SUBAB
CMPGT	MPYHU	B IRP	SUB	LDBU	SUBAH
CMPGTU	MPYHUS	B NRP	SUBU	LDH	SUBAW
CMPLT	MPYHSU	B reg	SUB2	LDHU	ZERO
CMPLTU	MPYHL	CLR	XOR	LDW	
LMBD	MPHLU	EXT	ZERO	LDB (15-bit offset)	
MV	MPYHULS	EXTU		LDBU (15-bit offset)	
NEG	MPYHSLU	MV		LDH (15-bit offset)	
NORM	MPYLH	MVC		LDHU (15-bit offset)	
NOT	MPYLHU	MVK		LDW (15-bit offset)	
OR	MPYLUHS	MVKH		MV	
SADD	MPYLSHU	MVKLH		STB	
SAT	SMPY	NEG		STH	
SSUB	SMPYHL	NOT		STW	
SUB	SMPYLH	OR			
SUBU	SMPYH				
SUBC					
XOR					
ZERO					

postincremented or postdecremented. All these options are shown in Table 1.4. Offsets are multiplied by 4, 2, or 1 (shifted left by 2, 1, or 0 bits) before being added or subtracted from the base address according to whether a word, half-word, or byte is being used . The addresses computed refer to the byte location of the data in memory. The BK0 and BK1 fields in the Addressing Mode Register (AMR) set the block sizes for circular addressing.

As an example, consider the instruction

$$\text{LDW} \quad \text{.D1} \quad *++\text{A4}[9], \text{A1}$$

This instruction loads a 32-bit word (LDW) using functional unit .D1 into register A1 from the memory byte address: *contents of* (A4) + 4 × 9. The offset, 9, is multiplied by 4 since there are 4 bytes per word.

Almost all of the programming for this course will be done in C, so you do not have to spend much time learning the details of assembly language programming using the instruction set. If a C program is written in a straight forward way without any special care, the

Table 1.3: Extra Instructions for the 'C67x

.L unit	.M Unit	.S Unit	.D Unit
ADDDP	MPYDP	ABSDP	ADDAD
ADDSP	MPYI	ABSSP	LDDW
DPINT	MPYID	CMPEQDP	
DPSP	MPYSP	CMPEQSP	
DPTRUNC		CMPGTDP	
INTDP		CMPGTSP	
INTDPU		CMPLTDP	
INTSP		CMPLTSP	
INTSPU		RCPDP	
SPINT		RCPSP	
SPTRUNC		RSQRDP	
SUBDP		RSQRSP	
SUBSP		SPDP	

Table 1.4: Addressing Modes

Addressing Type	No Modification of Address Register	Preincrement or Predecrement of Address Register	Postincrement or Postdecrement of Address Register
Register Indirect	*R	*++R *--R	*R++ *R--
Register Relative	*+R[ucst5] *-R[ucst5]	*++R[ucst5] *--R[ucst5]	*R++[ucst5] *R--[ucst5]
Register Relative with 15-bit Constant Offset	*+B14/B15[ucst15]	none	none
Base + Index	*+R[offsetR] *-R[offsetR]	*++R[offsetR] *--R[offsetR]	*R++[offsetR] *R--[offsetR]

Notes:
  ucst5 = 5-bit unsigned integer constant
  ucst15 = 15-bit unsigned integer constant
  R = base register
  offsetR = index register

TI C6000 C compiler generates reasonably efficient code that is adequate for many applications. Program efficiency can be improved significantly by using the compiler's optimization features.

It is important to know something about the DSP assembly instructions to understand how the DSP works and what happens when higher level instructions are turned into machine instructions by the compiler. Also, some DSP hardware capabilities like hardware circular buffering are not used by the compiler.

1.2.7   Parallel Operations and Pipelining

As mentioned earlier, the instruction word for each CPU functional unit is 32 bits long, and eight instructions are fetched at a time. The group of eight instructions or $8 \times 32 = 256$ bits is called a fetch packet. Fetch packets must start at an address that is a multiple of eight 32-bit words. Since the CPU has eight functional units, up to eight instructions can be executed in parallel. Each must use a different functional unit. However, not all functional units must be used. Each group of instructions operating in parallel is called an *execute packet*.

Bit 0 of each instruction is called the p-bit and determines if the instruction operates in parallel with another. The instructions in a fetch packet are scanned from the lowest address to the highest. If the p-bit of instruction i is 1, then instruction $i + 1$ is executed in parallel with instruction i. If it is 0, instruction $i + 1$ is executed one cycle after instruction i. The p-bit of the eighth instruction is always 0 because it cannot chain to the next fetch packet. If all the p-bits in a fetch packet are 0, the instructions are executed sequentially in time, one after the other, starting with the first instruction in the packet. If the first seven p-bits are 1, all eight instructions are executed in parallel. When some are 0 and some 1, the execute packets (chained instructions) are executed sequentially in time starting from the packet with the lowest address. If a branch occurs in the middle of an execute packet, all instructions in the entire fetch packet at lower addresses than the branch are ignored.

Instructions are processed in a multi-stage pipeline consisting of a program fetch, program decode, and execute stage as shown in Table 1.5. The fetch stage has four phases for all instructions: (1) program address generation (PG), (2) program address sent (PS), (3) program wait (PW), and (4) program data receive (PR). The decode stage has two phases for each instruction: (1) dispatch (DP) and (2) decode (DC). Instructions in the execute stage can pass through anywhere from 1 to 10 phases (E1–E10). Parallel instructions pass simultaneously through each pipeline phase. Serial instructions proceed through the pipeline with a fixed relative phase difference. There are no pipeline interlocks.

During the PG (program address generate) phase of the fetch stage, the program address is computed by the CPU. The address is sent to memory in the PS (program address send) phase and it is read in the PW (program access ready wait) phase. Finally, the CPU receives the fetch packet in the PR (program fetch packet receive) phase.

During the DP (instruction dispatch) phase of the decode stage, the fetch packet of eight instructions is partitioned into execute packets and the instructions are assigned to functional units. Then in the DC (instruction decode) phase, the source registers, destination registers, and required data paths are decoded for instruction execution by the functional units.

Table 1.5: TMS320C6x Pipeline Phases

Stage	Phase	Symbol
Program Fetch	Program Address Generation	PG
	Program Address Sent	PS
	Program Wait	PW
	Program Data Receive	PR
Program Decode	Dispatch	DP
	Decode	DC
Execute	Execute 1	E1
	⋮	⋮
	Execute 10	E10

Execute packets can take anywhere from 1 to 10 execute phases. Fixed-point instructions take at most 5 phases while floating-point instructions can take up to 10 phases. Most fixed-point instructions are single-cycle instructions and use only the execute phase E1. See [I.7] for details of the execution phases for all the different kinds of instructions.

An elementary example of the pipeline operation is shown in Figure 1.4. Here it is assumed that each fetch packet is one execute packet so all eight instructions are in parallel. The fetch packets flow in lockstep through each pipeline phase. It is also assumed that each execute packet requires all 10 execute phases. The pipeline is full at clock cycle 7 when fetch packet n reaches the E1 execute phase and fetch packet $n + 6$ is in the PG (program address generate) phase.

Since different instructions can require a different number of execute phases, the pipeline operation can be much more complicated than shown in Figure 1.4. Care must be taken so that the results of one instruction are ready when they are needed by another. Therefore, dummy *no operation* (NOP) instructions must be included where necessary to synchronize the computations. This makes hand assembly coding very difficult and tedious. Fortunately, TI provides an optimizing assembler that automatically assigns instructions to functional units and inserts NOP's where necessary. The assembler input source is a slightly higher level language than pure assembly instructions and is called *linear assembly*. The TI optimizing C compiler also automatically performs the scheduling. See the *TMS320C6000 Assembly Language Tools User's Guide* [I.5], *TMS320C6000 Optimizing Compiler User's Guide* [I.9], and *TMS320C6000 Programmer's Guide* [I.11] for complete details.

The concept of *delay slots* is useful in analyzing instruction execution. Each instruction type has a number of delay slots associated with it. A delay slot is a CPU cycle that occurs after the first execution phase (E1) of an instruction. Results for an instruction are not

| Clock | Fetch Packet | | | | | | | | | | |
Cycle	n	$n+1$	$n+2$	$n+3$	$n+4$	$n+5$	$n+6$	$n+7$	$n+8$	$n+9$	$n+10$
1	PG										
2	PS	PG									
3	PW	PS	PG								
4	PR	PW	PS	PG							
5	DP	PR	PW	PS	PG						
6	DC	DP	PR	PW	PS	PG					
7	E1	DC	DP	PR	PW	PS	PG				
8	E2	E1	DC	DP	PR	PW	PS	PG			
9	E3	E2	E1	DC	DP	PR	PW	PS	PG		
10	E4	E3	E2	E1	DC	DP	PR	PW	PS	PG	
11	E5	E4	E3	E2	E1	DC	DP	PR	PW	PS	PG
12	E6	E5	E4	E3	E2	E1	DC	DP	PR	PW	PS
13	E7	E6	E5	E4	E3	E2	E1	DC	DP	PR	PW
14	E8	E7	E6	E5	E4	E3	E2	E1	DC	DP	PR
15	E9	E8	E7	E6	E5	E4	E3	E2	E1	DC	DP
16	E10	E9	E8	E7	E6	E5	E4	E3	E2	E1	DC
17		E10	E9	E8	E7	E6	E5	E4	E3	E2	E1

Figure 1.4: Pipeline Operation Assuming One Execute Packet per Fetch Packet

available for use by another instruction until after the last delay slot. For example, single cycle instructions complete execution during the E1 pipeline phase and have no delay slots. The MPY instruction has one delay slot so its results are not available until the end of the E2 pipeline phase.

A store instruction uses the E1, E2, and E3 pipeline phases and writes its data to memory in the E3 phase but is considered to have no delay slots. The reason is that the results of a store must be accessed by a load instruction which uses phases E1 through E5. For a load instruction, the memory address is read during the E3 phase, is received at the CPU core boundary in the E4 phase and is written to the register in the E5 phase, so a load instruction has four delay slots. Since a store instruction writes its data to memory in the E3 phase, a load instruction following a store instruction finds the data from the store in memory when it reads the location in its E3 phase.

Branch instructions use just one execute phase. However, there are five delay slots between the execution of the branch and execution of the target code. This is illustrated in Figure 1.5. The branch target code is in the PG phase when the branch is in the E1 phase. There are then five delay lots until the target reaches the E1 phase.

1.3   The TMS320C6713 DSP Starter Kit (DSK)

The TMS320C6713 DSK is a low cost board designed to allow the user to evaluate the capabilities of the 'C6713 DSP and develop 'C6713-based products. It demonstrates how

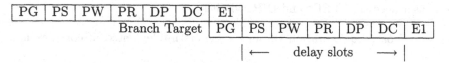

PG	PS	PW	PR	DP	DC	E1							
			Branch Target			PG	PS	PW	PR	DP	DC	E1	

$$| \longleftarrow \quad \text{delay slots} \quad \longrightarrow |$$

Figure 1.5: Branch Instruction Execution

the DSP can be interfaced with various kinds of memories and peripherals, and illustrates power, clock, JTAG and parallel peripheral interfaces. The board is approximately 5 inches wide and 8 inches long and is designed to sit on the desktop external to a host PC. It connects to the host PC through a USB port or an XDS510 emulator. A simplified block diagram of the DSK is shown in Figure 1.6.

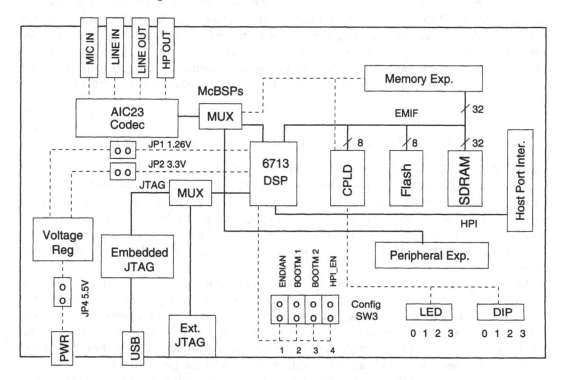

Figure 1.6: Block Diagram of the TMS320C6713 DSP Starter Kit (DSK)

The major DSK hardware features are:

- A TMS320C6713 DSP operating at 225 MHz.

- An AIC23 stereo codec with Line In, Line Out, MIC, and headphone stereo jacks

- 16 Mbytes of synchronous DRAM (SDRAM)

- 512 Kbytes of non-volatile Flash memory (256 Kbytes usable in default configuration)

- Four user accessible LEDs and DIP switches

- Software board configuration through registers implemented in complex logic device (CPLD)

- Configurable boot options

- Expansion connectors for daughter cards

- JTAG emulation through on-board JTAG emulator with USB host interface or external emulator

A printed technical reference manual comes with the DSK. Extensive documentation for the DSK and 'C6713 DSP is included on the CD that comes with the DSK. You can choose to have the documentation loaded on the PC hard drive when you install the DSK support software and Code Composer Studio. It includes a large collection of TI manuals. These TI manuals along with much other information can be conveniently accessed through Code Composer's "Help" menu. The version of Code Composer Studio delivered with the DSK is tailored to it.

1.3.1   The Audio Interface Onboard the TMS320C6713 DSK

The TMS320C6713 DSK uses a Texas Instruments AIC23 codec. In the default configuration, the codec is connected to the two serial ports, McBSP0 and McBSP1. McBSP0 is used as a unidirectional channel to control the codec's internal configuration registers. It should be programmed to send a 16-bit control word to the AIC23 in SPI format. The top 7 bits of the control word specify the register to be modified and the lower 9 bits contain the register value. Once the codec is configured, the control channel is normally idle while audio data is being transmitted.

McBSP1 is used as the bi-directional data channel for ADC input and DAC output samples. The codec supports a variety of sample formats. For the experiments in this course, the codec should be configured to use 16-bit samples in two's complement signed format. The codec should be set to operate in master mode so it supplies the frame sync and bit clocks at the correct sample rate to McBSP1. The preferred serial format is DSP mode which is designed specifically to operate with the McBSP ports on TI DSPs.

The codec has a 12 MHz system clock which is the same as the frequency used in many USB systems. The AIC23 can divide down the 12 MHz clock frequency to provide sampling rates of 8000, 16000, 24000, 32000, 44100, 48000, and 96000 Hz.

McBSP0 and McBSP1 can be individually disconnected through software from the AIC23 codec and routed to the Peripheral Expansion Connector. This allows commercially available and individually designed daughter cards to be plugged into the expansion sockets on the DSK. TI has published a daughter card standard that all its variety of DSK's follow. There are daughter cards for a significant number of codecs that are more capable than the AIC23. At the University of Maryland, we have designed a daughter card to convert RS232 serial port signal levels to TTL levels compatible with the McBSP signals.

More details about the hardware characteristics of the AIC23 and how to use it are presented in Chapter 2. Also look at the Code Composer "Help" menu for details. To see the details of how the AIC23 is included on the DSK, see the *TMS320C6713 DSK Technical Reference*, [I.2]. For complete details on the AIC23 chip itself, see *TLV320AIC23 Stereo Audio CODEC Data Manual*, [I.4].

1.4 Software Support for the DSK Board and 'C6x DSP's

1.4.1 The Board Support Library (BSL)

A special Board Support Library (BSL) is supplied with the TMS320C6713 DSK. The BSL provides C-language functions for configuring and controlling all the on-board devices. The library includes modules for general board initialization, access to the AIC23 codec, reading the DIP switches, controlling the LED's, and programming and erasing the Flash memory. The source code for this library is also included. The version of Code Composer supplied with the DSK is set up to automatically use the BSL. You can get complete documentation for the BSL by connecting the DSK to your PC, bring up Code Composer, and going to Help, Contents, TMS320C6713DSK, Software, Board Support Library.

The function for configuring the codec in the BSL sets McBSP1 to transmit and receive 16-bit words. The codec sends 16-bit left and right channel input samples to McBSP1 alternately and a program reading these samples from McBSP1's Data Receive Register (DRR1) would have to somehow figure out which is the right and which is the left channel sample. We have modified the code configuration function DSK6713_AIC23_openCodec() to send and receive data samples from the codec in DSP format using 32-bit words. The first word transmitted by the AIC23 codec is the left channel 16-bit sample and the right channel 16-bit sample is transmitted immediately after the left channel sample. The AIC23 generates a single frame sync at the beginning of the left channel sample. Therefore, a 32-bit word received by McBSP1 contains the left sample in the upper 16 bits and the right sample in the lower 16 bits. This solves the channel ambiguity problem. The reverse process takes place when sending samples from the DSP to the codec. The user's program should pack the left channel 16-bit sample in the upper 16 bits of an integer and the right channel 16-bit sample in the lower 16 bits and then write this word to the Data Transmit Register (DXR1) of McBSP1. We have replaced the original BSL codec configuration function with our modified function and renamed the file dsk6713bsl32.lib. The required header files as well as the source code are installed on our PC's. The codec configuration process is discussed in detail in Chapter 2.

We have also added the files intr.obj and intr_.obj, which are the compiled versions of intr.c and intr_.asm, to dsk6713bsl32.lib. The required header files have also been added to the "include" folder. These files provide a simple interrupt structure that allows interrupts to be dynamically hooked to or unhooked from interrupt service routines. The file intr.c includes functions to reset the interrupt registers to defaults, initialize the interrupt vector table and interrupt service routine jump table, assign interrupt sources to CPU inter-

rupt numbers, hook interrupt service routines to CPU interrupts, and set and clear bits in
the Interrupt Enable Register (IER) and Interrupt Flags Register (IFR). Chapter 2 discusses
this interrupt structure in much more detail. Complete documentation can be found in the
source files.

1.4.2 The Chip Support Library

TI has created a *Chip Support Library* (CSL) that contains C functions and macros for
configuring and interfacing with all the 'C6713 on-chip peripherals and CPU interrupt controller. Complete details are presented in the *TMS320C6000 Chip Support Library API
User's Guide* [I.6]. This library is loaded onto the PC when the DSK software is installed.
Each peripheral is covered by an individual API module. The CSL header files provide a
complete symbolic description of all peripheral registers and register fields.

The CSL provides two methods for initializing the registers of a peripheral. The symbolic
names of all peripherals are listed in Table 1 of [I.6]. Let PER be a peripheral name, for
example, McBSP. One method is to use the function PER_configArgs(reg0, reg1, ...).
You must construct all the register words before calling this function. The CSL has the
PER_REG_RMK macros to help set the fields in each register. The second method is to use the
function PER_config(&MyConfig) where MyConfig is a structure constructed by using the
CSL structure type PER_Config as shown in the following lines:

```
PER_Config MyConfig = {
    reg0,        .
    reg1,
    ...
};
```

You still need to construct the register values.

The CSL provides a graphical user interface (GUI) that is part of the DSP/BIOS Configuration Tool of Code Composer Studio to set the peripheral registers for most of the 'C6000
series DSP's. However, the *TMS320C6000 Chip Support Library API User's Guide* says the
GUI does not support the TMS320C6713.

1.5 Code Composer Studio

You will constantly use Texas Instruments' multi-function program, *Code Composer Studio*
(CCS), on the PC to generate programs for the TMS320C6713 DSP, load them into the DSP
memory, run them, and monitor program execution. See the *Code Composer Studio User's
Guide* [I.3] and online CCS help files for complete details.

1.5.1 Project Files and Building Programs

You can build a *project* in CCS to easily manage an application involving multiple source
files, libraries, memory maps, and special command files. The file containing all the project
information is given the extension pjt. By clicking on the Rebuild All or Incremental build task

bar buttons or by menu selections, you can create an executable module in COFF (common object format file) that can be loaded into the DSP's memory. The default behavior is to use the extension, out, for executable modules. To build a program, CCS invokes the program cl6x.exe with the appropriate command line options, input source file names, and output file names. The program cl6x.exe is a shell program that

1. compiles C source files using the TI optimizing C compiler, outputting assembly modules with the extension asm,

2. assembles assembly and linear assembly source files including those generated by the compiler using asm6x.exe creating *relocatable object modules* with the extension obj,

3. links the object modules and required library modules using lnk6x.exe according to information in a *linker command file* into an executable COFF file. Linker command files normally have the extension cmd.

The Incremental build option only compiles and assembles source files that have been modified since the last build, speeding up the building process. Figure 1.7 illustrates the process of building an executable module.

CCS includes a full featured editor with syntax highlighting for entering and modifying source code. When a file is compiled or assembled and errors are detected, CCS can jump from the error message window to the location of the error in the source file displayed in the editing window.

1.5.2   The Optimizing Compiler and Assembler

Code Composer Studio includes a C/C++ optimizing compiler that converts standard ANSI C source programs into C6000 assembly language source. An interlist facility can be invoked that creates an output file showing each C source statement followed by the assembly code generated to implement the statement. Several levels of optimization can be used. The compiler automatically schedules parallel use of the 'C6x execution units and inserts NOP's where necessary to account for delay slots. At the higher optimization levels, the compiler performs software pipelining which is a technique to make loops execute as efficiently as possible by making maximum parallel use of the execution units and pipeline stages. This would be extremely difficult and time consuming by hand. A smart compiler is a necessity to make the complicated hardware architecture of the 'C6000 family easy and productive to use.

The compiler has several extensions to ANSI C. Assembly statements can be included inline with the C source code. This is useful for manipulating registers in the DSP and using special hardware features that are not efficiently accessible thorough C. There are also a number of *intrinsics* that can be used like C functions and perform assembly instructions. An interrupt keyword has been added that can be used to declare a C function to be an interrupt handler. The compiler then generates the code necessary to save and restore the machine state on entry and exit. A volatile keyword can be used to type a variable so that the optimizer does not optimize it out of existence. For example, a status register in a

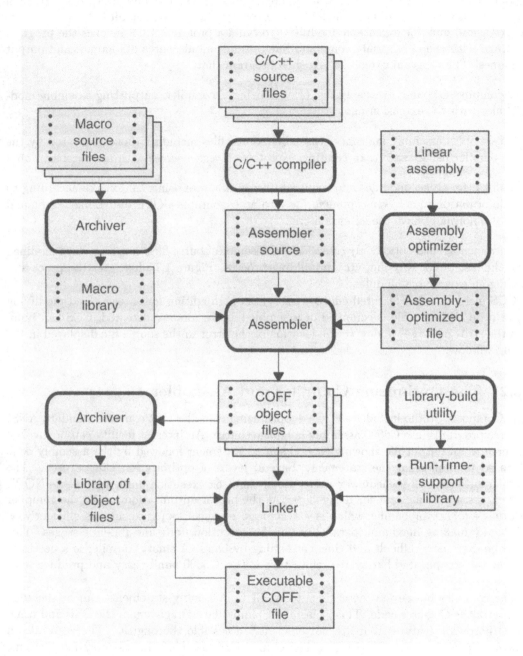

Figure 1.7: Building Programs

serial port may change when a word is received, but the compiler will not recognize this can happen and assume a reference to the register never changes.

Program efficiency can be improved significantly by using the compiler's optimization features, appropriate compiler directives, and good memory assignments. Extensive details about how to write optimized programs can be found in the *Programmer's Guide* [I.11]. Another jump in efficiency can be made by using the intrinsics. They are somewhat similar to inline assembly instructions and make the compiler use the DSP's hardware efficiently. A complete list of the intrinsics can be found in the *TMS320C6000 Optimizing Compiler User's Guide* [I.9]. By using all the optimization techniques, good programmers have been able to write C programs that, when compiled, are as efficient as hand optimized assembly code.

The trend in industry these days is to write almost an entire application in C because of the speed in writing the program, ease of reading by others, and its portability to new DSP types. Small portions of the program that are time intensive may be hand programmed and optimized. The TI code generation tools can profile running programs and produce statistics about the execution time of program segments.

TI has created a language called *linear assembly* that is part way between pure assembly language and C. Linear assembly source files have the extension sa. In linear assembly you do not have to be concerned with assigning registers or pipeline issues. Symbolic names can be used for registers. The assembly optimizer assigns registers and optimizes loops to generate highly parallel assembly code.

The assembly source code files generated by the compiler and optimizing assembler must then be passed through the assembler to generate relocatable object modules.

See the *TMS320C6000 Optimizing Compiler User's Guide* [I.9] and *TMS320C6000 Assembly Language Tools User's Guide* [I.5] for complete details on the compiler and assembler.

1.5.3 The Linker

The final step in building a program is to link all the relocatable modules together. The linker, lnk6x.exe, combines relocatable object modules to form an executable output program. The default extension for executable programs is out. In addition, the linker can generate a map file showing the absolute memory addresses of all global variables. A very important input to the linker is a *linker command file* which has the extension cmd. The command file can contain names of additional object modules to link, paths to libraries, names for the map and out files, a memory map for the target hardware system, and commands describing where to put specific program sections in memory. An example of a linker command file is presented early in Chapter 2. See the *TMS320C6000 Assembly Language Tools User's Guide* [I.5] for all the linker capabilities and options.

1.5.4 Building Programs from Command Line Prompts

The programs cl6x.exe, asm6x.exe, and lnk6x.exe can all be executed from a command line prompt. The general format for invoking the cl6x.exe shell is

```
cl6x [-compiler options] [filenames] [-z [link options]]
```

Items in rectangular brackets are optional. The entry [filenames] is a list of source files. Filenames that have no extension are automatically considered to have the extension, c, and to be C source code. Filenames with the extension, asm, are considered to be assembly language source code and are assembled and files with the extension, sa, are treated as linear assembly source and operated on by the optimizing assembler. Everything to the right of the -z option applies only to the linker. The assembler and linker can be executed in a similar manner. See the TI compiler and assembler manuals for complete details.

1.5.5   The Archiver

CCS includes an archiver program ar6x.exe that can be used to build libraries of relocatable object modules or source files. It can display a table of contents for an archive. The archiver can also insert modules into or extract modules from an archive. The archiver details are presented in the *TMS320C6000 Assembly Language Tools User's Guide* [I.5].

1.5.6   Additional Code Composer Studio Features

CCS has many additional features and some are described in the following list.

- Code Composer Studio has extensive capabilities for loading, running and monitoring program execution. It is the tool you will almost always use to load program into the target board memory and start the program running. It can single step through C or assembly instructions, stop at break points, display or change the contents of memory ranges and registers, watch selected C variables, and profile running programs.

- CCS can send data to the target board from a PC file or read data from probe points in the DSP program to a PC file.

- CCS can capture data from the target board and graph the results as a function of time, perform an FFT, or display a constellation diagram, eye diagram, or image.

- CCS has an interpretive general extension language (GEL) similar to C that allows you to extend CCS's capabilities. Through GEL functions CCS can access and change target memory locations including DSP registers and add options to the CCS menus.

- CCS facilitates building programs including a real-time operating system, DSP/BIOS. When CCS does normal file I/O, it halts the DSP during the data transfers so the program will not run in real-time. DSP/BIOS allows real-time data exchange (RTDX). DSP/BIOS can run multiple threads with different priorities. We will not use this facility in our experiments because it adds another level of complication and hides the basic DSP software and hardware issues. However, it certainly could be used in further independent projects.

1.6   Other Software

A variety of free-ware a commercial programs are available on the PC's. The free-ware programs include file compression archivers, a telnet terminal, an FTP program, and the GNUPLOT graphing program. Additional programs are listed below.

1.6.1   Digital Filter Design Programs

Filtering is a fundamental operation in communication systems. Several programs have been collected, modified, and written for designing digital filters for the experiments. They are all free-ware programs. These digital filter design programs are:

- `window.exe` is a program for designing lowpass, highpass, bandpass, bandstop, differentiation, and Hilbert transform FIR filters using the Fourier series and window function method. The basic program was taken from the classic IEEE Press book, *Programs for Digital Signal Processing* [II.C.7] and modified by the author.

- `remez87.exe` is a program for designing multiple passband/stopband, differentiation, and Hilbert transform FIR filters using the McClellan-Parks approach. The basic program was taken from [II.C.7] and modified by the author.

- `iir.exe` is a program for designing lowpass, highpass, bandpass, and band-reject IIR digital filters using the bilinear transformation to convert classical analog prototype filters into digital filters. Classical analog Butterworth, Chebyshev, inverse Chebyshev, and elliptic filters can also be designed. The basic program was downloaded from the TI bulletin board and modified by the author. The original program was written by S. Burrus at Rice University.

- `rascos.exe` is a program written by the author for designing FIR filters that approximate the raised cosine frequency response. The impulse response is separated into subfilters for an interpolation filter bank.

- `sqrtraco.exe` is a program written by the author for designing FIR filters that approximate the square-root of raised cosine frequency response. The impulse response is separated into subfilters for an interpolation filter bank.

The MATLAB signal processing package has filter design functions similar to some of the ones listed above.

1.6.2   Commercial Software

Commercial software available for the laboratory includes:

- The standard Microsoft Office suite including MS Word and Excel.

- Microsoft Visual C++

- MATLAB

- Anti-virus programs

1.7   Introductory Experiments

Code Composer Studio is the software utility you will almost always use to generate and edit source code, build executable DSP programs, and load the programs into the TMS320C6713 DSK. The goals of this introductory experiment are to become familiar with the audio connections to the DSK and to explore some of the capabilities of Code Composer Studio. No lab report is required for this experiment.

For your first lab session, perform the following tasks:

1. Check out the hardware. Find the three audio connectors for the DSK. They are MIC IN, LINE IN, and LINE OUT. The MIC IN jack is for low level signals from a microphone. You will be using the commercial signal generator for this course and should use only the LINE IN and LINE OUT connectors for these larger signal levels. Beware that a common mistake of lab students is to make the input too large and saturate the input amplifiers resulting in strange outputs.

2. If it has not already been done, connect the power supply to the DSK. Then connect the DSK to a USB port on the PC. You will hear the Microsoft "bing-bong" sound when the operating system detects the new USB Connection.

3. Start CCS by double clicking on the icon named C6713 DSK CCS on the desktop. The 'C6713 DSK version of CCS will not start unless the DSK is powered up and the USB connection has been made. The CCS splash screen will appear first for a few seconds and then the CCS workspace window will appear. Depending on the state of the PC, you may or may not see a message window in the lower right-hand corner with the message, "Waiting for USB Enumeration," and hear the USB connection sound. When the CCS workspace window is closed, you will hear the Microsoft "bong-bing bing-bong" sounds.

4. Work through as much of the Code Composer Studio IDE tutorial and other tutorials as possible during the first lab period. To get to the tutorial, click on Help on the CCS menu bar. Next select Tutorial and click on Code Composer Studio IDE in the table of contents on the left side of the window. Be sure to learn how to (1) create a project file, (2) build and run a program, (3) use break points and watch windows, (4) profile code execution, and (5) do file I/O and display graphs. Also browse through the extensive hardware and software documentation under the Help menu.

Please do not modify or work in the C:\TI, C:\CCStudio_v3.1, or C:\c6713 directories. Copy any files you will modify to a directory in your workspace on the network server or on the PC and do your work there. Almost all of the time you should use a directory in your workspace on the network server. Create files on the hard drive only in an emergency when the network is down. Finally, some words of wisdom learned by hard experience are, **"Always make backups of your programs before leaving the lab session!"**

Chapter 2

Learning to Use the Hardware and Software Tools by Generating a Sine Wave

The goal of this experiment is to learn how to use the hardware and software tools available at each station. The hardware includes a PC, TMS320C6713 DSK, a signal generator, and an oscilloscope. The main software tool you will use is Code Composer Studio (CCS) which contains an editor and a project building facility which automates calling the C compiler, assembler, and linker. You will also use CCS to load programs into the DSP boards, run them, and monitor their execution. You will gradually learn about the DSP's architecture including the McBSP serial ports, interrupt controller, and EDMA controllers by generating a sine wave by using polling, by using interrupts, and by using DMA from a table.

The experiments in this chapter assume that the TMS320C6713 DSK is being used along with the support library **dsk6713bsl32.lib** described in Section 1.4.1. The codec initialization function in this library configures McBSP1 to transmit and receive 16-bit left and right channel sample pairs packed into a single 32-bit word with the left channel sample in the upper 16 bits and the right channel sample in the lower 16 bits. Functions for setting up a simple interrupt environment and interfacing with the interrupt registers were also added to this library.

2.1 Getting Started with a Simple Audio Loop Through Program

2.1.1 A Linker Command File and Beginning C Program

First copy the linker command file, **dsk6713.cmd**, listed below to your working directory from the directory C:\c6713dsk. Linker command files are used to define how relocatable program sections are mapped into the physical system memory. They can also contain assembler options and lists of object programs to be included in the output modules. Additional object modules can also be included on the linker command line. The modules are loaded

29

in the order in which they appear in the command line list of .cmd and .obj files. Command files are convenient for saving definitions and operations that will be ordinarily used when linking programs for a particular project.

The -c line in `dsk6713.cmd` tells the linker to use the autoinitialization feature of C programs. The TI C compiler builds a table containing the data required to initialize all variables initialized in the C program. Code is included in the executable module to load the data values in the table into the variables when the program starts. The -heap and -stack lines allocate memory for the heap and stack. The number after these commands is the allocated memory size in bytes.

The -lrts6700.lib line tells the linker to search the C run-time library `rts6700.lib` for unresolved references. This library provides the standard functions the C compiler expects. The line -lcsl6713.lib tells the linker to search the Chip Support Library (CSL) `csl6713.lib`. CCS has been set to know the path to these libraries. They are almost always used by C programs. Including these lines in the linker command file automatically includes them in the linker search path without any further effort on your part.

The `MEMORY` portion of the command file is used to define the physical memory layout. For example, the line

```
IRAM : origin = 0x0,  len = 0x40000   /* 256 Kbytes */
```

defines the internal program memory to be a region called `IRAM` which starts at byte address 0x00000000, and has a length of 0x00040000 bytes which is 256 Kbytes.

The C compiler puts data and program code into named sections. Named sections can also be created by the programmer in assembly source code. The `SECTIONS` portion of the linker command file tells the linker how to place sections into defined memory regions. The standard conventions are to place program instructions in the `.text` section, initialized constants in the `.const` section, global and static variables in the `.bss` section, initialization tables for variables and constants in the `.cinit` section, local variables in the `.stack` section, and buffers for C I/O functions in the `.cio` section. Data from assembly programs can be put in the `.data` section. C does not use the `.data` section. There are many more options for linker command files that allow complex mappings of programs into physical memory. See the *TMS320C6000 Assembly Language Tools User's Guide* [I.5] for complete details.

Program 2.1 Linker Command File `dsk6713.cmd`

```
/***********************************************************/
/*  File dsk6713.cmd                                       */
/*    This linker command file can be used as the starting */
/*  point for linking programs for the TMS320C6713 DSK.    */
/*                                                         */
/*  This CMD file assumes everything fits into internal RAM.*/
/*  If that's not true, map some sections to the external  */
/*  SDRAM.                                                 */
/***********************************************************/
```

```
-c
-heap  0x1000
-stack 0x400
-lrts6700.lib
-lcsl6713.lib

MEMORY
{
  IRAM  : origin = 0x0,        len = 0x40000     /* 256 Kbytes */
  SDRAM : origin = 0x80000000, len = 0x1000000   /* 16 Mbytes SDRAM */
  FLASH : origin = 0x90000000, len = 0x40000     /* 256 Kbytes */
}

SECTIONS
{
  .vec:      load = 0x00000000 /* Interrupt vectors included */
                              /* by using intr_reset()  */
  .text:     load = IRAM /* Executable code */
  .const:    load = IRAM /* Initialized constants */
  .bss:      load = IRAM /* Global and static variables */
  .data:     load = IRAM /* Data from .asm programs */
  .cinit:    load = IRAM /* Tables for initializing */
                              /* variables and constants */
  .stack:    load = IRAM /* Stack for local variables */
  .far:      load = IRAM /* Global and static variables */
                              /* declared far */
  .sysmem:   load = IRAM /* Used by malloc, etc. (heap) */
  .cio:      load = IRAM /* Used for C I/O functions */
  .csldata   load = IRAM
  .switch    load = IRAM
}
```

Before executing the code that performs your desired signal processing algorithm, the DSK and DSP have to be initialized. This is partially taken care of when you start Code Composer. The version of CCS supplied with the 'C6713 DSK has been configured to automatically load the *general extension language* (GEL) file DSK6713.gel in the directory C:\ti\cc\gel for CCS v2.21 or C:\CCStudio_v3.1 for CCS v3.1. This file defines a memory map, creates some GEL functions for the CCS GEL menu, sets some CPLD registers to configure components on the DSK board, and initializes the EMIF for the memory on the board. Your program must do the remaining initialization.

Copy the file dskstart32.c from the directory C:\c6713dsk to your workspace. A listings of this file is shown below. You will add it to a project file shortly. This file can be used as the starting point for all your programs. It uses functions from the UMD modified Board Support Library (BSL), dsk6713bsl32.lib, in the directory C:\c6713dsk\dsk6713bsl32\lib.

The required header files and source files can be found in the parallel sub-directories, include and sources. You can find detailed documentation for the BSL by starting Code Composer and clicking on Help → Contents → TMS320C6713 DSK → Software → Board Support.

First, dskstart32.c calls DSK6713_init() to initialize the board support library. The source code for this function is in the BSL file dsk6713.c. The function initializes the chip's PLL, configures the EMF based on the DSK version, and sets the CPLD registers to a default state.

Next dskstart32.c initializes the interrupt controller registers and installs the default interrupt service routines by calling the function intr_reset() in the UMD added file intr.c. This function clears GIE and PGIE, disables all interrupts except RESET in IER, clears the flags in the IFR for the the maskable interrupts INT4 through INT15, resets the interrupt multiplexers, initializes the interrupt service table pointer (ISTP), and sets up the Interrupt Service Routine Jump Table. The object modules intr.obj and intr_.obj were added to BSL library so you should not include intr.c and intr_.asm in your project. See Section 2.5.5 and Table 2.7 for a list of the interrupt functions and macros. For complete details, see the source files intr.c and intr.h .

Next the codec is started by calling the function DSK6713_AIC23_openCodec(). This function configures serial port McBSP0 to act as a unidirectional control channel in the SPI mode transmitting 16-bit words. Then it configures the AIC23 stereo codec to operate in the DSP mode with 16-bit data words with a sampling rate of 48 kHz. Then McBSP1 is configured to send data samples to the codec or receive data samples from the codec in the DSP format using 32-bit words. The first word transmitted by the AIC23 is the left channel sample. The right channel sample is transmitted immediately after the left sample. The AIC23 generates a single frame sync at the beginning of the left channel sample. Therefore, a 32-bit word received by McBSP1 contains the left sample in the upper 16 bits and the right sample in the lower 16 bits. The 16-bit samples are in 2's complement format. Words transmitted from McBSP1 to AIC23 must have the same format. The codec and McBSP1 are configured so that the codec generates the frame syncs and shift clocks. See the text at the top of dskstart32.c for more details about the UMD modifications of DSK6713_AIC23_openCodec.c from the TI BSL version which sets McBSP1 to transmit and receive 16-bit words.

Finally, dskstart32.c enters an infinite loop that reads pairs of left and right channel samples from the codec ADC and loops them back out to the codec DAC. This loop should be replaced by the C code to achieve the goals of your experiments.

Program 2.2 dskstart32.c

```
/************************************************************/
/* Function DSK6713_AIC23_openCodec() in dsk6713_opencodec.c */
/* is a modification of the same function in the BSL module  */
/* DSK6713_AIC23_openCodec.c. It configures McBSP1 to trans- */
/* mit and receive 32-bit words rather than 16-bit words by  */
/* changing the RWDLEN1 value to 32BIT, XWDLEN1 to 32BIT,    */
/* RFRLEN1 to OF(0), and XFRLEN1 to OF(0) in structure       */
/* mcbspCfgData in dsk6713_opencodec.c. This causes McBSP1   */
```

```
/* to use a single phase frame consisting of one 32-bit word */
/* per frame. Words are sent to the codec by using the BSL    */
/* function DSK6713_AIC23_write() and read from the codec by */
/* using the function DSK6713_AIC23_read().                  */
/**************************************************************/

#include <stdio.h>
#include <stdlib.h>

#include <dsk6713.h>
#include <dsk6713_aic23.h>
#include <intr.h>

#include <math.h>

/* Codec configuration settings */
/* See dsk6713_aic23.h and the TLV320AIC23 Stereo Audio CODEC Data Manual */
/* for a detailed description of the bits in each of the 10 AIC23 control */
/* registers in the following configuration structure.               */

DSK6713_AIC23_Config config = { \
    0x0017,  /* 0 DSK6713_AIC23_LEFTINVOL  Left line input channel volume */ \
    0x0017,  /* 1 DSK6713_AIC23_RIGHTINVOL Right line input channel volume */\
    0x00d8,  /* 2 DSK6713_AIC23_LEFTHPVOL  Left channel headphone volume */ \
    0x00d8,  /* 3 DSK6713_AIC23_RIGHTHPVOL Right channel headphone volume */ \
    0x0011,  /* 4 DSK6713_AIC23_ANAPATH    Analog audio path control */     \
    0x0000,  /* 5 DSK6713_AIC23_DIGPATH    Digital audio path control */    \
    0x0000,  /* 6 DSK6713_AIC23_POWERDOWN  Power down control */            \
    0x0043,  /* 7 DSK6713_AIC23_DIGIF      Digital audio interface format */ \
    0x0081,  /* 8 DSK6713_AIC23_SAMPLERATE Sample rate control (48 kHz) */   \
    0x0001   /* 9 DSK6713_AIC23_DIGACT     Digital interface activation */   \
};


/**************************************************************/
/* Main program: Replace with your code                     */
/**************************************************************/

void main(void){

  DSK6713_AIC23_CodecHandle hCodec;
  Uint32 sample_pair = 0;

  /* Initialize the interrupt system */
```

```
intr_reset();

/* dsk6713_init() must be called before other BSL functions */
DSK6713_init(); /* In the BSL library */

/* Start the codec */
hCodec = DSK6713_AIC23_openCodec(0, &config);

/* Change the sampling rate to 16 kHz */
DSK6713_AIC23_setFreq(hCodec, DSK6713_AIC23_FREQ_16KHZ);

/* Read left and right channel samples from the ADC and loop */
/* them back out to the DAC.                                  */
for(;;){
  while(!DSK6713_AIC23_read(hCodec, &sample_pair));
  while(!DSK6713_AIC23_write(hCodec, sample_pair));
}
}
```

How Samples are Sent to the Codec

Left and right sample pairs are sent to the codec as 32-bit words with the left channel sample in the upper 16 bits and the right channel sample in the lower 16 bits. Each sample is in 16-bit two's complement format. These 32-bit words are sent to the codec by the BSL function DSK6713_AIC23_write(). This function polls the McBSP1 XRDY flag and returns immediately without sending the sample if it is false and also returns the value 0 (FALSE). It sends the sample word by writing it to the Data Transmit Register (DXR) of McBSP1 if XRDY is 1 (TRUE) and returns the value 1. The C code for this function is shown below.

```
#include <dsk6713.h>
#include <dsk6713_aic23.h>
Int16 DSK6713_AIC23_write(DSK6713_AIC23_CodecHandle hCodec, Uint32 val)
{ /* If McBSP not ready for new data, return false */
   if (!MCBSP_xrdy(DSK6713_AIC23_DATAHANDLE)) {
       return (FALSE);
   }
  /* Write 16 bit data value to DXR */
   MCBSP_write(DSK6713_AIC23_DATAHANDLE, val);
  /* Short delay for McBSP state machine to update */
   asm(" nop");
   asm(" nop");
   asm(" nop");
   asm(" nop");
   asm(" nop");
   asm(" nop");
```

```
    asm(" nop");
    asm(" nop");
    return(TRUE);
}
```

The lower level functions `McBSP_xrdy()` and `MCBSP_write()` are in TI's CSL library. `McBSP_xrdy()` tests the XRDY flag of the McBSP corresponding to the handle and `MCBSP_write()` writes the data word to the data transmit register (DXR) of the McBSP.

How Samples are Read from the Codec

Words are read from the codec by using the function `DSK6713_AIC23_read()`. This function polls the RRDY flag of McBSP1 and returns immediately if it is FALSE without reading a word and also returns the value FALSE. If RRDY is TRUE it reads a word from the Data Receive Register (DRR) of McBSP1 and returns the value TRUE. The source code for this function is shown below.

```
#include <dsk6713.h>
#include <dsk6713_aic23.h>
Int16 DSK6713_AIC23_read(DSK6713_AIC23_CodecHandle hCodec, Uint32 *val)
{/* If no new data available, return false */
    if (!MCBSP_rrdy(DSK6713_AIC23_DATAHANDLE)) {
        return (FALSE);
    }
    /* Read the data */
    *val = MCBSP_read(DSK6713_AIC23_DATAHANDLE);
    return (TRUE);
}
```

`MCBSP_rrdy()` and `MCBSP_read()` are in TI's CSL library. `MCBSP_rrdy()` tests the receive ready (RRDY) flag of the McBSP associated with the handle. `MCBSP_read()` reads a 32-bit word from the Data Receive Register (DRR) of the McBSP. Notice that the word read is typed as an unsigned int.

2.1.2 Properties of the AIC23 Codec

The TMS320C6713 DSK supplies a 12 MHz clock to the AIC23 codec which is divided down internally in the AIC23 to give the sampling rates shown in the table below. The codec can be set to these sampling rates by using the function `DSK6713_AIC23_setFreq(handle,freq ID)` from the BSL library. This function puts the quantity "Value" into AIC23 control register 8.

Some of the AIC23 analog interface properties are

- The ADC for the line inputs has a full-scale range of 1.0 V RMS.

- The microphone input is a high-impedance, low-capacitance input compatible with a wide range of microphones.

Table 2.1: AIC23 Sampling Rates

freq ID	Value	Frequency
DSK6713_AIC23_FREQ_8KHZ	0x06	8000 Hz
DSK6713_AIC23_FREQ_16KHZ	0x2c	16000 Hz
DSK6713_AIC23_FREQ_24KHZ	0x20	24000 Hz
DSK6713_AIC23_FREQ_32KHZ	0x0c	32000 Hz
DSK6713_AIC23_FREQ_44KHZ	0x11	44100 Hz
DSK6713_AIC23_FREQ_48KHZ	0x00	48000 Hz
DSK6713_AIC23_FREQ_96KHZ	0x0e	96000 Hz

- The DAC for the line outputs has a full-scale output voltage range of 1.0 V RMS.

- The stereo headphone outputs are designed to drive 16 or 32-ohm headphones.

- The AIC23 has an analog bypass mode that directly connects the analog line inputs to the analog line outputs.

- The AIC23 has a sidetone insertion mode where the microphone input is routed to the line and headphone outputs.

For complete details on the AIC23 codec including input and output filter frequency responses, see the *TLV320AIC23 Stereo Audio CODEC Data Manual* [I.4].

2.1.3 Creating a CCS Project for dskstart32.c

Now that you have gotten the starting C program and linker command file, it is time to make a project file and build the executable output file. Perform the following tasks:

1. The first time you use Code Composer Studio you need to save your **Workspace** in a place where you have write permission. To do this, start CCS, click on **File**, then **Workspace**, and then **Save Workspace As ...** and give it a valid name in your private workspace.

2. To start a project in CCS, click on **Project**, select **New**, and fill out the boxes as follows:

Project Name:	*give it a name*
Location:	*a directory in your private workspace*
Project type:	Executable (.out)
Target	TMS320C67xx

3. Copy C:\c6713dsk\dskstart32.c to your workspace and add the copied C source file to the project.

4. Next set the build options for Code Composer Studio. Click on **Project** and select **Build Options**. Enter the following options:

```
Compiler -> Basic
  Target Version:        671x (-mv6710)
  Generate Debug Info:   Full Symbolic Debug (-g)
  Opt Speed vs Size:     Speed Most Critical (no ms)
  Opt Level:             None
  Program Level Opt:     None

Compiler -> Preprocessor
  Include Search Path (-i):  .; c:\c6713dsk\dsk6713bsl32\include
  Define Symbols (-d):       CHIP_6713
  Preprocessing:             None

Compiler -> Files
  Asm Directory:  "a directory in your workspace"
  Obj Directory:  "a directory in your workspace"

Linker -> Basic
  Output Filename (-o): dskstart32.out (You can change this.)
  Map Filename (-m):    dskstart32.map (optional)
  Autoinit Model:       Run-time autoinitialization
  Library Search Path:
```

5. Add to the project the linker command file c:\c6713dsk\dsk6713.cmd and the library
   c:\c6713dsk\dsk6713bsl32\lib\dsk6713bsl32.lib

2.1.4   Experiment 2.1: Building and Testing dskstart32.c

The program, dskstart32.c, simply loops the input ADC samples back to the output DAC.
To check that your project builds and runs correctly, do the following:

1. Plug a stereo cable into the DSK Line Input and connect both channels to the same
   signal generator output. The program dskstart32 should set the codec to sample at
   16000 Hz, so set the signal generator to output a sine wave of less than 8000 Hz.

2. Plug a stereo cable into the DSK Line Output. Connect the left and right outputs
   to two different oscilloscope channels. You should use channels 1 and 4 on the HP
   oscilloscopes. NOTE: The right channel is the white plug and the left channel is the
   red plug.

3. Make sure the sampling rate is set to 16000 Hz in dskstart32.c.

4. After your project options have been set, build the executable module by clicking on
   the Rebuild All icon or Project → Rebuild All.

5. Load the program using File → Load Program

6. Start the program running and check that the sine waves appear on the scope. Make sure the input level is small enough so that there is no clipping.

7. Increase the input signal level until the clipping occurs and the output is distorted so you can recognize this problem in future experiments. Then reduce the level again so there is no clipping.

8. Vary the sine wave frequency. What happens when it is more than 8000 Hz? Why?

9. Measure the amplitude response of the system by varying the input frequency and dividing the output amplitude by the input amplitude. Plot the amplitude response. Use enough frequencies to get a smooth curve, particularly in regions where the amplitude response changes quickly. Your amplitude response results will be needed for Chapter 3 experiments.

2.2 More Details on the McBSP Serial Ports and Codecs

You will use the McBSP1 serial port whenever you send a signal sample to or get a sample from the codec. For complete McBSP details see [I.10]. A simplified block diagram of a McBSP is shown in Figure 2.1. The signals shown in the diagram are:

DX/DR	Serial transmit/receive data
FSX/FSR	Transmit/receive frame sync
CLKX/CLKR	Transmit/receive serial shift clock
XINT/RINT	Transmit/receive interrupt to CPU
XEVT/REVT	Transmit/receive interrupt to DMA
CLKS	External clock for Sample Rate Generator

The Sample Rate Generator (SRG) can be used to generate the frame syncs and shift clocks from and internal or external clock. The SRG is not used with the codecs because the codecs supply the frame syncs and shift clocks.

The Events/Interrupts block can send interrupt requests to the CPU and event notifications to the EDMA when words are received or transmitted by the serial port.

2.2.1 Basic McBSP Transmitter and Receiver Operation

A more detailed diagram of the McBSP transmitter is shown in Figure 2.2. The transmitter operates as follows:

- The CPU or DMA writes a 32-bit word in parallel into the Data Transmit Register (DXR) which is a 32-bit memory-mapped register. The XRDY flag is cleared whenever data is written to the DXR.

- When the transmit frame lynch (FSX) goes high, a word of the configured number of bits is serially shifted out of the Transmit Shift Register (XSR). After a word is shifted out of the XSR, a parallel transfer of the DXR into the XSR is performed. The XRDY

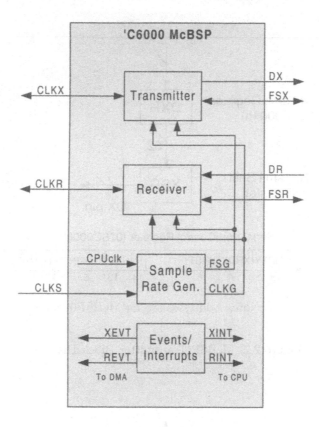

Figure 2.1: McBSP Block Diagram

flag is set when the transfer occurs. The CPU can test the XRDY flag to see if the DXR is empty and another word can be written to it. McBSP1 should be configured to transmit 32-bit words to the codec onboard the DSK.

- The serial port transmitter sends an interrupt request (XINT) to the CPU when the XRDY flag makes a transition from 0 to 1 if XINTM = 00b in the Serial Port Control Register (SPCR). It also sends a Transmit Event Notice (XEVT) to the DMA.

A McBSP receiver block diagram is shown in Figure 2.3. The receiver operation is essentially the reverse of the transmitter operation. It works as follows:

- When the receive frame synch (FSR) goes high, the received bits are shifted serially into the Receive Shift Register (RSR).

- When an element with the configured number of bits is received, the 32-bit RSR is transferred in parallel to the Receive Buffer Register (RBR) if it is empty.

- The RBR is then copied to the Data Receive Register (DRR) if it is empty. It is up to the programmer to select the element with the configured number of bits from the 32-bit DRR word. McBSP1 should be configured to receive 32-bit elements.

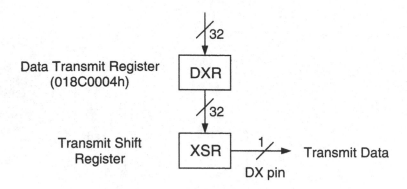

Note: Addresses are for McBSP0

Figure 2.2: McBSP Transmitter Block Diagram

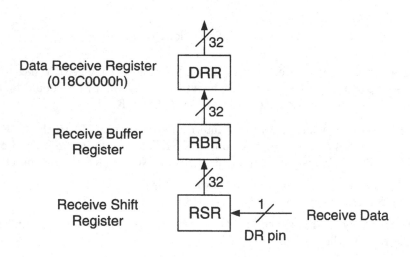

Note: Addresses are for McBSP0

Figure 2.3: McBSP Receiver Block Diagram

- The RRDY bit in the SPCR is set to 1 when the RBR is moved to the DRR, and it is cleared when the DRR is read.

- When RRDY transitions from 0 to 1, the McBSP generates a CPU interrupt request (RINT) if RINTM = 00b in the SPCR. A receive event (REVT) is also sent to the EDMA controller.

The TI chip support library (CSL) [I.6] contains a module with C functions and macros that supports every feature of the McBSP serial ports.

2.2.2   Example C Code for Reading from and Writing to the Codec

The following C code segments show one way of interfacing with the AIC23 stereo codec. It is assumed that earlier in the program the McBSP1 transmitter and receiver are configured for a one-phase frame containing one 32-bit word. The most significant 16-bits of each 32-bit word are the left channel sample and the least significant 16-bits are the right channel sample each in 2's complement format.

The following code segment shows how to get a left and right channel sample from the codec ADC. The sampling frequency is controlled by the frame syncs generated by the codec. The program sits in a `while` loop polling the RRDY flag until a sample arrives using the BSL function `DSK6713_AIC23_read(hCodec,&sample_pair)` described on page 35. When the RRDY flag becomes true, the word in the DRR is read, stored in `sample_pair`, and the RRDY flag is cleared. The right and left channel samples are then extracted from `sample_pair`. First `sample_pair` is cast into a signed int. The result is arithmetically shifted right 16 bits, so the right channel sample falls off the right end of the word leaving the left channel sample in the lower 16 bits with its sign extended through the upper 16 bits. The result is converted to a float when it is put in the float, `left`. The right channel sample is extracted by again casting `sample_pair` into a signed int, shifting it left 16 bits to knock off the 16 left channel bits and put the right channel sample sign bit in bit 31 which is the 32-bit integer sign bit, and then arithmetically shifting the result right 16 bits to extend the sign through the upper 16 bits and leave the right channel sample in the lower 16 bits. The result is converted to a float by setting it equal to the float, `right`.

Program 2.3 Example C Code for Stereo Read

```
DSK6713_AIC23_CodecHandle hCodec;
Uint32 sample_pair = 0;
float left, right;

    :

/* Poll RRDY. When TRUE, read DRR */
    while(!DSK6713_AIC23_read(hCodec,&sample_pair));

/* Extract left channel sample, sign extend, convert to float */
```

```
left  = ( (int) sample_pair) >> 16;
```

```
/* Extract right channel sample, sign extend, convert to float */
   right =( (int) sample_pair) << 16 >> 16;
```

Sending a pair of sample to the DAC involves the opposite operations. The following program segment assumes the samples are initially in floating-point format. The first step is to convert them to the integer words `ileft` and `iright`. Care must be taken so that the samples fit in 16-bit integers. Next the left sample is or-ed into the upper 16-bits of the output 32-bit integer word, `sample`, and the right channel sample into the lower 16 bits. Notice that the upper 16 bits of the right channel have been masked to 0. Then the program sits in a `while` loop until the transmit ready bit (XRDY) in the SPCR becomes true. When XRDY becomes true, the Data Transmit Register (DXR) has become empty and the new pair of samples is written to it. These last steps are performed by the function `DSK6713_AIC23_write(hCodec, sample)` described on page 34.

Program 2.4 Example C Code for Stereo Write

```
DSK6713_AIC23_CodecHandle hCodec;
float left,  right;
int   ileft, iright, sample;

       ⋮

/* Convert left and right values to integers */
   ileft  = (int) left;
   iright = (int) right;

/* Combine L/R samples into a 32-bit word */
   sample = (ileft<<16)|(iright & 0x0000FFFF);

/* Poll XRDY bit until true, then write to DXR*/
   while(!DSK6713_AIC23_write(hCodec, sample));
```

The method of sending or receiving a word by waiting for the XRDY or RRDY bits to become true is called *polling*.

2.3   The 'C6000 Timers

The TMS320C6713 has two 32-bit general purpose timers. They can be used to time events, count events, generate pulses, interrupt the CPU, and send synchronization events to the EDMA. The timers can be clocked by an internal or external signal. They have an input pin (TINP) and an output pin (TOUT) which can also be configured as general purpose I/O pins. See [I.10] for all the details about the timers.

The internal clock frequency for the 'C6713 timers is the CPU clock frequency divided by four. Therefore, the DSK timer clock frequency is $225Mhz/4 = 56.25$ MHz. You will use the external clock output as an external input to the McBSP1 serial port Sample Rate Generator (SRG) to create a bit clock for a bit-error rate tester in the RS232 experiments of Chapter 10.

When the external input pin, TINP, is selected as the timer clock source, the timer can count input pulses and interrupt the CPU when a preset number is reached.

Each timer has the three registers shown in Table 2.2. The CSL library [I.6] has C functions and macros for configuring and interfacing with the timers.

When the 'C6713 timers are driven by the internal clock source, the frequency of the signal on the TOUT pin is

$$f_{\text{out}} = \frac{\text{CPU clock frequency}}{N \times \text{Period Register value}} \tag{2.1}$$

where $N = 4$ for pulse mode and $N = 8$ for clock mode.

Table 2.2: Timer Registers

Name and Abbreviation	Description
Timer Control (CTL)	Sets the operating mode of the timer, monitors the timer status, and controls the function of the TOUT pin.
Timer Period (PRD)	Contains the number of input clock cycles to count or determines the period of the TOUT signal
Timer Counter (CNT)	A 32-bit register holding the current value of the incrementing counter.

2.4   Generating a Sine Wave by Polling XRDY

In the next three sections, you will use the simple task of generating a sine wave by three different methods to learn about various aspects of the DSP and DSK. In this section you will use the polling method described briefly above. The goal is to generate the continuous-time sine wave

$$s(t) = \sin 2\pi f_0 t$$

with frequency f_0 at the codec's line output jack. Let $f_s = 1/T$ be the desired sampling rate where T is the sampling period. Then the required signal samples are

$$s(nT) = \sin 2\pi f_0 nT = \sin 2\pi \frac{f_0}{f_s} n = \sin n\Delta \tag{2.2}$$

where $\Delta = 2\pi f_0/f_s$ is the change in the angle, $\theta(t) = 2\pi f_0 t$, between successive samples. The required sequence of angles can be generated recursively. Let

$$\theta[n] = n\Delta \tag{2.3}$$

Then
$$\theta[n+1] = (n+1)\Delta = n\Delta + \Delta = \theta[n] + \Delta \qquad (2.4)$$

and the desired sine wave samples are $\sin \theta[n]$. This algorithm will be used to generate sinusoidal carriers for modulators and demodulators in some of the following chapters. In some applications it is efficient to store an array of samples of one cycle of a sine wave taken at a sequence of angles sufficiently close together to meet the accuracy requirements of the application and use $\theta[n]$ as an index into the array to get the sine values. In this section, the C `sin()` function will be used to generate the values.

A sample program segment for generating a sine wave by polling is shown in Program Program 2.5. The initialization code is not shown. During initialization, the codec must be configured for the desired sampling rate. Two important items should be noted in the program:

1. The codec uses 16-bit two's complement integers which are in the range $\pm 2^{15} = \pm 32,768$. Also $|\sin x| \leq 1$ so if it is converted to an integer, it becomes zero almost everywhere. Therefore, `sin(angle)` which is a float is multiplied by the float 15,000.0 before it is converted to an integer to use a large portion of the dynamic range of the DAC.

2. If the angle θ is continually incremented by Δ it will grow without bound and eventually overflow. Also, some C implementations of $\sin x$ do not behave well for large x. Therefore, the program checks to see if θ becomes larger than 2π and subtracts 2π when it does. This works because $\sin(x) = \sin(x - 2\pi)$.

Program 2.5 Sample Program Segment for Polling

```
#include <math.h>
#define  pi 3.141592653589
int sample = 0;
float fs = 16000.;
float f0 = 1000.;
float delta = 2.*pi*f0/fs;
float twopi = 2.0*pi;
float angle = 0;
float left;

for(;;){                      /*  Infinite loop        */
   left = 15000.0*sin(angle);  /* Scale for DAC    */
   sample = ((int) left) <<16; /* Put in top half */

/* Poll XRDY bit until true, then write to DXR     */
   while(!DSK6713_AIC23_write(hCodec, sample));

   angle += delta;   /* Increment sine wave angle */
```

```
    if( angle >= twopi) angle -= twopi; /* Keep angle from overflowing */
}
```

2.4.1   Experiment 2.2: Instructions for the Polling Experiment

First, note the following important information:

- Remember to include `math.h` in your C program.

- The DSK has stereo **LINE IN** and **LINE OUT** jacks. The lab has cables to convert from the DSK stereo plug to an RCA mono connector for the left channel and an RCA mono connector for the right channel. The RCA connectors are plugged into RCA to BNC adapters so they can be connected to the oscillators and oscilloscopes.

- Cable Color Scheme

  - **Left Channel:** Red plug
  - **Right Channel:** White plug

For the polling experiment do the following:

1. Set the sampling rate to 16 kHz.

2. Generate a 1 kHz sine wave on the left channel and a 2 kHz sine wave on the right channel. Remember that $|\sin(x)| \leq 1$ and that floats less than 1 become 0 when converted to ints. Therefore, scale your floating point sine wave samples to make them greater than 1 and fill reasonable part of the DAC dynamic range before converting them to ints.

3. Combine the left and right channel integer samples into a 32-bit integer and write the resulting word to the McBSP1 DXR using polling of the XRDY flag.

4. Observe the left and right channel outputs on two oscilloscope channels.

5. Verify that the sine wave frequencies observed on the scope are the desired values by measuring their periods.

6. Use the signal generator to measure the frequencies. The HP oscilloscopes also can measure the frequencies.

7. When you have verified that your program is working, change the left channel frequency to 15 kHz and the right channel frequency to 14 kHz. Measure the DAC output frequencies. Explain your results by mathematical analysis. That is, give equations to show why you got what you did. (Hint: Look up "aliasing" in any reference on digital signal processing.)

2.5   Generating a Sine Wave Using Interrupts

Nearly all the time in the polling method is spent sitting in a loop waiting for the XRDY flag to get set and the DSP is doing almost nothing. In a real-world application the DSP would be performing many other tasks like running a CELP speech codec, modulating and demodulating data for a modem, or decoding a turbo code. A much more efficient approach is to let the DSP perform these desired tasks *in the background* and have the serial port interrupt the background tasks when it needs a sample to transmit or a sample has been received. The *interrupt service routine* is called a *foreground task*. See the *TMS320C6000 CPU and Instruction Set Reference Guide* [I.7] and *TMS320C6713B Floating-Point Digital Signal Processor* [I.12] for complete details on the interrupt controller.

2.5.1   The CPU Interrupt Priorities and Sources

TMS320C6713 DSP has a *vectored priority interrupt controller* that handles 16 different CPU interrupts. The highest priority interrupt is RESET which cannot be masked. The next priority interrupt is the Non-Maskable Interrupt (NMI) which is used to alert the DSP of a serious hardware problem. There are two reserved interrupts and 12 additional maskable CPU interrupts. The peripherals, such as, the timers, McBSP and McASP serial ports, EDMA controller, plus external interrupt pins sourced from the GPIO module present a set of many interrupt sources. The 16 CPU interrupts and their default sources are shown in Table 2.3. INT_00 has the highest priority and INT_15 the lowest.

The interrupt system includes a multiplexer to select the CPU interrupt sources and map them to the 12 maskable prioritized CPU interrupts. The complete list of C6713 interrupt sources is shown in the Table 2.4 along with the required Interrupt Selector values. The GPIO module can select external pins as interrupt sources. The mapping is shown in Table 2.5.

2.5.2   Interrupt Control Registers

The interrupt control registers and their purposes are shown in Table 2.6. Complete details about the layout of the registers are presented in the *TMS320C6000 CPU and Instruction Set Reference Guide* [I.7]. The CSL library provides functions for interfacing with these registers. The University of Maryland modified dsk6713bs132 library also contains a variety of macros and functions for interfacing with these registers and setting up an interrupt system with an interrupt service routine jump table that allows interrupt service routines to be hooked to or unhooked from CPU interrupts in a C program. A list of these functions is shown in Table 2.7.

The Control Status Register (CSR) bit 0 is the Global Interrupt Enable (GIE) bit. All maskable interrupts are disabled If GIE = 0, and maskable interrupts can be enabled if GIE = 1. Bit 1 is the Previous GIE (PGIE) which saves the value of GIE when an interrupt is taken.

Bits 0 through 15 of the Interrupt Enable Register (IER) correspond to the 16 CPU interrupts. An interrupt is enabled by setting its bit to 1 and disabled by clearing its bit

Table 2.3: Default CPU Interrupt Sources

CPU INTERRUPT NUMBER	BYTE OFFSET IN IST	INTERRUPT SELECTOR CONTROL REGISTER	DEFAULT SELECTOR VALUE (BINARY)	DEFAULT INTERRUPT EVENT
INT_00	000h	-	-	RESET
INT_01	020h	-	-	NMI
INT_02	040h	-	-	Reserved
INT_03	060h	-	-	Reserved
INT_04	080h	MUXL[4:0]	00100	GPINT4
INT_05	0A0h	MUXL[9:5]	00101	GPINT5
INT_06	0C0h	MUXL[14:10]	00110	GPINT6
INT_07	0E0h	MUXL[20:16]	00111	GPINT7
INT_08	100h	MUXL[25:21]	01000	EDMAINT
INT_09	120h	MUXL30:26]	01001	EMUDTDMA
INT_10	140h	MUXH[4:0]	00011	SDINT
INT_11	160h	MUXH[9:5]	01010	EMURTDXRX
INT_12	180h	MUXH[14:10]	01011	EMURTDXTX
INT_13	1A0h	MUXH[20:16]	00000	DSPINT
INT_14	1C0h	MUXH[25:21]	00001	TINT0
INT_15	1E0h	MUXH[30:26]	00010	TINT1

to 0. Bit 1 is the Nonmaskable Interrupt Enable (NMIE) bit. NMIE must be set, that is, NMIE = 1, for the maskable interrupts 4 through 15 to get serviced. If NMIE = 0, none of the maskable interrupts get serviced.

When an interrupt occurs, the corresponding bit gets set in the Interrupt Flags Register (IFR) . This happens whether or not the interrupt is enabled and allows pending interrupts to be serviced at a later time if they are not currently enabled. Interrupts are serviced in the order of their priority with a lower number interrupt having higher priority.

A bit in the IFR can be manually set by writing a 1 to the corresponding bit in the Interrupt Set Register (ISR).. A bit in the IFR can be cleared by writing a 1 to the corresponding bit in the Interrupt Clear Register (ICR) .

When an interrupt is serviced, the program jumps to the interrupt service routine. It must know where to return to after the ISR is completed. The Interrupt Return Pointer (IRP) saves the return address. The Nonmaskable Interrupt Return Pointer (NRP) serves the same purpose for nonmaskable interrupts.

The Interrupt Service Table Pointer (ISTP) holds the address of the Interrupt Service table. When the DSP is powered up, the default address is initialized to be address 0 at the start of memory.

Table 2.4: Interrupt Sources

INTERRUPT SELECTOR VALUE (BINARY)	INTERRUPT EVENT	MODULE
00000	DSPINT	HPI
00001	TINT0	Timer 0
00010	TINT1	Timer 1
00011	SDINT	EMIF
00100	GPINT4	GPIO
00101	GPINT5	GPIO
00110	GPINT6	GPIO
00111	GPINT7	GPIO
01000	EDMAINT	EDMA
01001	EMUDTDMA	Emulation
01010	EMURTDXRX	Emulation
01011	EMURTDXTX	Emulation
01100	XINT0	McBSP0
01101	RINT0	McBSP0
01110	XINT1	McBSP1
01111	RINT1	McBSP1
10000	GPINT0	GPIO
10001	Reserved	-
10010	Reserved	-
10011	Reserved	-
10100	Reserved	-
10101	Reserved	-
10110	I2CINT0	I2C0
10111	I2CINT1	I2C1
11000	Reserved	-
11001	Reserved	-
11010	Reserved	-
11011	Reserved	-
11100	AXINT0	McASP0
11101	ARINT0	McASP0
11110	AXINT1	McASP1
11111	ARINT1	McASP1

Table 2.5: External Interrupt Sources

PIN NAME	INTERRUPT EVENT	MODULE
GP[15]	GPINT0	GPIO
GP[14]	GPINT0	GPIO
GP[13]	GPINT0	GPIO
GP[12]	GPINT0	GPIO
GP[11]	GPINT0	GPIO
GP[10]	GPINT0	GPIO
GP[9]	GPINT0	GPIO
GP[8]	GPINT0	GPIO
GP[7]	GPINT0 or GPINT7	GPIO
GP[6]	GPINT0 or GPINT6	GPIO
GP[5]	GPINT0 or GPINT5	GPIO
GP[4]	GPINT0 or GPINT4	GPIO
GP[3]	GPINT0	GPIO
GP[2]	GPINT0	GPIO
GP[1]	GPINT0	GPIO
GP[0]	GPINT0	GPIO

Table 2.6: Interrupt Control Registers

	Name	Description
CSR	Control Status Register	Globally set or disable interrupts
IER	Interrupt Enable Register	Enable interrupts. Bit n corresponds to INT_n
IFR	Interrupt Flags Register	Shows status of interrupts. Bit n corresponds to INT_n
ISR	Interrupt Set Register	Manually set flags in IFR
ICR	Interrupt Clear Register	Manually clear flags in IFR
ISTP	Interrupt Service Table Pointer	Pointer to the beginning of the interrupt service table
NRP	Nonmaskable Interrupt Return Pointer	Return address used on return from a nonmaskable interrupt
IRP	Interrupt Return Pointer	Return address used on return from a maskable interrupt

2.5.3 What Happens When an Interrupt Occurs

In order for a maskable interrupt to occur, the following conditions must be true:

- The *global interrupt enable bit*(GIE) which is bit 0 in the control status register (CSR) is set to 1. When GIE = 0, no maskable interrupt can occur.

- The nonmaskable interrupt enable bit (NMIE) in the *interrupt enable register* (IER) is set to 1. No maskable interrupt can occur if NMIE = 0.

- The bit corresponding to the desired interrupt is set to 1 in the IER.

- The desired interrupt occurs, which sets the corresponding bit in the *interrupt flags register* (IFR) to 1 and no higher priority interrupt flags are 1 in the IFR

When CPU interrupt n occurs, program execution jumps to byte offset $4 \times 8 \times n = 32n$ in an *interrupt service table* (IST). The IST contains 16 *interrupt service fetch packets* (ISFP), each consisting of eight 32-bit instruction words. An ISFP may contain an entire interrupt service routine or may branch to a larger service routine. An example of an ISFP for RESET for C programs is shown below. The C compiler generates the C initialization code starting at address _c_int00, so the RESET ISFP simply branches to this address.

```
_RESET:  mvk       _c_int00,b0 ; load lower 16 bits of _c_int00
         mvkh      _c_int00,b0 ; load upper 16 bits of _c_int00
         b     .s2 b0          ; branch to C initialization
         nop       5           ; do 5 NOP's for branch latency 5
         nop                   ; add four words to fetch packet
         nop                   ;   to make a total of 8 words
         nop
         nop
```

We will normally start the interrupt service table (IST) at location 0. It can be relocated and the Interrupt Service Table Pointer register (ISTP) points to its starting address which must be a multiple of 256 words. The organization of the IST is shown in Table 2.3.

The DSP takes the following actions when an interrupt occurs:

- The corresponding flag in the interrupt flags register (IFR) is set to 1.

- If GIE = NMIE = 1 and no higher priority interrupts are pending, the interrupt is serviced:

  - GIE is copied to PGIE (previous global interrupt enable bit) and GIE is cleared to preclude other interrupts. GIE can be manually set to allow the interrupt service routine to be interrupted itself.

  - The flag bit in the IFR is cleared to show that the interrupt has been serviced.

  - The return address is put in the *interrupt return pointer* (IRP).

— Execution jumps to the corresponding fetch packet in the interrupt service table (IST).

— The service routine must save the CPU state on entry and restore it on exit.

— A return from a maskable interrupt is accomplished by the assembly instructions

```
B   IRP;  return, moves PGIE to GIE
NOP 5  ;  delay slots for branch
```

2.5.4   TI Extensions to Standard C Interrupt Service Routines

The Texas Instruments compiler includes extensions to standard C for interrupt service routines. To make the compiler use these extensions when writing an interrupt service routine in C, declare the function to be an ISR by using the `interrupt` keyword. For example, declare `your_isr_name()` to be an ISR by the statement

```
interrupt void your_isr_name(){...}
```

where the dots within the braces represent your source code. You can also use the *interrupt pragma* as shown on the following line:

```
#pragma INTERRUPT(your_isr_name)
```

The C compiler will then automatically generate code to:

1. Save the CPU registers used by the ISR on the stack. If the ISR calls another function, all registers are saved.

2. Restore the registers before returning with a `B IRP` instruction.

You cannot pass parameters to, or return values from an interrupt service routine[1]

2.5.5   Using the dsk6713bsl32 Library for Interrupts

A list of C functions and macros for interfacing with the interrupt system is shown in Table 2.7. To write and build programs using the TI C interrupt extensions and the `dsk6713bsl32` interrupt functions:

- Add the linker command file `C:\c6713dsk\dsk6713.cmd` to your project.

- Include the header file `C:\c6713dsk\dsk6713bsl32\include\intr.h` in your program. You should set the "Include Search Path" in your project, so it is only necessary to use the line "`include intr.h`" in your C program.

- Be sure to add the library `dsk6713bsl32.lib` to your project.

The interrupt service table will be generated in a section called `.vec`. The sample beginning linker command file `dsk6713.cmd` loads the `.vec` section starting at absolute address 0.

---

[1]When using the Chip Support Library (CSL) and DSP/BIOS the conventions for ISR's are different. The `interrupt` keyword should not be used and the ISR's can pass arguments. See [I.8] for details.

Table 2.7: Interrupt Functions Provided by `intr.h` and `intr.c` in `dsk6713bsl32`

MACRO FUNCTIONS

`INTR_GLOBAL_ENABLE()`	Sets GIE bit in CSR
`INTR_GLOBAL_DISABLE()`	Clears GIE bit in CSR
`INTR_ENABLE(bit)`	Sets bit in IER
`INTR_DISABLE(bit)`	Clears bit in IER
`INTR_CHECK_FLAG(bit)`	Returns value of bit in IFR
`INTR_SET_FLAG(bit)`	Set interrupt by writing to ISR bit
`INTR_CLR_FLAG(bit)`	Clears int. flag by writing 1 to ICR bit
`INTR_SET_MAP(cpu_intr, src,sel)`	Map interrupt source to CPU interrupt
`INTR_GET_ISN(cpu_intr,sel))`	Get ISN of selected interrupt
`INTR_MAP_RESET()`	Reset interrupt multiplexer map to defaults
`INTR_EXT_POLARITY(bit,val)`	Assign external interrupt's polarity

FUNCTIONS

`intr_hook(*fp,cpu_intr)`	Place function pointer into ISR jump table at location for `cpu_intr`
`intr_init()`	Initialize ISTP with base address of the interrupt service table (IST). Using this function causes default interrupt service fetch packets (ISFP) to be loaded into the Interrupt Service Table (IST) and a default interrupt service routine jump table to be created.
`intr_map(cpu_intr, isn)`	Maps interrupt source `isn` to the `cpu_intr`
`intr_isn(cpu_intr)`	Returns interrupt source number for CPU int.
`intr_reset()`	Reset interrupt registers to default values
`intr_get_cpu_intr(isn)`	Return CPU interrupt assigned to ISN. If isn not mapped, return -1

Installing a C Interrupt Service Routine

The following list shows the steps that should be taken to "install" a C interrupt service routine for CPU interrupt 15 as an example.

1. Use `intr_reset()` to set the interrupt control registers to their default values. It initializes the ISTP and causes the interrupt service table and interrupt service routine jump table to be installed.

2. Map the interrupt source number to a CPU interrupt number.
   `intr_map(CPU_INT15, ISN_XINT0);`

3. Clear the interrupt flag to make sure none is pending.
   `INTR_CLR_FLAG(CPU_INT15);`

4. Hook the ISR to the CPU interrupt. Let the ISR be `my_isr()`.
   `intr_hook(my_isr, CPU_INT15);`

5. Enable the NMI interrupt.
   `INTR_ENABLE(CPU_INT_NMI);`

6. Enable the CPU interrupt in the IE register.
   `INTR_ENABLE(CPU_INT15);`

7. Set the GIE bit in the CSR.
   `INTR_GLOBAL_ENABLE();`

An example of using these functions is shown in Program 2.6 on page 54.

2.5.6   Experiment 2.3: Generating Sine Waves by Using Interrupts

Repeat the instructions for generating a sine wave by polling in Section 2.4.1 but now use a C interrupt service routine to generate the sine wave samples and write them to the McBSP1 data transmit register (DXR1). No polling of the XRDY1 flag is needed because samples are transmitted only when interrupts occur at the codec's sampling rate and cause execution to jump into your interrupt service routine.

The `main()` function should:

- initialize McBSP0, McBSP1, and the codec with a 16 kHz sampling rate,

- map CPU INT15 to McBSP1 XINT1,
  Note: The choice of INT15 was arbitrary. Any of INT4 – INT15 can be used.

- hook CPU INT15 to your ISR,

- enable interrupts,

- and go into an infinite interruptible loop.

A partial C code segment for accomplishing these tasks is shown in the following program listing. An important point is to notice that `angle_left` and `angle_right` in the ISR must retain their values between ISR calls. This is done in the ISR in the program segment by declaring them to be *static*. They could also have been made global variables. Also, you should not use the polling McBSP write function since the program jumps to the ISR when the XRDY1 flag goes from FALSE to TRUE. Use the CSL non-polling write function `MCBSP_write()` as shown in the program segment.

Program 2.6 Sample C Code for Generating a Sine Wave by Interrupts

```
#include <stdio.h>
#include <stdlib.h>

#include <dsk6713.h>
#include <dsk6713_aic23.h>

#include <intr.h>

        ⋮

#define   sampling_rate 16000.
#define   freq_left   1000.
#define   freq_right  2000.
#define   scale   10000.0
#define   PI 3.141592653589

float twopi = 2.*PI;

/* phase increment left for sine wave */
float delta_left = 2.0*PI*freq_left/sampling_rate;

/* phase increment for right sine wave */
float delta_right = 2.0*PI*freq_right/sampling_rate;

interrupt void tx_isr(void); /* prototype the ISR */

void main(void){
DSK6713_AIC23_CodecHandle hCodec;

/**********************************************************************/
/* Initialize interrupt system with intr_reset()                    */
/*                                                                  */
/*   The default interrupt service routines are set up by calling the */
/* function intr_reset() in the UMD added file intr.c. This clears    */
/* GIE and PGIE, disables all interrupts except RESET in IER, clears  */
```

```
/* the flags in the IFR for the the maskable interrupts INT4 - INT15, */
/* resets the interrupt multiplexers, initializes the interrupt       */
/* service table pointer (ISTP), and causes the Interrupt Service     */
/* Table and Interrupt Service Routine Jump Table to be loaded.       */
/*********************************************************************/
  intr_reset();

  /* dsk6713_init() must be called before other BSL functions */
  DSK6713_init(); /* In the BSL library */

  /* Start the codec. McBSP1 uses 32-bit words */
  hCodec = DSK6713_AIC23_openCodec(0, &config);

  /* Change the sampling rate to 16 kHz */
  DSK6713_AIC23_setFreq(hCodec, DSK6713_AIC23_FREQ_16KHZ);

  /* Select McBSP1 transmit int for INT15    */
  intr_map(CPU_INT15, ISN_XINT1);

  /* Hook our ISR to INT15                    */
  intr_hook(tx_isr, CPU_INT15);

  /* Clear old interrupts                     */
  INTR_CLR_FLAG(CPU_INT15);

/* Enable interrupts                          */
  /* NMI must be enabled for other ints to occur */
  INTR_ENABLE(CPU_INT_NMI);

  /* Set INT15 bit in IER                     */
  INTR_ENABLE(CPU_INT15);

  /* Turn on enabled ints                     */
  INTR_GLOBAL_ENABLE();

/* Write a word to start transmission using CSL function */
  MCBSP_write(DSK6713_AIC23_DATAHANDLE, 0);

  for (;;); /* infinite interruptible loop */
}

interrupt void tx_isr(void){
    float x_left, x_right;
/*********************************************************************/
```

```
/* Note: angle_left and angle_right must retain their values between */
/* ISR calls. Do this by making them static as below or global.      */
/*******************************************************************/
    static float angle_left=0.;
    static float angle_right=0.;
    int output, ileft, iright;

/* Put your code here to do the following:
  1. Generate scaled left and right channel sine samples. Convert them
     to integers and combine them into a 32-bit output word.
  2. Increment phase angles of sines modulo 2*pi.
  3. There is no need to poll XRDY1 since its transition from false to
     true causes a jump to this ISR.  DSK6713_AIC23_DATAHANDLE is
     declared as a global variable in DSK6713_aic23_opencodec.c.  Just
     write the output sample to McBSP1 by the CSL library function
     MCBSP_write() as shown below. */

    MCBSP_write(DSK6713_AIC23_DATAHANDLE, output);
}
```

2.6   Generating a Sine Wave with the EDMA and a Table

In the polling and interrupt experiments, the sine wave sample values were numerically computed using the $\sin(\cdot)$ function and this consumes CPU cycles. In many real-time applications, a set of tasks is repeated periodically so the DSP has a limited amount of time to perform each task and computational efficiency is critical. An efficient method for generating a sine wave carrier signal in communication systems is to read the sample values out of a precomputed table. In this experiment, you will learn how to read the samples from a table and load them into the McBSP1 Data Transmit Register (DXR1) for transmission to the codec by using the Enhanced Direct Memory Access (EDMA) controller. The EDMA controller is another important internal TMS320C6713 peripheral. The EDMA controller handles all data transfers between the L2 cache/memory controller and peripherals in the 'C621x/'C671x/'C64x family of DSP's independently of the CPU operations. Earlier 'C6x series DSP's like the 'C6201 and 'C6701 have a simpler peripheral called the DMA controller. The architecture of the EDMA controller is quite different than that of the DMA controller. Enhancements of the EDMA include 16 channels for the 'C6713 with programmable priority, and the ability to link and chain data transfers.

2.6.1   EDMA Overview

The Enhanced Direct Memory Access Controller (EDMA) handles all data transfers between the L2 cache/memory controller and the peripherals. These include cache servicing, non-

cacheable memory access, user-programmed data transfers, and host access. The EDMA can move data to and from any addressable memory spaces including internal memory (L2 SRAM), peripherals, and external memory. The EDMA is quite complex and we will only touch on its operation. See the *TMS320C6000 Peripherals Reference Guide* [I.10, Chapter 6] and *TMS320C6713B Floating-Point Digital Signal Processor* [I.12] for complete details.
Some of the EDMA features are:

- The EDMA controller includes event and interrupt processing registers, an event encoder, a parameter RAM, and address generation hardware.

- The EDMA has 16 independent channels and they can be assigned priorities.

- Data transfers can be initiated by the CPU or events.

- When an event occurs, its transfer parameters are read from the Parameter RAM (PaRAM). These parameters are sent to address generation hardware.

- The EDMA can transfer elements that are 8-bit bytes, 16-bit halfwords, or 32-bit words.

- Very sophisticated block transfers can be programmed. The EDMA can transfer 1-dimensional and 2-dimensional data blocks consisting of multiple frames. See [I.10, Section 6.8] for details.

- After an element transfer, source and/or destination element addresses can stay the same, be incremented or decremented by one element, or incremented or decremented by the value in the index register ELEIDX for the channel. Arrays are offset by FRMIDX for the channel.

- After a programmed transfer is completed, the EDMA can continue data transfers by linking to another transfer programmed in the Parameter RAM for the channel or by chaining to a transfer for another channel.

- The EDMA can generate transfer completion interrupts to the CPU along with a programable transfer complete code. The CPU can then take some desired action based on the transfer complete code.

- The EDMA has a quick DMA mode (QDMA) that can be used for quick, one-time transfers.

2.6.2   EDMA Event Selection

The 'C6713 EDMA supports up to 16 EDMA channels. Channels 8 through 11 are reserved for chaining, leaving 12 channels to service peripheral devices. Data transfers can be initiated by the CPU or *events*. The default mapping of events to channels is shown in Table 2.8. The user can change the mapping of events to channels. The EDMA selector registers ESEL0, ESEL1, and ESEL2 control this mapping. Table 2.9 shows the events and selector codes.

Table 2.8: TMS320C6713 Default EDMA Events

EDMA CHAN.	EDMA SELECTOR CONTROL REGISTER	DEFAULT SELECTOR CODE (BINARY)	DEFAULT EDMA EVENT
0	ESEL0[5:0]	000000	DSPINT
1	ESEL0[13:8]	000001	TINT0
2	ESEL0[21:16]	000010	TINT1
3	ESEL0[29:24]	000011	SDINT
4	ESEL1[5:0]	000100	GPINT4
5	ESEL1[13:8]	000101	GPINT5
6	ESEL1[21:16]	000110	GPINT6
7	ESEL1[29:24]	000111	GPINT7
8	-	-	TCC8 (Chaining)
9	-	-	TCC9 (Chaining)
10	-	-	TCC10 (Chaining)
11	-	-	TCC11 (Chaining)
12	ESEL3[5:0]	001100	XEVT0
13	ESEL3[13:8]	001101	REVT0
14	ESEL3[21:16]	001110	XEVT1
15	ESEL3[29:24]	001111	REVT1

2.6.3   Registers for Event Processing

The EDMA controller contains four registers for event processing. They are described in the following list.

Event Register (ER)

   When event n occurs, bit n is set in the ER.

Event Enable Register (EER)

   Setting bit n of the EER enables processing of that event. Clearing bit n to 0 disables processing of event n. The occurrence of event n is latched in the ER even if it is disabled.

Event Clear Register (ECR)

   If an event is enabled in the EER and gets posted in the ER, the ER bit is automatically cleared when the EDMA processes the transfer for the event. If the event is disabled, the CPU can clear the event flag bit in the ER by writing a 1 to the corresponding bit in the ECR. Writing a 0 has no effect.

Event Set Register (ESR)

   Writing a 1 to a bit in the ESR causes the corresponding bit in the event register (ER) to get set. This allows the CPU to submit event requests and can be used as a good debugging tool.

Table 2.9: EDMA Event Selection

EDMA SELECTOR CODE (binary)	EDMA EVENT	MODULE	EDMA SELECTOR CODE (binary)	EDMA EVENT	MODULE
000000	DSPINT	HPI	100100	AREVTO	McASP0
000001	TINT0	Timer 0	100101	AREVT0	McASP0
000010	TINT1	Timer 1	100110	AXEVTE1	McASP1
000011	SDINT	EMIF	100111	AXEVTO1	McASP1
000100	GPINT4	GPIO	101000	AXEVT1	McASP1
000101	GPINT5	GPIO	101001	AREVTE1	McASP1
000110	GPINT6	GPIO	101010	AREVTO1	McASP1
000111	GPINT7	GPIO	101011	AREVT1	McASP1
001000	GPINT0	GPIO	101100	I2CREVT0	I2C0
001001	GPINT1	GPIO	101101	I2CXEVT0	I2C0
001010	GPINT2	GPIO	101110	I2CREVT1	I2C1
001011	GPINT3	GPIO	101111	I2CXEVT1	I2C1
001100	XEVT0	McBSP0	110000	GPINT8	GPIO
001101	REVT0	McBSP0	110001	GPINT9	GPIO
001110	XEVT1	McBSP1	110010	GPINT10	GPIO
001111	REVT1	McBSP1	110011	GPINT11	GPIO
010000–	Reserved		110100	GPINT12	GPIO
011111			110101	GPINT13	GPIO
100000	AXEVTE0	McASP0	110110	GPINT14	GPIO
100001	AXEVTO0	McASP0	110111	GPINT15	GPIO
100010	AXEVT0	McASP0	111000–	Reserved	
100011	AREVTE0	McASP0	111111		

2.6.4 The Parameter RAM (PaRAM)

The transfer parameter table for the EDMA channels and link information is stored in the Parameter RAM (PaRAM) which is a 2K-byte RAM block located within the EDMA. The table consists of six-word parameter sets for a total of 85 sets. Each set uses $6 \times 4 = 24$ bytes and contains the parameters for a transfer shown in the Table 2.10.

The OPT Field in the (PaRAM)

The meanings of all the fields in a transfer set are fairly obvious except for OPT which contains fields to set the priority to High or Low; set the element size to 8, 16, or 32 bits;

define the source and destination as 1 or 2-dimensional; set the source and destination address update modes; enable or disable the transfer complete interrupt; define the transfer complete code; enable or disable linking; and set the frame synchronization mode.

Table 2.10: Format of a Transfer Parameter Set Record

31	16	15	0	
Options (OPT)				Word 0
Source Address (SRC)				Word 1
Array/frame count (FRMCNT)		Element count (ELECNT)		Word 2
Destination Address (DST)				Word 3
Array/frame index (FRMIDX)		Element index (ELEIDX)		Word 4
Element count reload (ELERLD)		link address (LINK)		Word 5

Contents of the PaRAM

The PaRAM is organized as follows:

- The first 16 parameter sets are for the 16 EDMA events. Each set contains 24 bytes.

- The remaining parameter sets are used for linking transfers. Each set is 24 bytes.

- The remaining 8 bytes of unused RAM can be used as a scratch pad area. A part of or the entire PaRAM can be used as a scratch pad RAM when the events corresponding to this region are disabled.

When an event mapped to a particular channel occurs, say channel n with $n \in \{0, 1, \ldots, 15\}$, its parameters are read from parameter set n in the PaRAM and sent to the address generation hardware.

2.6.5   Synchronization of EDMA Transfers

The EDMA can make 1 or 2-dimensional transfers. We will only consider the 1-D case. A 1-D *block* transfer consists of FRMCNT + 1 *frames*. Each frame consists of the number of *elements* specified by the field ELECNT in the parameter set. The following two types of 1-D synchronized transfers are possible:

1. **Element Synchronized 1-D Transfer (FS = 0 in OPT)**
   When a channel sync event occurs, for example, a transition of a McBSP XRDY flag

from false to true, an element in a frame is transferred from its source to destination, the source and destination addresses are updated in the parameter set after the element is transferred, and the element count (ELECNT) is decremented in the parameter set.

When ELCNT = 1, indicating the final element in the frame, and a sync event occurs, the element is transferred. Then the element count is reloaded with the value of ELERLD in the parameter set and the frame count (FRMCNT) is decremented by 1. The EDMA continues transfers at sync events for a new frame if one still remains to be transferred.

2. **Frame Synchronized 1-D Transfers (FS = 1 in OPT)**
   A sync event causes all the elements in a frame to be transferred as rapidly as possible. Each new event causes another frame to be transferred as rapidly as possible until the requested number of frames has been transferred.

2.6.6   Linking and Chaining EDMA Transfers

Linking EDMA Transfers

When the LINK field, bit 1, in options parameter OPT is set to 0, the EDMA stops after a transfer is completed. When LINK = 1 and the requested transfer is completed, the transfer parameters are reloaded with the parameter set pointed to by the 16-bit link address, and the EDMA continues transfers with this new set. The entire parameter RAM is located in the memory area 01A0xxxxh, so a 16-bit link address is sufficient. The link address must be located on a 24-byte boundary. There is no limit to the number of transfers that can be linked. However, the final transfer should link to a NULL parameter set which is one with all its entries set to 0 (24 zero bytes). A transfer can be linked to itself to simulate the autoinitialization feature of the TMS320C6201 and TMS320C6701 DMA. This is useful for circular buffering and repetitive transfers. To eliminate timing problems resulting from the parameter reload time, the event register (ER) is not checked while the parameters are being reloaded. However, new events are registered in the ER. Any record in the PaRAM can be used for linking. However, a set in the first 16 should be used only if the corresponding event is disabled.

Chaining EDMA Channels

The EDMA chaining capability allows the completion of an EDMA channel transfer to trigger another EDMA channel transfer. EDMA chaining does not modify any channel parameters. It just gives a synchronization event to the chained channel. Linking and chaining are different. Linking reloads the current channel parameters with the linked parameters and transfers continue on the same channel with the linked parameters. Chaining does not modify or update any chained parameters. It simply gives a synchronization event to the chained channel. Channels 8, 9, 10, and 11 are reserved for chaining. Chaining is enabled by setting bit 8, 9, 10, or 11 in the channel chain enable register (CCER). The four-bit field, transfer complete code (TCC), in OPT for a channel must also be set to one of these four values to cause chaining to occur at the end of the transfer.

2.6.7   EDMA Interrupts to the CPU

When the TCINT bit is set to 1 in OPT for an EDMA channel and the event mapped to the channel occurs, the EDMA sets a bit in the channel interrupt pending register (CIPR) determined by the transfer complete code programmed in OPT. Then, if the bit corresponding to the channel is set in the channel interrupt enable register (CIER), the EDMA generates the interrupt EDMA_INT to the CPU. If the CPU interrupt EDMA_INT (default CPU_INT8) is enabled in the CPU IER, program execution jumps to the vectored interrupt service routine (ISR). The ISR can read the CIPR to check which EDMA events have been registered as completed and take the appropriate action.

2.6.8   Experiment 2.4: Generating a Sine Wave Using the EDMA Controller

To learn how to use the EDMA controller, write a program to do the following:

1. Configure the McBSP's and codec as in the polling and interrupt experiments and again use a 16 kHz sampling rate.

2. Generate a 512 word integer array, table[512], where the upper 16 bits are the samples for 32 cycles of a 1 kHz sine wave for the left channel, and the lower 16 bits are the samples for 64 cycles of a 2 kHz sine wave for the right channel. Of course, the left and right channel sine wave samples must be scaled to use a large part of the DAC's dynamic range and must be converted to 16-bit integers before being combined into 32-bit words.

3. Configure the EDMA controller to transfer the entire array of 512 samples to the Data Transmit Register (DXR) of McBSP1 which will send them to the codec. Synchronize the transfers with the XRDY1 event to get the 16 kHz sampling rate.

4. Link the channel parameter set back to itself so the sine waves are continuously sent.

Test your program by running it and observing the codec left and right channel outputs on the oscilloscope Verify that they are sine waves with the desired frequencies.

     The following example program segment does most of the work for you. It is on the PC's hard drive as C:\c6713dsk\edma_sines.c and on the class web site. It uses TI's Chip Support Library (CSL) to configure the EDMA. Detailed information about the CSL can be found in the *TMS320C6000 Chip Support Library API Reference Guide* [I.6]. You can also conveniently find CSL documentation by bringing up Code Composer and following the path:

Help → Contents → Chip Support Library → EDMA Module.

The program configures the EDMA to use element sync by the event XEVT1 which happens when XRDY1 makes a transition from 0 to 1 The default mapping of this event to EDMA channel 14 is used. The EDMA is set to transfer single frames containing 512 elements with

the elements being 32-bit words. The same 512-word sine wave sample frame is transmitted
repeatedly by linking back to the same parameter set at the end of each frame transfer.

The two lines

```
EDMA_link(hEdmaXmt,hEdmaReloadXmt);
EDMA_link(hEdmaReloadXmt, hEdmaReloadXmt);
```

require some clarification. The function EDMA_link() sets the LINK parameter in the parameter set record. In the structure gEdmaConfigXmt, the LINK parameter is set to NULL.
In the first code line above, the NULL for LINK is set in the PaRAM channel record to
point to the reload record. This completes the channel parameter set for the XEVT1 event.
When the last word in the frame is transmitted and a link occurs, the reload parameter set
is copied to the channel parameter set. To repeat the same transfer, the reloaded parameter
set should be identical to the initial set. Therefore, the second line above replaces the NULL
for LINK in the reload parameter set to its own address, that is, the address of that reload
parameter set. This makes the reload parameter set the same as the channel parameter set.

Program 2.7 Program for Generating Sine Waves Using the EDMA

```c
#include <stdio.h>
#include <stdlib.h>
#include <dsk6713.h>
#include <dsk6713_aic23.h>
#include <csl_edma.h>
#include <intr.h>
#include <math.h>

#define  sampling_rate 16000
#define SZ_TABLE 512
#define f_left   1000.
#define f_right 2000.
#define scale 15000.
#define pi 3.141592653589

int table[512];

/* Codec Configuration Settings
   See dsk6713_aic23.h and the TLV320AIC23 Stereo Audio CODEC Data
   Manual for a detailed description of the bits in each of the 10
   AIC23 control registers in the following configuration structure. */

DSK6713_AIC23_Config config = { \
    0x0017,  /* 0 DSK6713_AIC23_LEFTINVOL
                Left line input channel volume */ \
    0x0017,  /* 1 DSK6713_AIC23_RIGHTINVOL
                Right line input channel volume */\
    0x00d8,  /* 2 DSK6713_AIC23_LEFTHPVOL
```

```
                  Left channel headphone volume */  \
    0x00d8,  /* 3 DSK6713_AIC23_RIGHTHPVOL
                  Right channel headphone volume */ \
    0x0011,  /* 4 DSK6713_AIC23_ANAPATH
                  Analog audio path control */        \
    0x0000,  /* 5 DSK6713_AIC23_DIGPATH
                  Digital audio path control */       \
    0x0000,  /* 6 DSK6713_AIC23_POWERDOWN
                  Power down control */               \
    0x0043,  /* 7 DSK6713_AIC23_DIGIF
                  Digital audio interface format */ \
    0x0081,  /* 8 DSK6713_AIC23_SAMPLERATE Sample
                  rate control (48 kHz) */    \
    0x0001   /* 9 DSK6713_AIC23_DIGACT
                  Digital interface activation */   \
};

EDMA_Handle hEdmaXmt;          // EDMA channel handles
EDMA_Handle hEdmaReloadXmt;
Int16 gXmtChan;                // TCC code (see initEDMA())

/* Transmit side EDMA configuration */
EDMA_Config gEdmaConfigXmt = {
    EDMA_FMKS(OPT, PRI, HIGH)    |  // Priority
    EDMA_FMKS(OPT, ESIZE, 32BIT) |  // Element size
    EDMA_FMKS(OPT, 2DS, NO) |        // 1 dimensional source
    EDMA_FMKS(OPT, SUM, INC) |       // Src update mode
    EDMA_FMKS(OPT, 2DD, NO)  |       // 1 dimensional dest
    EDMA_FMKS(OPT, DUM, NONE)|       // Dest update mode
    EDMA_FMKS(OPT, TCINT, NO)|       // No EDMA interrupt
    EDMA_FMKS(OPT, TCC, OF(0))|      // Trans. compl. code
    EDMA_FMKS(OPT, LINK, YES)|       // Enable linking
    EDMA_FMKS(OPT, FS, NO),          // Use frame sync?

    (Uint32) table,                  // Src address

    EDMA_FMK(CNT, FRMCNT, NULL)  |  // Frame count
    EDMA_FMK(CNT, ELECNT, SZ_TABLE),// Element cnt

    EDMA_FMKS(DST, DST, OF(0)),      //Dest address

    EDMA_FMKS(IDX, FRMIDX, DEFAULT) | // Frame index
    EDMA_FMKS(IDX, ELEIDX, DEFAULT),  // Element index
    EDMA_FMK (RLD, ELERLD, NULL) |    // Reload element
```

```
    EDMA_FMK (RLD, LINK, NULL)          // Reload link
};

/* Function Prototypes  */
void initEdma(void);
void create_table(void);

void main(void){
  DSK6713_AIC23_CodecHandle hCodec;
  intr_reset(); /* Initialize interrupt system */

  /* dsk6713_init() must be called before other BSL functions */
  DSK6713_init(); /* In the BSL library */

  /* Start the codec. McBSP1 uses 32-bit words,
     1 phase, 1 word frame */
  hCodec = DSK6713_AIC23_openCodec(0, &config);

  /* Change the sampling rate to 16 kHz */
  DSK6713_AIC23_setFreq(hCodec, DSK6713_AIC23_FREQ_16KHZ);

  create_table();  /* You must write this function. */
  initEdma(); /* Initialize the EDMA controller (See below) */
   while(1);  /* infiniteloop */
} /* end of main()  */

/***************************************************/
/* Create a table where upper 16-bits are samples */
/* of a sine wave with frequency f_left, and the  */
/* lower 16 bits are samples of a sine wave with  */
/* frequency f_right.                             */
/***************************************************/

void create_table(void){
  PUT YOUR CODE TO GENERATE THE SINE TABLE HERE.
}

/***************************************************/
/* initEdma() - Initialize the EDMA controller.  */
/* Use linked transfers to automatically restart */
/* at beginning of sine table.                   */
/***************************************************/

void initEdma(void)
```

```
{
/* Configure transmit channel */

/* Get hEdmaXmt handle, Set channel event to XEVT1 */
    hEdmaXmt = EDMA_open(EDMA_CHA_XEVT1, EDMA_OPEN_RESET);

/* Get handle for reload table. This function reserves space for
   a 6-word link  parameter set in the PaRAM. The "-1" argument
   tells the function to choose an unused set.   */
    hEdmaReloadXmt = EDMA_allocTable(-1);

/* Get the address of DXR for McBSP1 and put it in the
   channel parameter configuration structure.  */
    gEdmaConfigXmt.dst = MCBSP_getXmtAddr(DSK6713_AIC23_DATAHANDLE);

/* Then configure the Xmt table. (LINK is set to NULL from
    structure gEdmaConfigXmt)  */
    EDMA_config(hEdmaXmt, &gEdmaConfigXmt);

/* Configure the Xmt reload table to be the same as the Xmt table
    (Link is set to NULL from structure gEdmaConfigXmt) */
    EDMA_config(hEdmaReloadXmt, &gEdmaConfigXmt);

/* Link back to table start  */
    /* Set LINK in channel parameter set to point to the reload
        parameter set */
    EDMA_link(hEdmaXmt,hEdmaReloadXmt);

    /* Set LINK in the reload parameter set to point to the reload
        parameter set */
    EDMA_link(hEdmaReloadXmt, hEdmaReloadXmt);

/* Enable EDMA channel  */
    EDMA_enableChannel(hEdmaXmt);

 /* Do a dummy write to generate the first McBSP transmit event */
    MCBSP_write(DSK6713_AIC23_DATAHANDLE, 0);
}
```

Chapter 3

Digital Filters

The major goal of this chapter is to learn how to implement discrete-time filters using a DSP and in in real-time. Discrete-time filters are often called digital filters. The theory is usually presented in a typical required Electrical Engineering undergraduate course on signals and systems and a Senior elective course on digital signal processing. Having to write a DSP program and getting every detail correct, and seeing filters work with live signals leads to a deeper understanding of digital filtering. In the process, you will learn more about the TMS320C6713 DSP and the TMS320C6713 DSK.

3.1 Discrete-Time Convolution and Frequency Responses

The output y[n] of a linear, time-invariant, discrete-time system (LTI) can be computed by convolving its input x[n] with its unit pulse response h[n]. The equation for this discrete-time convolution is

$$y[n] = \sum_{k=-\infty}^{\infty} x[k]h[n-k] = \sum_{k=-\infty}^{\infty} h[k]x[n-k] \tag{3.1}$$

The z-transform of the discrete-time convolution of two signals is the product of the two transforms, that is,

$$Y(z) = \sum_{n=-\infty}^{\infty} y[n]z^{-n} = X(z)H(z) \tag{3.2}$$

where

$$H(z) = \sum_{n=-\infty}^{\infty} h[n]z^{-n} \quad \text{and} \quad X(z) = \sum_{n=-\infty}^{\infty} x[n]z^{-n} \tag{3.3}$$

The response of an LTI system to a sinusoid after the transients have become negligible is called its sinusoidal steady-state response. To determine this response, let the input be the sampled complex sinusoid

$$x[n] = Ce^{j\omega nT}$$

According to (3.1), the output is

$$y[n] = \sum_{k=-\infty}^{\infty} h[k]Ce^{j\omega(n-k)T} = Ce^{j\omega nT} \sum_{k=-\infty}^{\infty} h[k]e^{-j\omega kT} = x[n]H(z)|_{z=e^{j\omega T}} \qquad (3.4)$$

Thus, the output is a sinusoid at the same frequency as the input but with its amplitude scaled by the complex number

$$H^*(\omega) = H(z)|_{z=e^{j\omega T}} \qquad (3.5)$$

The function $H^*(\omega)$ is called the *frequency response* of the system. The function $A(\omega) = |H^*(\omega)|$ is called the *amplitude response* of the system and the angle $\theta(\omega) = \arg H^*(\omega)$ is called its *phase response*. Notice that all of these responses are periodic as functions of ω with period equal to the sampling rate $\omega_s = 2\pi/T$. In polar form

$$H^*(\omega) = A(\omega)e^{j\theta(\omega)} \qquad (3.6)$$

so, according to (3.4), the output can be expressed as

$$y[n] = CA(\omega)e^{j[\omega nT+\theta(\omega)]} \qquad (3.7)$$

When the input is the real sinusoid

$$x[n] = C\cos(\omega nT + \phi) = \Re e\{Ce^{j\phi}e^{j\omega nT}\}$$

the output is

$$y[n] = \Re e\{H^*(\omega)Ce^{j\phi}e^{j\omega nT}\} = CA(\omega)\cos[\omega nT + \theta(\omega) + \phi]$$

In other words, the system scales the magnitude of the sinusoidal input by the amplitude response and shifts its phase by the phase response. This is the basis for digital filtering.

3.2 Finite Duration Impulse Response (FIR) Filters

3.2.1 Block Diagram for Most Common Realization

If the unit pulse response is identically zero outside the set of integers $\{0, 1, \cdots, N-1\}$, the convolution (3.1) becomes

$$y[n] = \sum_{k=0}^{N-1} h[k]x[n-k] = \sum_{k=n-N+1}^{n} x[k]h[n-k] \qquad (3.8)$$

A filter of this type is called an N-tap *finite duration impulse response* (FIR) filter, *non-recursive* filter, *transversal* filter, or *moving average* filter. A block diagram for the most common method of implementing FIR filters is shown in Fig. 3.1 on page 69. It consists of a delay line represented by the chain of blocks labeled z^{-1} and a set of taps into the delay line with weights equal to the unit pulse response samples.

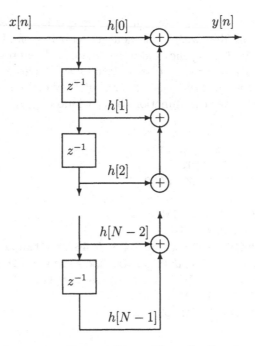

Figure 3.1: Type 1 Direct Form Realization

The block diagram of an FIR filter shown in Fig. 3.1 could represent the physical layout of a hardware implementation or just the structure of a software algorithm. In a software implementation, the delay line would just be an array in memory. Entering a new sample into an array by shifting the entire array is inefficient. In the exercises described below, you will learn how to implement the "shift register" by forming a circular array or buffer in C and also by using the hardware circular addressing capabilities of the TMS320C6x DSP's.

3.2.2 Two Methods for Finding the Filter Coefficients to Achieve a Desired Frequency Response

Two programs for designing digital filters are included in the directory C:\DIGFIL. Both methods design filters with exactly linear phase which is a reason FIR filters are sometimes preferred over IIR filters.

Historically, the first method for designing digital filters was the *Fourier series and window function* method. See [II.C.17, Chapter 8] for the theory. The program window.exe implements this method. It was taken from the IEEE Press book, *Programs for Digital Signal Processing* [II.C.7] and modified to make it more user friendly. The program presents a selection of seven different window types. The Hamming window (3) and Kaiser window (6) are the ones you will most likely find best. Six different filter types are available: (1) lowpass, (2) highpass, (3) bandpass, (4) bandstop, (5) bandpass Hilbert transform, and (6) bandpass differentiator. To use this program, first copy it to your working directory and then type **window**. As an example of how to use the program, suppose the sampling rate is 8 kHz, and a 21 tap bandpass filter with a lower cutoff frequency of 1 kHz and an upper cutoff frequency of 3 kHz is desired using the Hamming window. The program screen output and typical responses are shown below. The text written by the program is in capital letters and the user responses are shown in lower case letters but they can be entered as either capital or lower case letters. The lower case functions after window types 3, 4, and 5 are displayed by the program.

```
ENTER NAME OF LISTING FILE: junk.lst
ENTER FILENAME FOR COEFFICIENTS: junk.cof
ENTER SAMPLING FREQUENCY IN HZ: 8000
        WINDOW TYPES
        1       RECTANGULAR WINDOW
        2       TRIANGULAR WINDOW
        3       HAMMING WINDOW    0.54 + 0.46 cos(theta)
        4       GENERALIZED HAMMING WINDOW alpha + (1-alpha) cos(theta)
        5       HANNING WINDOW    0.5 + 0.5 cos(theta)
        6       KAISER (IO-SINH) WINDOW
        7       CHEBYSHEV WINDOW


        FILTER TYPES
        1       LOWPASS FILTER
        2       HIGHPASS FILTER
        3       BANDPASS FILTER
        4       BANDSTOP FILTER
        5       BANDPASS HILBERT TRANSFORM
        6       BANDPASS DIFFERENTIATOR

ENTER FILTER LENGTH, WINDOW TYPE, FILTER TYPE: 21,3,3
SPECIFY LOWER, UPPER CUTOFF IN HZ: 1000,3000

CREATE (FREQUENCY,RESPONSE) FILE (Y OR N)?  y
ENTER FILENAME: junk.dat
LINEAR (L) OR DB (D) SCALE ?: d
```

The **LISTING FILE** is where the number of taps, filter type, window type, sampling frequency, and filter coefficients are written. The **FILENAME FOR COEFFICIENTS** has one entry per line. The first line is the number of coefficients. The remaining lines are the coefficients in order of increasing index and in floating point format. This file is useful for using the coefficients in another program. The **(FREQ, RESPONSE)** file is a listing of the amplitude response of the filter on a linear or dB scale. Each line contains a pair of numbers consisting of the frequency and corresponding amplitude response. The frequency increment is automatically selected to show the ripples in the amplitude response.

The second program for designing digital filters is **remez87.exe**. It is a modified version of the program in the IEEE book, *Programs for Digital Signal Processing* [II.C.7]. The program was developed by J. McClellan and T. Parks who were at Rice University at the time. It uses the Remez exchange algorithm to design filters that are optimum in the Chebyshev sense, that is, the maximum absolute error is minimized and causes the error to be equal ripple. The program can design (1) multiple passband/stopband filters, (2) differentiators, and (3) Hilbert transform filters.

Use of `remez87.exe` will be demonstrated by an example. Suppose the sampling rate is 8 kHz and a 21 tap bandpass filter with a passband extending from 1000 to 3000 Hz is desired. Using the conventions of this program, three bands must be specified. They are: (1) a lower stopband, (2) the passband, and (3) an upper stopband. Let the lower stopband edges extend from 0 to 500 Hz, the passband edges extend from 1000 to 3000 Hz, and the upper stopband edges extend from 3500 to 4000 Hz. Values must be specified for the amplitude in each of the bands. Let the values in the two stopbands be 0 and the value in the passband be 1. Also, weights for each band must be specified. The weight values scale the error in each band. Since the algorithm generates an equal ripple weighted error, larger weights result in bands with smaller unweighted ripple. For this example let the bands be equally weighted with the value 1.

The program `remez87` uses a variable GRID DENSITY to determine the frequency increment for computing the frequency response error with larger numbers corresponding to closer spaced frequencies. Values in the range of 16 to 32 seem to work well with little difference observed in the results. The smaller number requires less computation and uses smaller arrays. The program computes the frequency response of the resulting filter and asks the user to enter the lower and upper frequency limits to use for the response. Program screen prompts are in capital letters. User responses can be in either upper or lower case. The prompts and responses for this example are shown below.

```
ENTER LISTING FILENAME: junk.1st
ENTER COEFFICIENT STORAGE FILENAME: junk.cof
LINEAR OR DB AMPLITUDE SCALE FOR PLOTS? (L OR D): d
ENTER SAMPLING FREQUENCY (HZ): 8000
ENTER START AND STOP FREQUENCIES IN HZ FOR
   RESPONSE CALCULATION (FSTART,FSTOP): 0,4000

FILTER TYPES AVAILABLE:
  1 MULTIPLE PASSBAND/STOPBAND FILTER
  2 DIFFERENTIATOR
  3 HILBERT TRANSFORM

ENTER: FILTER LENGTH, TYPE, NO. OF BANDS, GRID DENSITY: 21,1,3,32
ENTER THE BAND EDGES (FREQUENCIES IN HERTZ)
0,500,1000,3000,3500,4000
SPECIAL USER DEFINED AMPLITUDE RESPONSE(Y/N)? n
SPECIAL USER DEFINED WEIGHTING FUNCTION(Y/N)? n

ENTER (SEPARATED BY COMMAS):
 1. VALUE FOR EACH BAND FOR MULTIPLE PASS/STOP BAND FILTERS
```

```
  2. SLOPES FOR DIFFERENTIATOR (GAIN = Ki*f -> SLOPE = Ki
       WHERE Ki = SLOPE OF i-TH BAND, f IN HERTZ)
  3. MAGNITUDE OF DESIRED VALUE FOR HILBERT TRANSFORM
0,1,0
ENTER WEIGHT FOR EACH BAND. (FOR A DIFFERENTIATOR
 THE WEIGHT FUNCTION GENERATED BY THE PROGRAM FOR THE i th
  BAND IS WT(i)/f WHERE WT(i) IS THE ENTERED BAND WEIGHT AND
  f IS IN HERTZ.)
1,1,1

STARTING REMEZ ITERATIONS
DEVIATION =      .159436E-03
  .

  .

  .

CALCULATING IMPULSE RESPONSE

CALCULATING FREQUENCY RESPONSE

CREATE (FREQ,RESPONSE) FILE (Y OR N)? y
ENTER FILENAME: junk.dat
```

The files requested by `remez87` are essentially the same as for `window`. However, the `LISTING` file contains more information, such as, the frequencies where the peak errors occur, a frequency response listing in linear and dB form, and a crude plot of the response used in the days when only line printers without graphics capabilities were available.

The program `remez87` asks if you want a SPECIAL USER DEFINED AMPLITUDE RESPONSE or a SPECIAL USER DEFINED WEIGHTING FUNCTION. You can write your own special subroutines for a special desired amplitude response and/or weighting function and link them into the main program. No special functions are included in this version of the program.

Similar filter design programs can be found in the MATLAB signal processing package.

3.3 Using Circular Buffers to Implement FIR Filters in C

For an N-tap FIR filter with coefficients nonzero only for indices in the set $\{0, \ldots, N-1\}$, the convolution sum (3.1) becomes

$$y[n] = \sum_{k=0}^{N-1} h[k]x[n-k] = h[0]x[n] + h[1]x[n-1] + \cdots + h[N-1]x[n-N+1] \qquad (3.9)$$

Notice that the oldest input sample $x[n - (N - 1)]$ is multiplied by the impulse response sample $h[N - 1]$ with the largest index and the newest sample $x[n]$ is multiplied by the impulse response sample $h[0]$ with the smallest index. This equation is shown schematically in Figure 3.1 where the N required signal samples are shown stored in a delay line. In a software implementation, the delay line represents an array in memory. Entering the newest sample into the "delay line" by shifting the elements in the entire array is inefficient for a software implementation and a better approach is to use *circular buffers*. Circular buffers can be implemented in C. Also, the 'C6000 DSP's have hardware support that can be accessed by assembly instructions to implement circular buffers even more efficiently.

The concept of a circular buffer is illustrated in Figure 3.2. The filter coefficients are stored in the N element array h[ ]. A variable, *newest*, points to the location in the

Array Index	Filter Coefficient Array h[]	Circular Buffer Array xcirc[]
0	$h[0]$	$x[n - newest]$
1	$h[1]$	$x[n - newest + 1]$
⋮	⋮	⋮
		$x[n - 1]$
newest		$x[n]$
oldest		$x[n - N + 1]$
		$x[n - N + 2]$
⋮	⋮	⋮
$N - 2$	$h[N - 2]$	$x[n - newest - 2]$
$N - 1$	$h[N - 1]$	$x[n - newest - 1]$

Figure 3.2: Contents of Coefficient Array and Circular Buffer

circular buffer array that contains the most recently entered sample. When a new sample is received at time n, it is written over the sample at location *oldest = newest + 1* modulo N and *newest* is incremented modulo N. In a physical implementation using a shift register, the overwritten sample would be shifted out of the end of the shift register when the new one is shifted in. Notice that when *newest* initially has the value $N - 1$, it becomes 0 when incremented modulo N. Thus, data samples are written into the array in a circular fashion moving down the array one element at a time to the bottom at location $N - 1$ and then jumping back up to the top of the array at location 0. Finally, the filter output can be calculated as

$$y[n] = \sum_{k=0}^{N-1} \text{h}[k]\text{xcirc}[\text{mod}(newest - k, N)] \tag{3.10}$$

where $\text{mod}(newest - k, N)$ is the integer in the set $\{0, \ldots, N-1\}$ formed by adding multiples of N to $newest - k$ until it falls in this set.

A segment of a C program for implementing the FIR filter with a circular buffer is shown below.

Program 3.1 C Program Segment for an FIR Filter with Circular Buffer

```
main()
{
    int x_index = 0;
    float y, xcirc[N];
        ...

/*-------------------------------------------*/
/* circularly increment newest               */
    ++newest;
    if(newest == N) newest = 0;
/*-------------------------------------------*/
/* Put new sample in delay line.             */
    xcirc[newest] = newsample;
/*-------------------------------------------*/
/* Do convolution sum                        */
    y = 0;
    x_index = newest
    for (k = 0; k < N; k++)
      {
        y += h[k]*xcirc[x_index];
        /*-----------------------------------*/
        /* circularly decrement x_index      */
        --x_index;
        if(x_index == -1) x_index = N-1;
        /*-----------------------------------*/
      }
    ...
}
```

Warning: DSK6713_AIC23_read() and MCBSP_read() each return a 32-bit unsigned int. Convert the returned value to an int before shifting right 16 bits to knock off the right channel and get the left channel with sign extension. Shifting an unsigned int right fills the MSB's with 0's so the sign is not extended.

Note: C has the mod operator, %, but its implementation by the compiler is very inefficient because the compiler must account for all general cases. Therefore, you should implement the mod operation as shown in the code segment above.

3.4 Circular Buffers Using the 'C6000 Hardware

The TMS320C6000 family of DSP's has built-in hardware capability for circular buffers. The eight registers, A4–A7 and B4–B7, can be used for linear or circular indirect addressing. The Address Mode Register (AMR) contains 2-bit fields, as shown in Table 3.1, for each register that determine the address modes as shown in Table 3.2. The number of words in the buffer is called the *block size*. The block size is determined by either the BK0 or BK1 5-bit fields in the AMR. The choice between them is determined by the 2-bit mode fields. Let Nblock be the value of the BK0 or BK1 field. Then the circular buffer has the size BUF_LEN = $2^{Nblock+1}$ bytes. So, the circular buffer size can only be a power of 2 bytes.

Table 3.1: Address Mode Register (AMR) Fields

31 26	25 21	20 16	15 14	13 12	11 10
Resvd	BK1	BK0	B7 mode	B6 mode	B5 mode

9 8	7 6	5 4	3 2	1 0
B4 mode	A7 mode	A6 mode	A5 mode	A4 mode

Table 3.2: AMR Mode Field Encoding

Mode	Addressing Option
00	Linear Mode
01	Circular Mode Using BK0 Size
10	Circular Mode Using BK1 Size
11	Reserved

The buffer must be aligned on a byte boundary that is a multiple of the block size BUF_LEN. Therefore, the Nblock+1 lsb's of the buffer base address must all be 0. This can be done in a C program by using the DATA_ALIGN pragma. Suppose the buffer is an array x[ ]. The alignment command is:

```
#pragma DATA_ALIGN(x, BUF_LEN)
```

The array x[ ] must be a global array. The alignment can also be done by creating a named section in the assembly program and using the linker to align the section properly.

3.4.1 How the Circular Buffer is Implemented

Circular addressing is implemented by inhibiting carries or borrows between bits Nblock and Nblock+1 in the address calculations. Therefore, bits Nblock+1 through 31 do not change as the address is incremented or decremented by an amount less than the buffer size.

3.4.2   Indirect Addressing Through Registers

Hardware circular addressing cannot be performed in C. It must be carried out by assembly instructions. Circular addressing is accomplished by indirect addressing through one of the eight allowed registers using the auto-increment/decrement and indexed modes. A typical circular buffering instruction is

```
LDW   *A5--, A8
```

where the A5 field in the AMR has been set for circular addressing. LDW is the mnemonic for "load a word." The word is loaded into the destination register A8 from the address pointed to by A5 and the address is decremented by 4 bytes according the mode in the AMR after being used (post decremented).

3.5   Interfacing C and Assembly Functions

Because of the tremendous advances in DSP hardware capabilities and software code generation tools, it is becoming standard practice to implement applications almost entirely in a higher level language like C. Some advantages are:

- Rapid software development using a high level language.

- Can use powerful optimizing compilers.

- Application can be easily ported to different DSP's.

- Profiling tools can find time intensive code segments which can then be written in optimized assembly code.

Generating efficient assembly code for the 'C6000 family by hand is very difficult because there are the multiple execution units, there is a multi-level pipeline, and different instructions take different times to execute.

Another reason for using assembly routines is that some hardware capabilities of the DSP, such as, hardware circular buffering, cannot be directly accessed by C. Therefore, we will learn the fundamentals of how to call assembly functions from C in this section. Look at Sections 8.4 and 8.5 of the *TMS320C6000 Optimizing Compiler User's Guide*, [I.9] for the complete details.

The C compiler has a specific set of conventions for register usage and argument passing. The register usage conventions are shown in Tables 3.3 and 3.4. In particular, notice that register B15 is used as the stack pointer, B3 contains the return address, register A4 is used to return 32-bit results, and the A5:A4 pair is used to return 64-bit results like long ints and doubles.

3.5.1   Responsibilities of the Calling and Called Function

The following steps must be taken by the calling function (parent) which can be a C or assembly routine. The called function (child) can be a C or assembly function also.

1. Passed arguments are placed in registers or on the stack. By convention, argument 1 is the left most argument.

   - The first ten arguments are passed in A and B registers as shown in Tables 3.3 and 3.4
   - Additional arguments are passed on the stack.

2. The calling function (parent) must save A0 through A9 and B0 through B9 if needed after the call, by pushing them on the stack.

3. The caller branches to the function (child).

4. Upon returning, the caller reclaims stack space used for arguments.

Table 3.3: "A" Side Register Usage

Register	Preserved By	Special Uses
A0	Parent	
A1	Parent	
A2	Parent	
A3	Parent	Structure register
A4	Parent	Argument 1 or return value
A5	Parent	Argument 1 or return value with A4 for doubles and longs
A6	Parent	Argument 3
A7	Parent	Argument 3 with A6 for doubles and longs
A8	Parent	Argument 5
A9	Parent	Argument 5 with A8 for doubles and longs
A10	Child	Argument 7
A11	Child	Argument 7 with A10 for doubles and longs
A12	Child	Argument 9
A13	Child	Argument 9 with A12 for doubles and longs
A14	Child	
A15	Child	Frame pointer (FP)

The called function or child must do the following:

1. The called function allocates space on the stack for local variables, temporary storage, and arguments to functions this function might call. The frame pointer (FP) is used to access arguments on the stack.

Table 3.4: "B" Side Register Usage

Register	Preserved By	Special Uses
B0	Parent	
B1	Parent	
B2	Parent	
B3	Parent	Return address
B4	Parent	Argument 2
B5	Parent	Argument 2 with B4 for doubles and longs
B6	Parent	Argument 4
B7	Parent	Argument 4 with B6 for doubles and longs
B8	Parent	Argument 6
B9	Parent	Argument 6 with B8 for doubles and longs
B10	Child	Argument 8
B11	Child	Argument 8 with B10 for doubles and longs
B12	Child	Argument 10
B13	Child	Argument 10 with B12 for doubles and longs
B14	Child	Data page pointer (DP)
B15	Child	Stack pointer (SP)

2. If the called function calls another, the return address must be saved on the stack. Otherwise it is left in B3.

3. If the called function modifies A10 through A15 or B10 through B15, it must save them in other registers or on the stack.

4. The called function code is executed.

5. The called function returns an int, float, or pointer in A4. Double or long double are returned in the A5:A4 pair.

6. A10–A15 and B10–B15 are restored if used.

7. The frame and stack pointers are restored.

8. The function returns by branching to the value in B3.

3.5.2 Using Assembly Functions with C

To write assembly functions that can be called from C, the following items must be kept in mind:

- C variable names are prefixed with an underscore by the compiler when generating assembly code. For example, a C variable named x is called _x in the assembly code.

- The caller must put the arguments in the proper registers or on the stack for arguments beyond number 10.

- A10–A15 and B10–B15, B3 and, possibly, A3 must be preserved by the called function. It can use all other registers freely.

- The called function must pop everything it pushed on the stack before returning to the caller.

- Any object or function declared in the assembly function that is accessed or called from C must be declared with a .def or .global directive in the assembly code. This allows the linker to resolve references to it.

3.6 Linear Assembly Code and the Assembly Optimizer

Writing efficient assembly code is very difficult and time consuming. The TI code generation tools allow you to write in a language called *linear assembly code* which is very similar to full assembly code but you do not have to worry about many of the hardware and software details. Linear assembly files should be given the extension **sa**. Linear assembly code does not include information about parallel instructions or instruction latencies. Register usage is usually not assigned in the source code but may be. Also, symbolic names can be used for registers. The *assembly optimizer* operates on linear assembly files and converts them to regular assembly code. The optimizer assumes files with the **sa** extension include linear assembly code. Some of the tasks it performs are:

- finding instructions that can operate in parallel

- handling pipeline latencies

- assigning register usage

- defining which execution units to use

- optimizing execution time by software pipelining

- creating entry and exit assembly code for functions to be called by C.

See [I.9] and [I.11] for complete details on linear assembly code, and how to use the assembly optimizer and interpret its diagnostic reports.

A C-callable linear assembly function must declare its entry point to be global and include .cproc and .endproc directives to mark the assembly code region to be optimized. An example of a C-callable linear assembly function for performing one convolution iteration using a hardware circular sample buffer is shown in Section 3.6.1 in the file convol1.sa. You will find the following lines in the program:

```
          .global _convolve
_convolve .cproc  x_addr, h_addr, Nh, Nblock, newest
          .reg   sum, prod, x_value, h_value
             .

             .

             .

          .return sum   ; By C convention, put sum in A4
          .endproc
```

In this example, the entry point is _convolve. The names following .cproc are the function's arguments. The .reg line lists symbolic variable names that the assembly optimizer will assign to registers or the stack, if necessary. Finally, the .return directive causes the assembly optimizer to return sum to the caller by putting it in A4, which is the C convention.

Invoking the Assembly Optimizer

The linear assembly file can be processed by the assembly optimizer by using the command prompt shell command

```
cl6x -mv6713 -o3 -k convol1.sa
```

The items on the command line have the following meanings:

- -mv6713 specifies the 'C6713 DSP.

- -o3 specifies optimization level 3. The 3 can be replaced by 0, 1, or 2. The -o option can be left out for no optimization.

- -k specifies that the .asm output should be kept

- convol1.sa is the input file to be optimized.

You can also use Code Composer Studio to process the .sa file by adding it to a project, clicking on Project and then Options, selecting the Compiler tab and setting the desired optimization level, and building the project.

3.6.1   A Linear Assembly Convolution Function that Uses a Circular Buffer and Can be Called from C

A C callable linear assembly program for computing one output sample of an FIR filter by a convolution sum using a hardware circular buffer is shown in Program 3.2. The source file `convol1.sa` contains the function `convolve()` and can be found on the class web site and in the directory `C:\c6713dsk`. A segment of a main C program for calling this function is shown in Program 3.5 later. See the *TMS320C6000 CPU and Instruction Set Reference Guide* [I.7] for complete descriptions of all the assembly instructions. Documentation for what the instructions are doing is included as comments in the linear assembly code.

Partial output from the assembly optimizer for no optimization and the highest level of optimization, -o3, is shown in Program 3.3 on page 84 and Program 3.4 on page 85, respectively. Notice that the assembly optimizer uses the C conventions and assigns the stack pointer (SP) to register B15, the data page pointer (DP) to B14, and the frame pointer (FP) to A15. The input arguments are found by the function in registers following the C conventions and were placed there by the calling C program. Observe how the loop is implemented by a conditional branch based on the contents of register B0. The final convolution sum is returned in A4 using the C convention.

The resulting assembly code for no optimization contains many NOP's and essentially no parallel scheduling of the execution units. The result for -o3 optimization contains a significant amount of parallelization. For example, consider the lines

```
   [!A2]   ADDSP   .L1     A0,A3,A3 ;  ^ |92|  sum of products
||         MPYSP   .M1X    B5,A5,A0 ; @|91|  h[k]*x[n-k]
```

Repetitions of these lines form most of the convolution sum. The product of a filter coefficient and data sample is computed and placed in A0 and in parallel the previous product from A0 is added to the accumulated sum in A3 and put back into A3. The instruction `ADDSP` stands for "floating-point add single precision" and `MPYSP` for "floating-point multiply single precision." The assembly optimizer reports a variety of statistics showing some bounds on the optimization and how the execution units are used. In the -o3 example, it reports that

```
;*   ii = 4 Schedule found with 4 iterations in parallel
```

This means that the convolution loop kernel requires `ii` = 4 cycles and four iterations of the convolution summation loop are being processed in the pipeline at the same time.

Program 3.2   convol1.sa

```
;*********************************************************
;  File: convol1.sa
;   By: S.A. Tretter
;
;   Compile using
;
;     cl6x -mv6713 -o3 convol1.sa
```

```
;
;    or by using Code Composer Studio with these options.
;
;    This is a C callable assembly function for computing
;    one convolution iteration. The circular buffering
;    hardware of the C6000 is used. The function
;    prototype is:
;
;    extern float convolve( float x[ ], float h[ ], int Nh,
;                           int Nblock, int newest );
;
;        x[ ]     circular input sample buffer
;        h[ ]     FIR filter coefficients
;        Nh       number of filter taps
;        Nblock   circular buffer size in bytes is
;                 2^{Nblock+1} and in words is 2^{Nblock-1}
;        newest   index pointing to newest sample in buffer
;
;    According to the TI C Compiler conventions, the
;    arguments on entry are found in the following
;    registers:
;
;        &x[0]    A4
;        &h[0]    B4
;        Nh       A6
;        Nblock   B6
;        newest   A8
;
;    WARNING:  The C calling function must align the
;    circular buffer, x[ ], on a boundary that is a
;    multiple of the buffer size in bytes, that is, a
;    multiple of BUF_LEN = 2^{Nblock+1} bytes.  This can
;    be done by a statement in the C program of the form
;     #pragma DATA_ALIGN(x, BUF_LEN)
;    Note: x[] must be a global array.
;***********************************************************

            .global _convolve
_convolve .cproc  x_addr, h_addr, Nh, Nblock, newest
            .reg  sum, prod, x_value, h_value

;  Compute address of x[newest] and put in x_addr
;  Note: The instruction ADDAW shifts the second argument,
;        newest, left 2 bits, i.e., multiplies it by 4,
```

```
;            before adding it to the first argument to form
;            the actual byte address of x[newest].

     ADDAW   x_addr, newest, x_addr ; &x[newest]
;------------------------------------------------------------
;  Set up circular addressing
;  Load Nblock into the BK0 field of the Address Mode
;     Register (AMR)

     SHL Nblock, 16, Nblock ; Shift Nblock to BK0 field

;   Note: The assembly optimizer will assign x_addr to
;         some register it likes.  You will have to
;         manually look at the assembled and optimized
;         code to see which register it picked and then
;         set up the circular mode using BK0 by writing
;         01 to the field for that register in AMR.
;         The assembler will give you a warning that
;         changing the AMR can give unpredictable
;         results but you can ignore this.
;
;         Suppose B4 was chosen by the optimizer.
;
     set Nblock, 8,8, Nblock; Set mode circular, BK0, B4
;    set Nblock, 10,10, Nblock; Use this for B5.
     MVC Nblock, AMR         ; load mode into AMR
;------------------------------------------------------------

;  Clear convolution sum registers

     ZERO  sum

;  Now compute the convolution sum.

loop:    .trip 8, 500  ; assume between 8 and 500 taps
     LDW *x_addr--, x_value ; x[newest-k] -> x_value
     LDW *h_addr++, h_value ; h[k] -> h_value
     MPYSP  x_value, h_value, prod ; h[k]*x[n-k]
     ADDSP prod, sum, sum ; sum of products

 [Nh] SUB Nh, 1, Nh   ; Decrement count by 1 tap
 [Nh] B  loop    ; Continue until all taps computed

     .return sum  ; By C convention, put sum in A4
```

```
        .endproc
```

Program 3.3 Part of Assembly Optimizer Output for No Optimization

```
        .asg    A15, FP
        .asg    B14, DP
        .asg    B15, SP


        .global _convolve
        .sect   ".text"


;**************************************************************************
;* FUNCTION NAME: _convolve                                               *
;*                                                                        *
;*   Regs Modified    : A0,A3,A4,B0,B4,B5,B6                              *
;*   Regs Used        : A0,A3,A4,A6,A8,B0,B3,B4,B5,B6                     *
;**************************************************************************
_convolve:
;            .reg  sum, prod, x_value, h_value
             MV     .S2X    A8,B5          ;  |47|

             MV     .S2X    A4,B4          ;  |47|
||           MV     .S1X    B4,A0          ;  |47|

             MV     .S2X    A6,B0          ;  |47|
        .line   10
             ADDAW  .D2     B4,B5,B4       ;  |56|  &x[newest]
        .line   17
             SHL    .S2     B6,0x10,B6     ;  |63|  Shift Nblock to BK0 field
        .line   31
             SET    .S2     B6,0x8,0x8,B6  ;  |77|  Set mode circular, BK0, B4
        .line   33
             MVC    .S2     B6,AMR         ;  |79|  load mode into AMR
             NOP            1
        .line   38
             ZERO   .D1     A4             ;  |84|
        .line   42

loop:
        .line   43
             LDW    .D2T2   *B4--,B5       ;  |89|  x[newest-k] -> x_value
             NOP            4
        .line   44
             LDW    .D1T1   *A0++,A3       ;  |90|  h[k] -> h_value
             NOP            4
        .line   45
             MPYSP  .M1X    B5,A3,A3       ;  |91|  h[k]*x[n-k]
```

```
              NOP          3
         .line   46
              ADDSP    .L1    A3,A4,A4          ; |92|  sum of products
              NOP          3
         .line   48
   [ B0]    ADD      .D2    0xffffffff,B0,B0  ; |94|  Decrement count by 1 tap
         .line   49
   [ B0]    B        .S1    loop              ; |95|  Continue until done
              NOP          5
              ; BRANCH OCCURS                  ; |95|
;** ------------------------------------------------------------------------*
         .line   51
         .line   52
              B        .S2    B3                ; |98|
              NOP          5
              ; BRANCH OCCURS                  ; |98|
         .endfunc        98,000000000h,0
```

Program 3.4 Part of Assembly Optimizer Output for -o3 Optimization

```
         .asg    A15, FP
         .asg    B14, DP
         .asg    B15, SP


        .global _convolve
        .sect  ''.text''
;****************************************************************************
;* FUNCTION NAME: _convolve                                                 *
;*                                                                          *
;*   Regs Modified     : A0,A1,A2,A3,A4,A5,B0,B4,B5                         *
;*   Regs Used         : A0,A1,A2,A3,A4,A5,A6,A8,B0,B3,B4,B5,B6             *
;****************************************************************************
_convolve:
;*--------------------------------------------------------------------------*
;*   SOFTWARE PIPELINE INFORMATION
;*
;*      Loop label : loop
;*      Known Minimum Trip Count          : 8
;*      Known Maximum Trip Count          : 500
;*      Known Max Trip Count Factor       : 1
;*      Loop Carried Dependency Bound( )  : 4
;*      Unpartitioned Resource Bound      : 1
;*      Partitioned Resource Bound(*)     : 1
;*      Resource Partition:
;*                              A-side   B-side
;*      .L units                  1*        0
;*      .S units                   0        1*
```

```
;*          .D units                      1*       1*
;*          .M units                      1*       0
;*          .X cross paths                1*       0
;*          .T address paths              1*       1*
;*          Long read paths               0        0
;*          Long write paths              0        0
;*          Logical  ops (.LS)            0        0       (.L or .S unit)
;*          Addition ops (.LSD)           0        1       (.L or .S or .D unit)
;*          Bound(.L .S .LS)              1*       1*
;*          Bound(.L .S .D .LS .LSD)      1*       1*
;*
;*          Searching for software pipeline schedule at ...
;*             ii = 4   Schedule found with 4 iterations in parallel
;*          done
;*
;*          Epilog not entirely removed
;*          Collapsed epilog stages     : 2
;*
;*          Prolog not entirely removed
;*          Collapsed prolog stages     : 2
;*
;*          Minimum required memory pad : 0 bytes
;*
;*          For further improvement on this loop, try option -mh8
;*
;*          Minimum safe trip count     : 1
;*-------------------------------------------------------------------------------*
L1:    ; PIPED LOOP PROLOG
           NOP              1
           MV       .S2X    A6,B0
           MV       .S2X    A8,B5


           MV       .S2X    A4,B4
||         MV       .S1X    B4,A4


       .line   10
           ADDAW    .D2     B4,B5,B5           ; |56|   &x[newest]
       .line   17
           SHL      .S2     B6,0x10,B4         ; |63|   Shift Nblock to BK0 field
       .line   31
           SET      .S2     B4,0x8,0x8,B4      ; |77|   Set mode circular, BK0, B4
       .line   33
           MVC      .S2     B4,AMR             ; |79|   load mode into AMR
       .line   38
           NOP              1
           ZERO     .D1     A3                 ; |84|
       .line   42
```

```
          MV      .D2     B5,B4
||        B       .S2     loop              ; (P) |95|  Continue until done
          SUB     .L1X    B0,1,A1
||        MVK     .S1     0x2,A2            ; init prolog collapse predicate
||        LDW     .D2T2   *B4--,B5          ; (P) |89|  x[newest-k] -> x_value
||        LDW     .D1T1   *A4++,A5          ; (P) |90|  h[k] -> h_value

;** -------------------------------------------------------------------*
loop:     ; PIPED LOOP KERNEL

   [!A2]  ADDSP   .L1     A0,A3,A3          ;      |92|  sum of products
||        MPYSP   .M1X    B5,A5,A0          ; @|91|  h[k]*x[n-k]

   [ B0]  ADD     .D2     0xffffffff,B0,B0  ; @|94|  Decrement count by 1 tap

   [ A2]  SUB     .D1     A2,1,A2           ;
|| [ B0]  B       .S2     loop              ; @|95|  Continue until done

   [ A1]  SUB     .S1     A1,1,A1           ;
|| [ A1]  LDW     .D2T2   *B4--,B5          ; @@@|89|  x[newest-k] -> x_value
|| [ A1]  LDW     .D1T1   *A4++,A5          ; @@@|90|  h[k] -> h_value

;** -------------------------------------------------------------------*
L3:       ; PIPED LOOP EPILOG
          ADDSP   .L1     A0,A3,A3          ; (E) @@@   |92|  sum of products
          .line   52
          .line   51
          B       .S2     B3                ; |98|
          NOP             2
          MV      .D1     A3,A4             ; |97|
          NOP             2
          ; BRANCH OCCURS                   ; |98|
          .endfunc        98,000000000h,0
```

Segment of a C Program for Calling the .asm Convolution Function

A segment of a main C function for calling the assembly function convolve() is shown in Program 3.5. Suppose we want to do an Nh = 25 tap filter. The circular buffer must be 32 words or BUF_LEN = 4 × 32 = 128 bytes. Since BUF_LEN = $2^{Nblock+1}$, we need Nblock = 6. Notice that the program uses the DATA_ALIGN pragma to align the circular buffer on a multiple of the block length. Also, the circular modification of the buffer input pointer, newest, is performed in the main routine rather than in the convolution function since it occurs only once for each new input sample.

Program 3.5 C Program to Call convolve()

```
    ...
#define Nh 25      /* number of filter taps*/
#define Nblock  6 /*length field in AMR */
#define BUF_LEN 1<<(Nblock+1) /* circular buffer */
                              /* size in bytes   */
#define BUF_LEN_WORDS 1<<(Nblock-1) /* BUF_LEN/4 */
/*** NOTE: x[ ] must be a global array *******/
   float x[BUF_LEN_WORDS]; /* circular buffer */
/* Align circ. buf. on multiple of block length */
   #pragma DATA_ALIGN(x, BUF_LEN)
   ...
main(){
   ...
   Uint32 sample_pair = 0;
   int newest = 0; /* Input pointer for buffer */
   float y = 0; /* filter output sample */
   int iy = 0; /* int output for codec */
   int ix;  /* new input sample */
   float h[Nh] = { Put your filter coefficients here
                   separated by commas };
/* Prototype the convolution function. */
 extern float convolve(float x[], float h[],
         int N_taps, int N_block, int newest);

 /*  Configure McBSP's and codec      */
   ...
   for(;;){
    /* Send last filter output to codec.  */
    while(!DSK6713_AIC23_write(hCodec, iy));

/* NOTE: DSK6713_AIC23_read() returns unsigned int.*/
/*  Convert returned value to an ''int'' before   */
/*  shifting right to extend sign.                 */

   /* Get a new sample pair   */
    while(!DSK6713_AIC23_read(hCodec, &sample_pair));

    ix = ( (int) sample_pair) >> 16;/* Extend sign. Eliminate right
                                      channel (16 LSB's).*/
    newest++;  /* Increment input pointer modulo buffer size */
    if(newest==BUF_LEN_WORDS) newest = 0;
```

```
    x[newest] = ix; /* Put new sample in buffer    */
    y = convolve(x, h, Nh, Nblock, newest); /* Compute new output */
    iy = ( (int) y) << 16; /* Convert to int and put in left channel */
  }
```

3.7   Infinite Duration Impulse Response (IIR) Filters

A filter with an impulse response, $h(n)$, that has infinite duration is known as an IIR filter. When $h(n)$ is the sum of damped exponentials, its z-transform, $H(z)$, which is also called its transfer function, is a rational function of z. That is, it is the ratio of two finite degree polynomials. We will use a rational function of the form

$$H(z) = \frac{b_0 + b_1 z^{-1} + b_2 z^{-2} + \cdots + b_N z^{-N}}{1 + a_1 z^{-1} + a_2 z^{-2} + \cdots + a_M z^{-M}} = \frac{B(z)}{A(z)} \tag{3.11}$$

3.7.1   Realizations for IIR Filters

Rational transfer functions can be realized in many ways. Three common realizations will be described below. The first realization will be called a *type 0 direct form*. The ratio of the z-transforms of the filter output and input is

$$\frac{Y(z)}{X(z)} = H(z) = \frac{B(z)}{A(z)} \tag{3.12}$$

Cross multiplying gives

$$Y(z)A(z) = X(z)B(z) \quad \text{or} \quad Y(z)\left(1 + \sum_{k=1}^{M} a_k z^{-k}\right) = X(z) \sum_{k=0}^{N} b_k z^{-k} \tag{3.13}$$

Taking all except the $Y(z)$ term to the righthand side yields

$$Y(z) = \sum_{k=0}^{N} b_k X(z) z^{-k} - \sum_{k=1}^{M} a_k Y(z) z^{-k} \tag{3.14}$$

The time domain equivalent is the difference equation

$$y[n] = \sum_{k=0}^{N} b_k x[n-k] - \sum_{k=1}^{M} a_k y[n-k] \tag{3.15}$$

This equation shows how to compute the current filter output from the current and N past inputs and M past outputs. A filter implemented in this way is also called a *recursive* filter since past outputs are used to calculate the current output. It is called a direct form because the coefficients in the transfer function appear directly in the difference equation.

Another realization which we will call a *type 1 direct form* is based on observing that (3.12) can be rearranged into the cascade form

$$Y(z) = \frac{X(z)}{A(z)}B(z) = V(z)B(z) \tag{3.16}$$

where

$$V(z) = X(z)\frac{1}{A(z)} \tag{3.17}$$

This is illustrated in Figure 3.3. The intermediate signal $v[n]$ can be computed using the

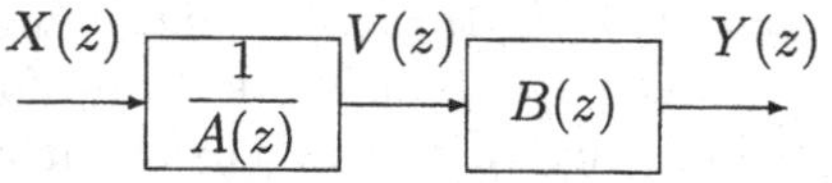

Figure 3.3: First Step in Finding Type 1 Direct Form Realization

direct form 0 realization

$$v[n] = x[n] - \sum_{k=1}^{M} a_k v[n-k] \tag{3.18}$$

Then, the output can be computed as

$$y[n] = \sum_{k=0}^{N} b_k v[n-k] \tag{3.19}$$

A block diagram for these equations is shown in Figure 3.4 where it is assumed that $M = N$. This form requires less storage than the type 0 direct form.

The contents of the delay elements, $s_1[n], \ldots, s_N[n]$, are *state variables* for the filter. The current output and next state can be computed from the current input and state. The following sequence of steps can be used to compute the filter outputs and states:

Step 1: Compute $v[n]$

$$v[n] = x[n] - \sum_{k=1}^{N} a_k s_k[n]$$

Step 2: Compute the output $y[n]$

$$y[n] = b_0 v[n] + \sum_{k=1}^{N} b_k s_k[n]$$

Step 3: Update the state variables

$$
\begin{aligned}
s_N[n+1] &= s_{N-1}[n] \\
s_{N-1}[n+1] &= s_{N-2}[n] \\
&\vdots \\
s_2[n+1] &= s_1[n] \\
s_1[n+1] &= v[n]
\end{aligned}
$$

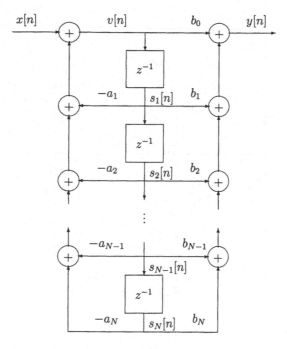

Figure 3.4: Type 1 Direct Form Realization

Another realization called the *type 2 direct form* can be found by rearranging (3.13). For simplicity, let $M = N$. Then

$$Y(z) = b_0 X(z) + \sum_{k=1}^{N} [b_k X(z) - a_k Y(z)] z^{-k} \tag{3.20}$$

A block diagram for this realization is shown in Figure 3.5. It requires essentially the same storage and arithmetic as a type 1 direct form.

The sequence of steps for computing the output of the type 2 direct form and updating its state is:

Step 1: Compute the output $y[n]$

$$y[n] = b_0 x[n] + s_1[n]$$

Step 2: Update the state variables

$$\begin{aligned}
s_1[n+1] &= b_1 x[n] - a_1 y[n] + s_2[n] \\
s_2[n+1] &= b_2 x[n] - a_2 y[n] + s_3[n] \\
&\vdots \\
s_{N-1}[n+1] &= b_{N-1} x[n] - a_{N-1} y[n] + s_N[n] \\
s_N[n+1] &= b_N x[n] - a_N y[n]
\end{aligned}$$

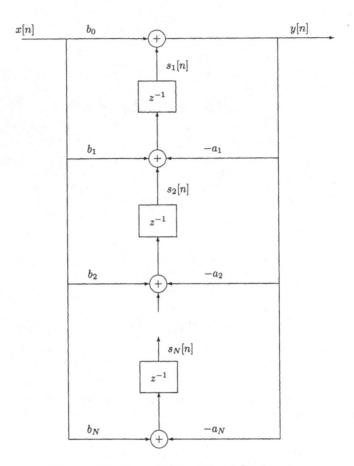

Figure 3.5: Type 2 Direct Form Realization

3.7.2   A Program for Designing IIR Filters

The program, `C:\digfil\iir\iir.exe`, designs IIR filters by using the bilinear transformation [II.C.17, pp. 212-219] with a Butterworth, Chebyshev, inverse Chebyshev, or elliptic analog prototype filter. It can design lowpass, highpass, bandpass, or bandstop analog and digital filters. The form of the resulting filter is a cascade (product) of sections, each with a second order numerator and denominator with the leading constant terms normalized to 1, possibly a first order section normalized in the same way, and an overall scale factor. These second order sections are also know as *biquads*. The sections can be realized by any of the three direct forms described above or other structures that can be found in DSP books. MATLAB has a similar IIR filter design package.

Care must be taken to prevent overflows and underflows when digital filters are implemented with fixed point DSP's. This problem is significantly reduced with floating point DSP's. Sometimes the overall scale factor generated by `iir.exe` is quite small and to maintain numerical accuracy it should be split among the different sections.

An example of how to use iir.exe is shown below. The program prompts are shown in upper case letters and the user responses in lower case letters or numbers. In this example, a bandpass filter is designed based on an elliptic analog prototype filter. The nominal lower stopband extends from 0 to 600 Hz, the passband extends from 1000 to 2000 Hz, and the upper stopband extends from 3000 to 4000 Hz. The questions and answers are explained more fully after the dialog.

```
SAVE RESULTS IN A FILE (Y OR N): y
ENTER LISTING FILENAME: junk.lst
ENTER 1 FOR ANALOG, 2 FOR DIGITAL: 2
ENTER SAMPLING RATE IN HZ: 8000
ENTER NUMBER OF FREQS TO DISPLAY: 100
ENTER STARTING FREQUENCY IN HZ: 0
ENTER STOPPING FREQUENCY IN HZ: 4000
ENTER 1 FOR BW, 2 FOR CHEBY, 3 FOR ICHEBY, 4 FOR ELLIPTIC: 4
ENTER 1 FOR LOWPASS, 2 FOR HP, 3 FOR BP, OR 4 FOR BR: 3
ENTER F1,F2,F3,F4 FOR BP OR BR FREQS: 600,1000,2000,3000
ENTER PASSBAND RIPPLE AND STOPBAND ATTENUATION IN +DB: 0.2,40

ELLIPTIC FILTER ORDER =        4

CREATE FREQ, LINEAR GAIN FILE (Y,N)? n
CREATE FREQ, DB GAIN FILE (Y,N)? Y
ENTER FILENAME: junkdb.dat
CREATE FREQ, PHASE FILE (Y,N)? n
CREATE FREQ, DELAY FILE (Y,N)? y
ENTER FILENAME: JUNKDEL.DAT
```

The first line of the dialog asks if you want to save the results in in a disk file. If the answer is Y or y, you are prompted for the name of a file. If the answer is N or n, the results appear on the screen (usually too fast to be read). The program computes the frequency response of the designed filter at the number of points specified which are equally spaced over the range of frequencies selected. You are then prompted for the type of analog prototype filter desired and the frequency selectivity type of the digital filter. In the case of a bandpass (BP) filter, four critical frequencies, $F1 < F2 < F3 < F4$, must be entered. The frequency $F1$ is the upper edge of the lower stopband, $F2$ is the lower edge of the passband, $F3$ is the upper edge of the passband, and $F4$ is the lower edge of the upper stopband. In the case of an elliptic filter, you are then prompted for the desired maximum passband ripple and the minimum stopband attenuation. The program then computes the order of the required analog lowpass prototype filter which in this example is 4. The actual order of the digital filter is double this number for bandpass and band reject filters. The user is given the option of choosing the filter order or letting iir.exe choose the order for some of the other prototype filters. Finally you are prompted for the types of frequency response files you wish to generate which can then be plotted with your favorite graphing program.

The RESULTS file for this example is shown below. First, the z-plane zeros and poles are displayed in rectangular form. Then they are shown in polar form. The radius is the magnitude of the pole or zero and the frequency is $f_s\theta/(2\pi)$ where θ is the angle and f_s is the sampling frequency. Notice, that for this bandpass filter, the zeros are all exactly on the unit circle with frequencies in the stop bands. The pole frequencies are in the passband.

The coefficients of the numerators and denominators of the second order sections are given and they can be realized by the direct forms. It is shown in many DSP books that it is computationally better to realize an IIR filter by splitting it into low order sections rather than by one high order section.

Finally, the amplitude response on a linear scale, the amplitude response on a dB scale, the phase response, and the envelope delay are listed for the chosen range. This data also appears in separate files if selected in the dialog.

```
              DIGITAL BANDPASS ELLIPTIC FILTER

                     FILTER ORDER =  8

                         Z PLANE


         ZEROS                          POLES

      .977149 +- j  .212554      .173365 +- j  .761580
      .902015 +- j  .431705     -.028463 +- j  .919833
     -.538154 +- j  .842847      .683010 +- j  .651915
     -.873779 +- j  .486323      .482595 +- j  .656484


      RADIUS       FREQUENCY      RADIUS       FREQUENCY

   .100000E+01    .272712E+03   .781063E+00   .171502E+04
   .100000E+01    .568352E+03   .920273E+00   .203939E+04
   .100000E+01    .272351E+04   .944190E+00   .970348E+03
   .100000E+01    .335335E+04   .814782E+00   .119288E+04


         4 CASCADE STAGES, EACH OF THE FORM:

F(z) = ( 1 + B1*z**(-1) + B2*z**(-2) ) / ( 1 + A1*z**(-1) + A2*z**(-2) )


       B1            B2            A1            A2
    -1.954298     1.000000     -.346731      .610059
    -1.804029     1.000000      .056927      .846903
     1.076307     1.000000    -1.366019      .891495
     1.747559     1.000000     -.965191      .663870


 SCALE FACTOR FOR UNITY GAIN IN PASSBAND:   1.8000479016654E-002
```

FREQUENCY RESPONSE

FREQUENCY	GAIN	GAIN (dB)	PHASE	DELAY (SEC)
.0000	2.1048E-03	-5.3536E+01	.00000	.13458E-03
40.0000	2.0557E-03	-5.3741E+01	-.03385	.13493E-03
80.0000	1.9093E-03	-5.4382E+01	-.06789	.13600E-03
120.0000	1.6681E-03	-5.5556E+01	-.10228	.13780E-03

.
.
.

3.7.3   Two Methods for Measuring a Phase Response

You will be asked to measure the phase response of an IIR filter in the laboratory experiments below. Two methods for measuring phase differences, (1) Lissajous figures and (2) relative time delays, are explained in this section. Suppose the input to a system is

$$x(t) = A \sin \omega_0 t \tag{3.21}$$

and the output is

$$y(t) = B \sin(\omega_0 t + \theta) \tag{3.22}$$

Measuring Phase Differences by Lissajous Figures

If $x(t)$ is applied to the horizontal input of an oscilloscope and $y(t)$ is applied to the vertical input, the ellipse

$$\left(\frac{y}{B}\right)^2 - 2\left(\frac{x}{A}\right)\left(\frac{y}{B}\right)\cos\theta + \left(\frac{x}{A}\right)^2 = \sin^2\theta \tag{3.23}$$

will be observed. If $\theta = 0$ the ellipse degenerates into a straight line with positive slope, if $\theta = \pi$ it becomes a line with negative slope, and if $\theta = \pi/2$ or $3\pi/2$ its principal axes become aligned with the x and y axes.

From (3.21), it can be seen that the maximum value for x is $x_{max} = A$. The ellipse crosses the x-axis when $y = 0$ or $\omega_0 t + \theta = \pi$. The corresponding value for x is

$$x_0 = A \sin(\pi - \theta) = A \sin \theta \tag{3.24}$$

Thus

$$\frac{x_0}{x_{max}} = \sin \theta \tag{3.25}$$

and so

$$\theta = \sin^{-1} \frac{x_0}{x_{max}} \tag{3.26}$$

A similar equation can be derived for y measurements.

Measuring Phase Differences by Relative Time Delay

You will find it difficult to accurately measure x_{max} and x_0 using the oscilloscope and also to determine what value to use for the arcsine. An easier and more accurate method for finding the phase difference is to measure the time delay between the positive zero crossings of $x(t)$ and $y(t)$. The output can also be expressed as

$$y(t) = B\sin[\omega_0(t + d)] = B\sin(\omega_0 t + \theta) \tag{3.27}$$

where

$$\theta = \omega_0 d = 2\pi\frac{d}{T_0} \quad \text{radians} \tag{3.28}$$

Therefore, the phase difference can be easily found by multiplying the relative time delay by the radian frequency. When using this method, make sure the oscilloscope traces for $x(t)$ and $y(t)$ are synchronized to the same time reference.

3.8   Laboratory Experiments for Digital Filters

3.8.1   Experiment 3.1: FIR Filters Entirely in C

1. Initialize McBSP0, McBSP1, and the AIC23 codec as before and set the sampling rate to 16000 Hz.

2. If you have not done this already, measure the amplitude response of the DSK left channel analog path. We will assume the right channel is the same. Apply a sine wave from the signal generator to the left channel of the line input and loop the samples internally in the DSP back to the line output. Vary the frequency and record the values of the output amplitude divided by the input amplitude. Use enough frequencies to get an accurate plot of the response. In particular, be sure to use enough points in the transition region from the passband to the stopband. Plot the response using your favorite plotting program. You should use the set of frequencies chosen here in the rest of Chapter 3.

3. Design a 25-tap bandpass FIR filter for a sampling rate of 16 kHz using WINDOW.EXE, REMEZ87.EXE, or MATLAB. The passband should extend from 2,000 Hz to 5,000 Hz. Plot the theoretical amplitude response in dB.

4. Write a C program to implement the filter using a circular sample buffer. Convert the input samples to floating point format before putting them into the circular buffer. The left channel is the upper 16 bits. So, arithmetically shift the received word 16 bits right to extend the sign and lop off the lower 16 bits (right DAC channel) and then convert the result to a float.

   The start of each iteration should be controlled by synchronizing it to the McBSP1 XRDY flag. Each time a sample is transmitted, a new input sample can be read because the transmit and receive frame syncs are identical.

5. First compile your program without optimization. Look at the assembly code generated by the compiler to get some idea of how the C source code is implemented by the 'C6713. Use the profiling capabilities of Code Composer Studio to measure the number of cycles required to generate one output sample. (Do not include the time spent polling the XRDY flag!)

6. Browse through Chapter 3 Optimizing Your Code in the *TMS320C6000 Optimizing Compiler User's Guide* [I.9]. Then compile your program using the four optimization levels o0, o1, o2, and o3. Look at the assembly code generated for each optimization level. Measure and record the number of cycles required to generate one output sample for each optimization level.

7. Measure the amplitude response of the filtering system from the line input to line output jack and plot the results on a dB scale after correcting for the DSK response. Compare your measured result with the theoretical response.

8. Increase the number of filter taps from 25 to find the largest number of taps that can be used without running out of time and report the result. (Hint: Do not re-design your filter for each new length. Simply append zeros to your 25-tap design. Multiplying by zero takes as long as multiplying by any other number because the compiler does not use the fact multiplying any finite number by zero gives zero.)

3.8.2 Experiment 3.2: FIR Filters Using C and Assembly

1. Complete the C program that calls the assembly function `convolve()` in the file `convol1.sa`. Use the 25-tap filter you designed for Experiment 3.1. The file `convol1.sa` can be found in the directory `C:\c6713dsk` and on the class web site.

2. Build the complete executable module using level -o3 optimization for both the C and linear assembly programs.

3. Attach the signal generator to the input jack and observe the output on the oscilloscope. Sweep the input frequency to check that the frequency response is correct. You do not have to do a detailed frequency response measurement.
   Note: You may have to click on **Debug** → **Reset CPU** to get the program to run properly.

4. Use the profiling capabilities of Code Composer Studio to measure the number of cycles required for one call to the convolution function with and without optimization. Compare the results to those for the Experiment 3.1 implementation totally in C.

5. Get the file `convolve.sa` from our web site or the directory `C:\c6713dsk`. It unrolls the convolution sum loop once to compute the contributions from two taps in each iteration of the summation loop. The number of filter taps must be an even number. However, a filter with an odd number of taps can be implemented by adding one dummy tap which is zero. The idea is to improve efficiency by eliminating branching

overhead and by allowing the optimizer to schedule use of the execution units more optimally.

Rebuild your FIR filter implementation using this new assembly function and level -o3 optimization. Compare the execution time for one call this convolution routine with that of the function in `convol1.sa`

The variable, ii, reported by the assembly optimizer indicates the number of cycles required by the convolution loop kernel. With level -o2 or -o3 optimization it reports ii = 4 for `convol1.sa` and `convolve.sa`, and that 4 instructions are executing in parallel. Therefore, the kernel for `convol1.sa` requires 4 cycles per tap while the kernel for `convolve.sa` requires only 2 cycles per tap. Notice the `convol1.asm` only uses multiplier .M1 while `convolve.asm` use both .M1 and .M2.

3.8.3   Experiment 3.3: Implementing an IIR Filter

In these experiments you will design a bandpass IIR filter, plot its theoretical amplitude and phase responses, and compare them with measured responses. Perform the following tasks for IIR filters:

1. Design an IIR bandpass filter based on an elliptic lowpass analog prototype. Use a 16 kHz sampling rate. The lower stopband should extend from 0 to 800 Hz, the passband from 2000 to 5000 Hz, and the upper stopband from 7000 to 8000 Hz. The passband ripple should be no more than 0.3 dB and the stopband attenuation should be at least 40 dB.

   Plot the theoretical amplitude response generated by the filter design program on a dB scale. Plot the phase response also. Explain any discontinuities in the phase response.

2. Write a program to implement your filter on the DSK. Use a cascade of second-order and, possibly, a first-order type 1 direct form filter sections.

3. Use the signal generator and oscilloscope to measure the amplitude response from the input to output jacks and plot it in dB. Also measure the phase response and plot the results. Be sure to adjust the measured responses for the responses of the analog paths in the DSK. Compare your theoretical and measured responses.

4. Use the profiling capability of Code Composer Studio to measure the number of clock cycles and time required to process one sample, and record the result. Do this for the two cases where the program is compiled without optimization and with level -o3 optimization.

3.9   Additional References

There are many excellent books covering the theoretical and practical aspects of digital signal processing. A few of them are included in Section II.C of the list of references at the end of this text. The initial books that dealt with discrete-time systems focused on sampled-data control systems and appeared in the late 1950's. They did not discuss methods for designing digital filters with the demanding frequency responses required in many communications and signal processing applications. The first complete college textbook that focused on digital signal processing, Oppenheim and Schafer [II.C.12], appeared in 1975 and was followed a year later by Tretter [II.C.17]. The theory has advanced very little since these books were published, but some of the newer books include sections on DSP chips.

See Proakis and Manolakis [II.C.15, Chapter 5] for a more extensive discussion of the frequency responses of FIR and IIR digital filters including a variety of examples. Also see Oppenheim and Schafer [II.C.13, Chapter 5].

Detailed discussions of structures for realizing FIR and IIR filters can be found in Oppenheim and Schafer [II.C.13, Chapter 6] and Proakis and Manolakis [II.C.15, Chapter 7]. They discuss the direct forms presented in this experiment and additional structures like lattice filters.

All the books listed in Section II.C of the references discuss FIR and IIR digital filter design techniques to meet frequency domain requirements. For example, see Oppenheim and Schafer [II.C.13, Chapter 7], Proakis and Manolakis [II.C.15, Chapter 8], and Tretter [II.C.17, Chapter 8]. The first two of these references present the theory for the method used in the REMEZ87 FIR filter design program and filter design examples. An excellent source for FORTRAN digital filter design programs and other signal processing programs is the classic IEEE Press book [II.C.7].

Important topics not discussed in this experiment are the effects of quantization in A/D conversion and finite word length arithmetic. See Oppenheim and Schafer [II.C.13, Sections 6.7 – 6.10], Proakis and Manolakis [II.C.15, Sections 6.2 and 7.6 – 7.8], and Tretter [II.C.17, Chapter 9] for discussions of these topics.

Chapter 4

The FFT and Power Spectrum Estimation

In this chapter, you will review and implement some important techniques for digital signal processing and data transmission. In particular, you will learn about the Fast Fourier Transform (FFT) and build a spectrum analyzer using the FFT. A technique called Orthogonal Frequency Division Multiplexing (OFDM) has become very popular for broadband wireline and wireless data transmission. OFDM uses an inverse FFT for modulation and an FFT for demodulation. Therefore, it is important for people specializing in communications and signal processing to have a strong understanding of the FFT. OFDM is the subject of Chapter 17

It is assumed that the reader is taking or has had a course on the theory of digital signal processing, so the presentation is brief. It sets the notation and summarizes important results. Comprehensive developments of the theory can be found in the books on digital signal processing listed in the references. References for specific topics are suggested at the end of this chapter.

4.1   The Discrete-Time Fourier Transform

Suppose that a continuous-time signal $x(t)$ is sampled with period T or sampling frequency $\omega_s = 2\pi/T$ to obtain the discrete-time signal $x[n] = x(nT)$. We will define the discrete-time Fourier transform of $x[n]$ to be the following sum, if it exists:

$$X(\omega) = \sum_{n=-\infty}^{\infty} x[n]e^{-j\omega nT} \qquad (4.1)$$

The z-transform of the signal is obtained by making the substitution $z = e^{j\omega T}$. Notice that $X(\omega)$ has period ω_s since the sum is a Fourier series.

The discrete-time signal can be determined from its discrete-time Fourier transform by the inversion integral

$$x[n] = \frac{1}{\omega_s} \int_{-\omega_s/2}^{\omega_s/2} X(\omega)e^{j\omega nT} d\omega \qquad (4.2)$$

Thus, $x[n]$ can be considered to be the sum of sampled sine waves at a continuum of frequencies in the Nyquist band $-\omega_s/2 < \omega \leq \omega_s/2$ with complex amplitudes given by $X(\omega)$. This suggests calling $X(\omega)$ the frequency spectrum of the signal.

4.2   Data Window Functions

The observed data sequence must be limited to a finite duration to compute the transform summation in practice. The most obvious approach is to simply truncate the summation to a finite range, for example, $0 \leq n \leq N-1$. This is equivalent to forming a new data sequence by multiplying the original signal $x[n]$ by the rectangular N-point data window function

$$h_1[n] = \begin{cases} 1 & \text{for } n = 0, 1, \ldots, N-1 \\ 0 & \text{elsewhere} \end{cases} \tag{4.3}$$

The effect of the truncation on the spectrum can be determined by the following product theorem. Let $h[n]$ and $x[n]$ be discrete-time signals with discrete-time Fourier transforms $H(\omega)$ and $X(\omega)$, respectively. Then, it can be shown that $y[n] = h[n]x[n]$ has the transform

$$Y(\omega) = \frac{1}{\omega_s} \int_{-\omega_s/2}^{\omega_s/2} X(\lambda) H(\omega - \lambda) \, d\lambda \tag{4.4}$$

Thus, the transform of a product in the time-domain becomes a convolution of the transforms in the frequency domain.

The discrete-time Fourier transform of the rectangular window is

$$H_1(\omega) = \sum_{n=0}^{N-1} e^{-j\omega nT} = e^{-j\omega(N-1)T/2} \frac{\sin(\omega NT/2)}{\sin(\omega T/2)} \tag{4.5}$$

The transform of a data window is often called a *spectral window*. A plot of $|H_1(\omega)|$ for $N = 10$ is shown in Figure 4.1. This function has a peak magnitude of height N at the origin and is zero at the frequencies $k\omega_s/N$ that are not multiples of ω_s. Thus, the main lobe centered about the origin has width $2\omega_s/N$. From (4.4) we see that the transform of the truncated sum is a smoothed version of the true spectrum, $X(\omega)$, obtained by convolving $X(\omega)$ with $H_1(\omega)$. The value at frequency ω is predominantly an average of values in the vicinity of ω weighted by $H_1(\omega - \lambda)$ over its main lobe which extends from $\lambda = \omega - (\omega_s/N)$ to $\lambda = \omega + (\omega_s/N)$. However, the maximum sidelobe magnitude of $H_1(\omega)$ is down only about 13 dB from the main lobe peak, so the value of $X(\omega)$ estimated by the truncated summation can be significantly distorted by large values away from ω "leaking through" the spectral window.

The spectral leakage problem can be reduced by using a data window that has smaller sidelobes in its transform. To obtain unbiased power spectral density estimates for a flat spectrum, a data window $h[n]$ should be normalized so that

$$\frac{1}{N} \sum_{n=0}^{N-1} h^2[n] = 1 \tag{4.6}$$

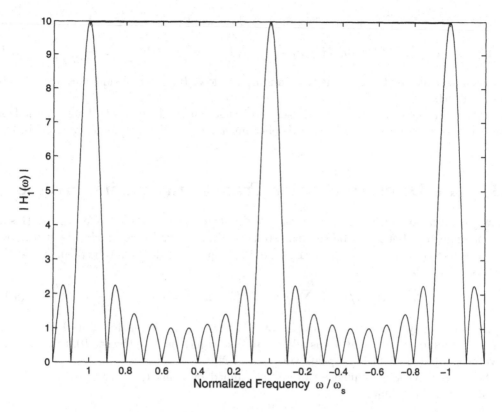

Figure 4.1: Magnitude of the Rectangular Window Transform for $N = 10$

Note that this normalization is different from that used for windows in digital filter design.

One popular data window is the Hanning window

$$h_2[n] = \begin{cases} c_2 \, 0.5 \left\{ 1 + \cos\left[\left(n - \frac{N-1}{2} \right) \frac{2\pi}{N} \right] \right\} & \text{for } n = 0, 1, \ldots, N-1 \\ 0 & \text{elsewhere} \end{cases} \tag{4.7}$$

where $c_2 = (3/8)^{-1/2}$ provides the proper normalization. The corresponding spectral window is

$$H_2(\omega) = c_2 e^{-j\omega(N-1)T/2} \left[0.5 H_1(\omega) + 0.25 H_1 \left(\omega - \frac{\omega_s}{N} \right) + 0.25 H_1 \left(\omega + \frac{\omega_s}{N} \right) \right] \tag{4.8}$$

The maximum sidelobe amplitude is down by 37.5 dB for this window. However, the mainlobe has double the width of that for the rectangular window.

Another popular data window is the Hamming window

$$h_3[n] = \begin{cases} c_3 \left\{ 0.54 + 0.46 \cos\left[\left(n - \frac{N-1}{2} \right) \frac{2\pi}{N} \right] \right\} & \text{for } n = 0, 1, \ldots, N-1 \\ 0 & \text{elsewhere} \end{cases} \tag{4.9}$$

where $c_3 = (0.3974)^{-1/2}$. The corresponding spectral window is

$$H_3(\omega) = c_3 e^{-j\omega(N-1)T/2} \left[0.54 H_1(\omega) + 0.23 H_1 \left(\omega - \frac{\omega_s}{N} \right) + 0.23 H_1 \left(\omega + \frac{\omega_s}{N} \right) \right] \quad (4.10)$$

This window is almost the same as the Hanning window. Its spectral sidelobes are down by at least 40 dB.

Other good windows such as the Kaiser and Blackman windows can be found in the DSP reference books. In general, reduced sidelobes must be traded against increased main lobe width.

4.3   The Discrete Fourier Transform and its Inverse

Let $x[n]$ be a signal which is zero for n outside the set $\{0, 1, \ldots, N-1\}$. We will call this an N-point sequence. Let $X(\omega)$ be the discrete-time Fourier transform of $x[n]$ defined above. Then, the *discrete Fourier transform* (DFT) of this sequence is defined to be the new N-point sequence

$$X_k = X(k\omega_s/N) = \sum_{n=0}^{N-1} x[n] e^{-j\frac{2\pi}{N}nk} \quad \text{for} \quad k = 0, 1, \ldots, N-1 \quad (4.11)$$

The DFT is simply the set of N samples of $X(\omega)$ taken at frequencies spaced by ω_s/N in the Nyquist band. Notice that if k is allowed to take values outside the set $\{0, 1, \ldots, N-1\}$, the value computed by (4.11) repeats with period N.

The original N-point sequence can be determined by using the inverse discrete Fourier transform (IDFT) formula

$$x[n] = \frac{1}{N} \sum_{k=0}^{N-1} X_k e^{j\frac{2\pi}{N}nk} \quad \text{for} \quad k = 0, 1, \ldots, N-1 \quad (4.12)$$

4.4   The Fast Fourier Transform

Direct computation of a single DFT point using (4.11) requires $N-1$ additions and N multiplications ignoring the fact that for some k the exponentials are 1 or -1. Thus, direct computation of all N points requires $N(N-1)$ complex additions and N^2 complex multiplications. The computational complexity can be reduced to the order of $N \log_2 N$ by algorithms known as *fast Fourier transforms* (FFT's) that compute the DFT indirectly. For example, with $N = 1024$ the FFT reduces the computational requirements by a factor of

$$\frac{N^2}{N \log_2 N} = 102.4$$

The improvement increases with N.

One FFT algorithm is called the *decimation-in-time* algorithm. A brief derivation is presented below for reference. To simplify the notation, let $W_N = e^{-j2\pi/N}$ so (4.11) becomes

$$X_k = \sum_{n=0}^{N-1} x[n] W_N^{nk} \quad (4.13)$$

This algorithm assumes that N is a power of 2. Splitting the sum into a sum over even n and one over odd n gives

$$X_k = \sum_{n=0}^{\frac{N}{2}-1} x[2n]W_N^{2nk} + \sum_{n=0}^{\frac{N}{2}-1} x[2n+1]W_N^{(2n+1)k} \quad \text{for} \quad k = 0, 1, \ldots, N-1 \qquad (4.14)$$

Let the even numbered points be the $N/2$ point sequence

$$a[n] = x[2n] \quad \text{for} \quad n = 0, 1, \ldots, \frac{N}{2} - 1 \qquad (4.15)$$

and the odd numbered points be the $N/2$ point sequence

$$b[n] = x[2n+1] \quad \text{for} \quad n = 0, 1, \ldots, \frac{N}{2} - 1 \qquad (4.16)$$

Also observe that $W_N^2 = W_{N/2}$. Thus, (4.14) can be written as

$$X_k = \sum_{n=0}^{\frac{N}{2}-1} a[n]W_{N/2}^{nk} + W_N^k \sum_{n=0}^{\frac{N}{2}-1} b[n]W_{N/2}^{nk} \quad \text{for} \quad k = 0, 1, \ldots, N-1 \qquad (4.17)$$

Let A_k and B_k be the $N/2$-point DFT's of $a[n]$ and $b[n]$ so that these DFT's have period $N/2$. With these definitions, (4.17) becomes

$$X_k = A_k + W_N^k B_k \quad \text{for} \quad k = 0, 1, \ldots, N-1 \qquad (4.18)$$

The next step results in the key equations for the decimation-in-time FFT. First observe that $W_N^{N/2} = -1$. Then, the previous equation can be separated into the two equations

$$X_k = A_k + W_N^k B_k \quad \text{for} \quad k = 0, 1, \ldots, \frac{N}{2} - 1 \qquad (4.19)$$

$$X_{k+\frac{N}{2}} = A_k - W_N^k B_k \quad \text{for} \quad k = 0, 1, \ldots, \frac{N}{2} - 1 \qquad (4.20)$$

Equations (4.19) and (4.20) show how to compute an N-point DFT by combining a pair of $N/2$-point DFT's. A flowgraph for this pair of equations is shown in Figure 4.2. This computation is called an FFT butterfly because of the shape of the flowgraph. A complete flowgraph for this first step with $N = 8$ is shown in Figure 4.3.

Assuming that the $N/2$-point DFT's, A_k and B_k, are known, it requires $N/2$ complex multiplications to compute $B_k W_N^k$, $N/2$ complex additions to compute $X_k = A_k + W_N^k B_k$, and $N/2$ complex subtractions to compute $X_{k+\frac{N}{2}} = A_k - W_N^k B_k$ for $k = 0, 1, \ldots, \frac{N}{2} - 1$. Addition and subtraction can be considered to be the same in terms of computational complexity. Thus, the entire N-point DFT can be computed with $N/2$ complex multiplications and N complex additions from the pair of $N/2$-point DFT's.

The same procedure can be used to compute the $N/2$-point DFT's from $N/4$-point DFT's. Computation of A_k by this method would require $N/4$ complex multiplications and $N/2$ complex additions. Finding B_k would require the same amount of computation. Thus the

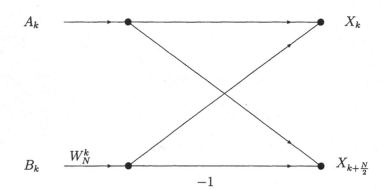

Figure 4.2: Flowgraph for an FFT Butterfly

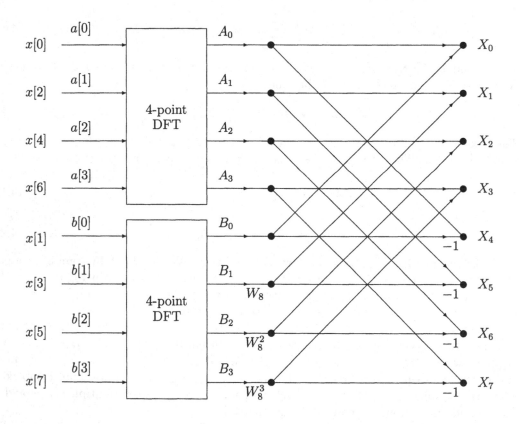

Figure 4.3: First Step in an 8-Point Decimation-in-Time FFT

total amount of computation to compute both A_k and B_k is $N/2$ multiplications and N additions, which is the same as for computing X_k from A_k and B_k.

The reduction by 2 procedure can be repeated until one-point DFT's are reached. A one-point DFT of a point is just the point itself. This requires $\log_2(N)$ stages. Therefore, the entire amount of computation required to compute the N-point DFT is $\frac{N}{2}\log_2(N)$ complex multiplications and $N\log_2(N)$ complex additions.

A complete flowgraph for an 8-point decimation-in-time FFT is shown in Figure 4.4. Notice that the input points are arranged in a scrambled order while the output DFT is in its natural order. It can be shown that the successive separation into even and odd numbered sequences puts the input sequence in bit-reversed order for any N that is a power of 2. The bit-reversed order is obtained by reversing the bits of the indexes for the original input array elements. .

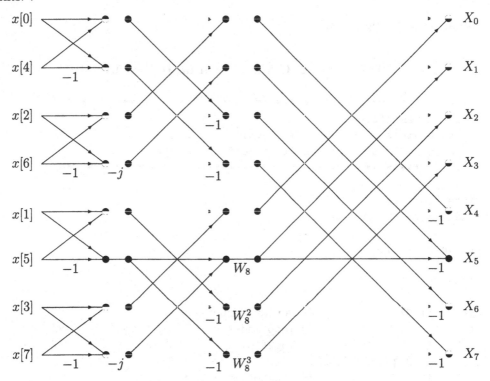

Figure 4.4: Complete Flowgraph for an 8-Point Decimation-in-Time FFT

A C function for computing a complex, radix-2, decimation-in-time FFT is included below. You can find the sources in `c:\digfil\fft`. It takes its complex input array in natural order and then rearranges it into bit reversed order. The output is in natural order. The computations are performed *in-place* with the output array written over the input array. The complex exponentials W^k are computed recursively. The program could be made more efficient by precomputing and storing a cosine/sine table at angle increments needed for the largest N to be used and addressing the table appropriately for smaller N. A header file defining a complex structure type and an example main function that computes the 16-point FFT of $\cos(5 \times 2\pi n/N)$ is also shown. The output of this program should be all 0 except for $X_5 = X_{11} = 8$.

Program 4.1 Header File Defining Complex Data Structure

```
/*  Header File complex.h                                 */
struct cmpx
  {
  float real;
  float imag;
  };
  typedef struct cmpx complex;
```

Program 4.2 C Main Program to Test fft.c

```
/*************************************************************/
/*  Program testfft.c                                      */
/*     An example of how to use function fft.c             */
/*  Compile by: gcc testfft.c fft.c -o fft.out -lm         */
/*************************************************************/

#include "complex.h"
#include <math.h>

extern void fft();

main()
  {
  complex X[16];         /* Declare input array         */
  int i;                 /* loop index                  */
  int M = 4;             /* log2(16)                    */
  float pi = 3.141592653589;
  int N = 16;            /* Number of FFT points        */
/*---------------------------------------------------------*/
/*   Initialize input array                               */
/*  Generate spectral lines at k = 5 and 11 of height 8.  */

  for(i=0; i<N; i++)
    {
    (X[i]).real = cos(i*5*2.0*pi/N);
    (X[i]).imag = 0.0;
    }
```

```
/*-------------------------------------------------------------*/
/* Perform FFT                                                 */

  fft(X,M);

/*  Display results on screen                                  */

  for(i=0; i<N; i++)
 printf("%4d%15.5f\t%15.5f\n",i,(X[i]).real, (X[i]).imag);

  }
```

Program 4.3 C Function for Radix-2 Decimation-in-Time FFT

```
/*****************************************************************/
/*  Function fft(complex *X, int M)                            */
/*                                                             */
/*   This is an elementary, complex, radix 2, decimation in    */
/*  time FFT.  The computations are performed "in place" and   */
/*  the output overwrites the input array.                     */
/*****************************************************************/

#include "complex.h"/* Definition of complex variable structure */
#include <math.h>   /* Definitions for math library             */

void fft(complex *X, int M)
        /* X is an array of N = 2**M complex points. */
  {
  complex temp1;           /* temporary storage complex variable */
  complex W;               /* e**(-j 2 pi/ N)                    */
  complex U;               /* Twiddle factor W**k                */
  int i,j,k;               /* loop indexes                       */
  int id;                  /* Index of lower point in butterfly  */
  int N = 1 << M;          /* Number of points for FFT           */
  int N2 = N/2;
  int L;                   /* FFT stage                          */
  int LE;              /* Number of points in sub DFT at stage L, */
                       /* and offset to next DFT in stage        */
  int LE1;                 /* Number of butterflies in one DFT at*/
                           /*  stage L. Also is offset to lower  */
                           /*  point in butterfly at stage L     */
  float pi = 3.1415926535897;

/*===============================================================*/
```

```
/*  Rearrange input array in bit-reversed order                    */
/*                                                                  */
/*    The index j is the bit reversed value of i.  Since 0 -> 0 */
/*  and N-1 -> N-1 under bit-reversal, these two reversals are  */
/*  skipped.                                                         */

  j = 0;
  for(i=1; i<(N-1); i++)
    {

/*++++++++++++++++++++++++++++++++++++++++++++++++++++++++++++++++++*/
/*  Increment bit-reversed counter for j by adding 1 to msb and */
/*   propagating carries from left to right.                        */

    k = N2;                    /* k is 1 in msb, 0 elsewhere        */

/*----------------------------------------------------------------*/
/*  Propagate carry from left to right                             */

    while(k<=j)                /* Propagate carry if bit is 1       */
      {
      j = j - k;               /* Bit tested is 1, so clear it.     */
      k = k/2;                 /* Set up 1 for next bit to right.   */
      }
    j = j+k;                   /* Change 1st 0 from left to 1       */
/*----------------------------------------------------------------*/
/*++++++++++++++++++++++++++++++++++++++++++++++++++++++++++++++++++*/

/*  Swap samples at locations i and j if not previously swapped.*/

    if(i<j)                    /* Test if previously swapped.       */
      {
      temp1.real = (X[j]).real;
      temp1.imag = (X[j]).imag;
      (X[j]).real = (X[i]).real;
      (X[j]).imag = (X[i]).imag;
      (X[i]).real = temp1.real;
      (X[i]).imag = temp1.imag;
      }
/*++++++++++++++++++++++++++++++++++++++++++++++++++++++++++++++++++*/
    }

/*================================================================*/
/*  Do M stages of butterflies                                     */
```

```
  for(L=1; L<= M; L++)
    {
    LE = 1 << L;            /*  LE = 2**L = points in sub DFT   */
    LE1 = LE/2;             /* Number of butterflies in sub-DFT */
    U.real = 1.0;
    U.imag = 0.0;           /* U = 1 + j 0                      */
    W.real = cos(pi/LE1);
    W.imag = - sin(pi/LE1); /* W = e**(-j 2 pi/LE)              */
/*----------------------------------------------------------------*/
/*   Do butterflies for L-th stage                               */

    for(j=0; j<LE1; j++)    /* Do the LE1 butterflies per sub DFT*/
      {
/*................................................................*/
/*    Compute butterflies that use same W**k                     */

      for(i=j; i<N; i += LE)
        {
        id = i + LE1;       /* Index of lower point in butterfly */
        temp1.real = (X[id]).real*U.real - (X[id]).imag*U.imag;
        temp1.imag = (X[id]).imag*U.real + (X[id]).real*U.imag;

        (X[id]).real = (X[i]).real - temp1.real;
        (X[id]).imag = (X[i]).imag - temp1.imag;

        (X[i]).real  = (X[i]).real + temp1.real;
        (X[i]).imag  = (X[i]).imag + temp1.imag;
        }
/*................................................................*/
/*   Recursively compute W**k as W*W**(k-1) = W*U                */

      temp1.real = U.real*W.real - U.imag*W.imag;
      U.imag = U.real*W.imag + U.imag*W.real;
      U.real = temp1.real;
/*................................................................*/
      }
/*----------------------------------------------------------------*/
    }
  return;
  }
```

4.5   Using the FFT to Estimate a Power Spectrum

One method for estimating power spectral densities is based on using a function called the *periodogram*. The periodogram of an N-point sequence $y[n]$ is defined to be

$$I_N(\omega) = \frac{1}{N}|Y(\omega)|^2 \tag{4.21}$$

where

$$Y(\omega) = \sum_{n=0}^{N-1} y[n]e^{-j\omega nT} \tag{4.22}$$

is the discrete-time Fourier transform of $y[n]$. It can be shown that the inverse transform of the periodogram is the *sample autocorrelation function*

$$R(n) = \begin{cases} \dfrac{1}{N}\displaystyle\sum_{k=0}^{N-1} y[n+k]\bar{y}[k] & \text{for } |n| \le N-1 \\ 0 & \text{elsewhere} \end{cases} \tag{4.23}$$

The variable, n, in the autocorrelation function is called the *lag*. For zero lag

$$R(0) = \frac{1}{N}\sum_{k=0}^{N-1} |y[k]|^2 = \frac{1}{\omega_s}\int_{-\omega_s/2}^{\omega_s/2} I_N(\omega)\,d\omega \tag{4.24}$$

is the average power in the sequence. This equation provides some justification for interpreting the periodogram as a function that shows how the power is distributed in the frequency domain.

At first glance, it is natural to assume that as N increase, the periodogram becomes a better estimate of the power spectral density for a stationary random process. However, this is not true. Actually, the mean of the periodogram converges to the true spectral density but its variance remains large. As N increases, the periodogram tends to oscillate more and more rapidly. See references [II.C.12] or [II.C.17] for details of this property and, more generally, estimation of power spectral densities.

A solution to this problem is to average the periodograms of different N-point sections of the observed data sequence. Let $x[n]$ be an observed data sequence with duration $M = LN$ and form the L windowed N-point data sections

$$y_k[n] = \begin{cases} h[n]x[n+kN] & \text{for } n = 0, 1, \ldots, N-1 \\ 0 & \text{elsewhere} \end{cases} \quad \text{for } k = 0, 1, \ldots, L-1 \tag{4.25}$$

where h[n] is a desired data window function. Designate the periodogram formed from the k-th widowed section by $I_{N,k}(\omega)$. Then the desired power spectral density estimator is

$$\hat{S}(\omega) = \frac{1}{L}\sum_{k=0}^{L-1} I_{N,k}(\omega) \tag{4.26}$$

When the data sections are statistically independent, averaging L sections reduces the variance by a factor of L. Additional gains can be achieved by overlapping the sections to some degree.

The periodograms can be computed at the uniformly spaced frequencies $\{k\omega_s/N; k = 0, 1, \ldots, N-1\}$ by using an N-point FFT. When the observed data sequence is real and the FFT program is designed to accept complex inputs, the computation time can be reduced by almost a factor of two by using the following identity. Let $a[n]$ and $b[n]$ be two real N-point sequences and form the complex sequence $c[n] = a[n] + jb[n]$. Then

$$|A_k|^2 + |B_k|^2 = \frac{|C_k|^2 + |C_{N-k}|^2}{2} \tag{4.27}$$

Thus, the sum of the periodograms of the two real sequences can be computed from the FFT of the single complex sequence.

4.6 Laboratory Experiments

4.6.1 Experiment 4.1: FFT Experiments

You can use the C FFT function starting on page 109 for these experiments. The files testfft.c, complex.h, and fft.c can be found in C:\DIGFIL\FFT. To test and extend your understanding of FFT's, perform the following tasks:

1. Let the sampling rate be 16 kHz and the sequence length be $N = 1024$ points. Theoretically find the DFT of the sequence

$$x_n = \sin(2\pi \times 2000 \times n/16000) \text{ for } n = 0, \ldots, 1023$$

2. Generate a program for the TMS320C6713 to compute the FFT of the sequence x_n defined above by doing the following:

   (a) First copy the linker command file, dsk6713.cmd from C:\c6713dsk to your project directory and increase -stack from 0x400 to 0x1000 so the stack does not overflow. Be sure to use this modified command file in your project.

   (b) Fill an N-point complex array with the real and imaginary parts of the sequence x_n defined above.

   (c) Compute the FFT of x_n by calling the function fft().

   (d) Fill a separate N-point real array with the squared complex magnitudes of the FFT values.

3. Use Code Composer Studio to:

   (a) Check your answer by displaying the complex FFT array in a Code Composer Studio watch window. Alternatively, you can use the C function printf() to write the FFT values to the CCS display window.

   (b) Read the squared magnitude array from the DSP into a disk file by using a probe point. Compare the disk file with the theoretical result to further check your program. You can also use the C function fprintf() to write the array to a PC disk file.

(c) Plot the squared magnitude array using the CCS graphing capability.

(d) Find the time required to compute the FFT by using the CCS profiling capability.

4. Repeat steps 1, 2, and 3 except multiply the input sequence by a Hamming window. When checking your results be sure to examine the FFT in the vicinity of 2000 and 14000 Hz.

5. Change the input sequence to:

$$x_n = \sin\left[2\pi\left(2000 + 0.5\,\frac{16000}{1024}\right)\frac{n}{16000}\right]$$

Repeat steps 1 through 4 for this new signal. Explain the results.

4.6.2   Experiment 4.2: Making a Spectrum Analyzer

Now you will make an elementary spectrum analyzer. The DSK will be used to collect blocks of $N = 1024$ samples taken at a 16 kHz rate. The DSP will compute and average the periodograms. The results will be displayed on the PC by using Code Composer Studio's animation and graphing capabilities.

Suggestions on How to Structure the Spectrum Analyzer Program

The power spectral density estimates will be based on periodograms of 1024-point blocks of input samples taken at a 16 kHz rate. The technique described by (4.27) on page 113 to compute the sum of pairs of periodograms should be used to efficiently utilize the 1024-point FFT. The following list suggests a method of data collection for your program and the tasks it should perform.

1. Initialize the DSK as usual.

2. Set up a 1024-word array that contains the floating-point samples of the Hamming window.

3. Set up an external 513-word array of floats, spectrum[], for the spectrum estimates at frequencies from 0 to 8000 Hz ($k = 0, \ldots, 512$). (Values from 8000 to 16000 Hz are the mirror images of the ones from 0 to 8000 Hz.)

4. **Ping-Pong Buffers:** Use the technique of *ping-pong buffers*. Set up two external complex floating-point arrays named ping[] and pong[] each of size 1024 complex words. (See the header file complex.h.) One array will be used to collect new samples from the ADC using RRDY interrupts from the McBSP1 receiver while an FFT and periodogram averaging are being performed on the other. The samples should be read, converted to floating-point words, and stored in the ping or pong array in an interrupt service routine. The first 1024 samples should be stored in the real part of the array and the next 1024 samples in the imaginary part of the array.

5. While one array is being filled with new samples through interrupts:

   (a) **Hamming Windowing:** Multiply the real and imaginary parts of the previously filled array by the Hamming window and leave the results in the same array.

   (b) Perform a complex 1024-point FFT on the windowed array.

   (c) Use equation (4.27) on page 113 to compute the sum of the squared magnitudes of the FFT's of the real and imaginary parts of the array. Remember that when x_n is real, $X_k = \bar{X}_{N-k}$ so that the second half of the FFT is totally redundant. Thus, it is only necessary to compute the sum of the squared magnitudes for $n = 0, \ldots, 512$. As each value is computed, add it to the corresponding element of `spectrum[]`.

   (d) Once the FFT and additions to `spectrum[]` have been completed, wait for the array collecting new samples to be filled. You'll have to devise a way that the main function can determine when arrays are filled. Then switch arrays, allow the array just processed to begin collecting new samples, and begin processing the array that was newly filled.

   (e) Continue to accumulate FFT squared magnitudes in `spectrum[]` until $L = 8$ have been added and then divide the elements by 8×1024 or whatever is appropriate depending on any previous normalizations. You can experiment later with using larger or smaller values for L.

   (f) Add a dummy line to your main function after averaging $L = 8$ periodograms, that is, when one spectrum estimate has been completed, and before going back to begin a new spectrum estimate, as a place for a Code Composer probe point. For example, the line might be

   $$\texttt{dummy = ping[0];}$$

   (g) Your program should then loop back, clear `spectrum[]`, and compute a new averaged periodogram, etc.

6. Compile your program using -o3 optimization.

Important Note: Using the "volatile" Declaration to Stop the Optimizer from Breaking Things

Consider the C source code lines:

```
int insample;
insample = MCBSP_read(DSK6713_AIC23_DATAHANDLE);
```

The optimizer looks at `MCBSP_read()` and sees a function with a constant argument. It thinks the returned value of this function will never change. It does not know the serial port DRR contents can be different each time the function is executed. Therefore, it creates code to set "`insample`" just once and never do it again. The declaration "volatile" informs the optimizer that "`insample`" can actually change and should be updated every time it is encountered. So, to make sure the optimizer does the correct thing, use the code:

```
volatile int insample;
insample = MCBSP_read(DSK6713_AIC23_DATAHANDLE);
```

WARNING: Remember to increase -stack to 0x1000 in the linker command file. Otherwise the stack might overflow and overwrite some variables causing strange answers.

Another Method of Averaging Periodograms

Periodograms can be averaged by using a one-pole IIR lowpass filter. A filter of this type has the transfer function

$$H(z) = (1 - \alpha)/(1 - \alpha z^{-1}) \text{ where } 0 < \alpha < 1$$

The closer α is to 1 the more lowpass the filter is and the slower its output changes. The impulse response of the filter is

$$h(n) = (1 - \alpha)\alpha^n u(n)$$

and it is sometimes called an *exponential averager*. Let the current averaged spectrum estimate at DFT slot k be $S_k(n)$ and the current periodogram at slot k be $I_k(n)$. Then the exponential averager output is computed by the formula:

$$S_k(n) = (1 - \alpha)I_k(n) + \alpha S_k(n - 1) \tag{4.28}$$

This computation can be performed each time the sum of a new pair of periodograms is computed.

Testing Your Spectrum Analyzer

1. Initial Testing Using a Known Synthesized Input

As a first test of your spectrum analyzer, replace the input samples in your interrupt service routine by samples of a sine wave generated in your program at one of the FFT bin frequencies. For example, for bin $k = 100$ your synthesized input could be

$$10000 \cos(n \times 100 \times 2\pi/1024)$$

The scale factor 10000 is to model the dynamic range of the integer samples arriving from the ADC. In your program, the angle inside the cosine function should be generated recursively by adding the constant $100 \times 2\pi/1024$ to the old angle each time the interrupt routine is entered. Also, limit the size of the angle by subtracting 2π when it exceeds 2π as you did in Chapter 2. Perform the following two exercises:

(a) Temporarily replace the Hamming window by a rectangular window, that is, all 1's, and fill the imaginary parts of the sample arrays with all 0's. Prove that the FFT values for $n = 100$ and $1024 - 100 = 924$ should be $10000 \times 1024/2$ and should be 0 for all other k. Determine the theoretical values your spectrum analyzer should produce. Check that your analyzer is giving the theoretical answer and correct your program if it isn't. Try cosines at a few other bin frequencies.

(b) Re-enable the Hamming window. Theoretically determine what the analyzer output should be. Check that your analyzer is giving the correct results.

2. **Displaying the Spectrum Using Code Composer Studio**

To display the spectrum estimates using Code Composer Studio's break point, probe point, graphing, and animation capabilities:

(a) Load your executable program.

(b) On the menu bar, select "Debug".

(c) On the "Debug" menu select "Go Main".

(d) Scroll down in the source code window to the line "dummy = ping[0];" and click on it to put the cursor there.

(e) Click on the break point icon (hand) to set a break point at this line. A red circle should appear next to the line in the margin on the left to indicated that a break point is set there.

(f) Click on the probe point icon (scope probe) to set a probe point at this line. A blue circle should be added to the left margin.

(g) On the menu bar, click "View".

(h) Go down the list and click "Graph".

(i) On the Graph menu, select "Time/Frequency".

(j) On the Time/Frequency menu enter:
Start Address: spectrum
Acquisition Buffer Size: 513
Display Data Size: 513
DSP Data Type: 32-bit floating point
Data Plot Style: Bar

(k) Click "OK" and the graph should appear.

(l) Click on the "animate" button. The program should begin running and the graph should get updated with each new spectrum estimate.

3. **Testing with External Inputs**

Now attach the signal generator output to the DSK line input and an oscilloscope. Set your program to use the actual input samples rather than the synthesized cosine wave. Perform the following experiments:

(a) Set the signal generator to generate a 2 kHz sine wave and observe the output of your spectrum analyzer using CCS as described above. Compute which FFT bin, that is, value of k, corresponds to 2 kHz and check that your spectrum analyzer display is correct.

(b) Derive a formula for the Fourier coefficients of a non-symmetric square-wave with one period given by the following equation.

$$x(t) = \begin{cases} A/2 & \text{for } |t| < \tau/2 \\ -A/2 & \text{for } \tau/2 \leq |t| \leq T_0/2 \end{cases}$$

    i. Set the signal generator to generate a 200 Hz square-wave with a 50% duty cycle and compare the measured and theoretical spectra.

    ii. Set the duty cycle to 20% and compare the measured and theoretical spectra.

(c) Test your spectrum analyzer with amplitude modulated signals. These have the form

$$A_c[1 + m(t)] \cos 2\pi f_c t$$

See Chapter 5 for a detailed discussion of AM. The waveform $m(t)$ is called the modulating signal and f_c the carrier frequency. The function $|A_c[1 + m(t)]|$ is called the signal envelope. In particular, let $f_c = 4$ kHz and do the following:

    i. Let $m(t) = A_m \cos 2\pi 500t$ and derive the theoretical spectrum for $x(t)$. A_m is called the modulation index for the AM signal. Set the function generator to generate a signal of this type with a modulation index of 100%. Compare the theoretical and measured spectra.

    ii. Repeat the previous item but change $m(t)$ to a 200 Hz square-wave with a 20% duty cycle.

(d) Experiment with the FM signals of the signal generator. FM is discussed in detail in Chapter 8. In particular, see Equation 8.12 of Section 8.1.2 for the spectrum. It is much more complex than the AM spectrum. Change the modulation index and observe how the carrier component can disappear and how the spectrum spreads out as the modulation index is increased.

4. **Testing with an Exponential Averager**

If time permits and for extra credit, modify your spectrum analyzer program to use exponential periodogram averaging instead of the arithmetic average of L pairs of periodograms. Experiment with different values of α and see how your display responds when the input signal is changed. For example, you could change the carrier frequency of an AM or FM signal, or the modulation index of an FM signal.

4.7 Additional References

The theory and properties of the discrete-time Fourier transform can be found in all the DSP books listed in the references. In particular, see Oppenheim and Schafer [II.C.13, Sections 2.6–2.9] and Proakis and Manolakis [II.C.15, Sections 3.2–3.4].

For discussions of data windows see Oppenheim and Schafer [II.C.13, Sections 7.4–7.5, 11.2–11.3], Rabiner and Gold [II.C.16, Sections 3.8–3.16], and Tretter [II.C.17, Sections 8.8 and 11.4–11.5].

For detailed presentations of the discrete-Fourier transform and its properties see Oppenheim and Schafer [II.C.13, Chapter 8], Proakis and Manolakis [II.C.15, Chapter 9], and Tretter [II.C.17, Chapter 10].

Extensive coverage of the FFT can be found in Burrus and Parks [II.C.3], Oppenheim and Schafer [II.C.13, Chapter 9], Proakis and Manolakis [II.C.15, Chapter 9], Rabiner and Gold [II.C.16, Chapter 6], and Tretter [II.C.17, Chapter 10].

There are many subtleties to the estimation of the power spectral density of a random signal. Very complete coverage of this topic can be found in the books by Jenkins and Watts [II.C.9] and Kay [II.C.10]. More concise presentations can be found in Oppenheim and Schafer [II.C.12 Chapter 11][II.C.13, Chapter 11]; Proakis and Manolakis [II.C.15, Chapter 12]; Manolakis, Ingle, and Kogon [II.C.11]; and Tretter [II.C.17, Chapter 11].

Chapter 5

Amplitude Modulation

A very common method of transmitting information known as *amplitude modulation* (AM) will be examined in this chapter. AM was the first widespread technique used in commercial radio broadcasting. The approaches presented here are particularly suited for implementation by digital signal processors. More complete discussions of AM and analog implementations can be found in the textbooks on communication systems suggested at the end of this chapter.

5.1 Theoretical Description of Amplitude Modulation

5.1.1 Mathematical Formula for an AM Signal

The purpose of modulation is to transform a message $m(t)$ into another signal $s(t)$ that can be transmitted through some medium like a radio link or telephone cable. The transformation must be reversible so that $m(t)$ can be recovered exactly from $s(t)$ at the receiver. The original message $m(t)$ is often called the *baseband* signal and is usually a lowpass signal. The transmitted signal $s(t)$ is usually a bandpass signal with its spectrum centered in the passband of the communication channel, so it is often called the *passband* signal.

An AM signal has the mathematical form

$$s(t) = A_c[1 + k_a m(t)] \cos \omega_c t \qquad (5.1)$$

where

$$c(t) = A_c \cos \omega_c t \qquad (5.2)$$

is called the *carrier* wave and has amplitude A_c and frequency $f_c = \omega_c/(2\pi)$ Hz. The carrier frequency, f_c, should be larger than the highest spectral component in $m(t)$. The parameter k_a is a positive constant called the *amplitude sensitivity* of the modulator.

The signal

$$e(t) = A_c|1 + k_a m(t)| \qquad (5.3)$$

is called the envelope of $s(t)$. When f_c is large relative to the bandwidth of $m(t)$, the envelope is a smooth signal that passes through the positive peaks of $s(t)$ and it can be

121

viewed as modulating (changing) the amplitude of the carrier wave in a way related to $m(t)$. In standard AM broadcasting, $k_a m(t)$ is adjusted so that

$$1 + k_a m(t) \geq 0 \quad \text{for all } t \tag{5.4}$$

In this case, the envelope is

$$e(t) = A_c[1 + k_a m(t)] \tag{5.5}$$

so $m(t)$ can be recovered from the envelope to within a scale factor and constant offset. An envelope detector is called a *noncoherent* demodulator because it makes no use of the carrier phase and frequency.

5.1.2 Example for Single Tone Modulation

In the special case of the sinusoidal message, $m(t) = A_m \cos \omega_m t$, the transmitted signal has the form

$$s(t) = A_c(1 + \mu \cos \omega_m t) \cos \omega_c t \tag{5.6}$$

where $\mu = k_a A_m$ is called the *modulation index*.

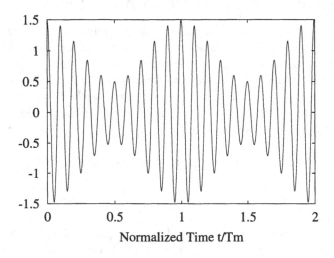

Figure 5.1: Single Tone AM Example with $\mu = 0.5$ and $\omega_c = 10\,\omega_m$

For $0 \leq \mu \leq 1$, the modulation index can be measured by observing that the envelope has the maximum value

$$e_{max} = A_c(1 + \mu) \tag{5.7}$$

and minimum value

$$e_{min} = A_c(1 - \mu) \tag{5.8}$$

Taking the ratio of these two equations and solving for μ gives the following formula for easily computing the modulation index from a display of the modulated signal.

$$\mu = \frac{1 - \frac{e_{min}}{e_{max}}}{1 + \frac{e_{min}}{e_{max}}} \tag{5.9}$$

When $\mu = 1$ the AM signal is said to be 100% modulated and the envelope periodically reaches 0. The signal is said to be overmodulated when $\mu > 1$. In general, the AM signal is said to be overmodulated when $1 + k_a m(t)$ is negative some of the time and then $m(t)$ cannot be determined from the envelope without distortion.

Using the trigonometric identity for the product of cosines, the transmitted signal can be expressed as

$$s(t) = A_c \cos \omega_c t + 0.5 A_c \mu \cos(\omega_c + \omega_m)t + 0.5 A_c \mu \cos(\omega_c - \omega_m)t \qquad (5.10)$$

The first term is a sinusoid at the carrier frequency and carries no message information. The other two terms are called sidebands and carry the information in $m(t)$. The total power in $s(t)$ is

$$P_s = 0.5 A_c^2 + 0.25 A_c^2 \mu^2 \qquad (5.11)$$

while the power in the sidebands due to the message is

$$P_m = 0.25 A_c^2 \mu^2 \qquad (5.12)$$

and their ratio is

$$\eta = \frac{P_m}{P_s} = \frac{\mu^2}{2 + \mu^2} \qquad (5.13)$$

This ratio increases monotonically from 0 to 1/3 as μ increases from 0 to 1. Since the carrier component carries no message information, the modulation is most efficient for 100% modulation.

5.1.3   The Spectrum of an AM Signal

Suppose the baseband message $m(t)$ has a Fourier transform $M(\omega)$ and $M(\omega) = 0$ for $|\omega| \geq W$. The message is said to be a lowpass band limited signal with cutoff frequency W. The Fourier transform of the transmitted signal, $s(t)$, is

$$S(\omega) = A_c \pi \delta(\omega + \omega_c) + A_c \pi \delta(\omega - \omega_c) + \frac{A_c}{2} k_a M(\omega + \omega_c) + \frac{A_c}{2} k_a M(\omega - \omega_c) \qquad (5.14)$$

An example is shown in Figure 5.2. Notice that the modulation adds a spectral line at the carrier frequency and translates the baseband spectrum so that it is centered about the carrier frequency.

5.2   Demodulating an AM Signal by Envelope Detection

Commercial radios use a simple, inexpensive, analog circuit known as an *envelope detector* to demodulate AM signals. This circuit employs a diode, capacitor, and resistors to follow the positive peaks of the AM wave with the assumption that the signal is less than 100% modulated. Two methods for envelope detection that are particularly suited to digital signal

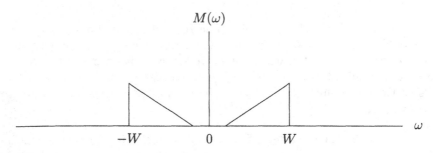

(a) Fourier Transform of Baseband Signal

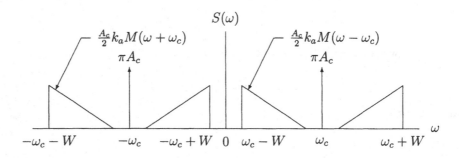

(b) Fourier Transform of Transmitted AM Signal

Figure 5.2: Spectrum of an AM Signal

processing will be studied in this chapter. The first method is called *square-law detection* and the second method uses the Hilbert transform to create something called the *complex envelope*. As a designer, you should always evaluate whether the analog envelope detector or DSP implementation is the most cost effective method for accomplishing your task.

5.2.1   Square-Law Demodulation of AM Signals

The block diagram of a square-law envelope detector is shown in Figure 5.3. The input $s(t)$ has the form of the AM signal given by (5.1). It will be assumed that the baseband message $m(t)$ is a lowpass signal with cutoff frequency W. The first block in the detector squares the input resulting in the signal

$$s^2(t) = A_c^2[1 + k_a m(t)]^2 \cos^2 \omega_c t = 0.5 A_c^2[1 + k_a m(t)]^2 + 0.5 A_c^2[1 + k_a m(t)]^2 \cos 2\omega_c t \quad (5.15)$$

The first term on the right-hand side of (5.15) is a lowpass signal except that the cutoff frequency has been increased to $2W$ by the squaring operation. The second term has a spectrum centered about $\pm 2\omega_c$. For positive frequencies, this spectrum is confined to the interval $(2\omega_c - 2W, 2\omega_c + 2W)$. For the square-law detector to work properly, the spectra for these two terms must not overlap. This requirement is met if

$$2W < 2\omega_c - 2W \quad \text{or} \quad \omega_c > 2W \quad (5.16)$$

The sampling rate required to implement the square-law detector by digital signal processing techniques will now be examined. We saw in the previous paragraph that the squared AM signal is band limited with upper cutoff frequency $2(\omega_c + W)$. Therefore, the input $s(t)$ must be sampled at a rate of at least $4(\omega_c + W)$ to prevent aliasing and the lowpass filter $H(\omega)$ must operate on samples of $s^2(t)$ taken at this rate. The output of the lowpass filter is band limited in the Nyquist band with a cutoff of $2W$. Thus, if $H(\omega)$ is implemented by an FIR filter with tap spacing corresponding to the required fast input sampling rate, computation can be reduced by computing the output only at times resulting in an output sampling rate of at least $4W$. This technique is called *skip sampling* or *decimation*.

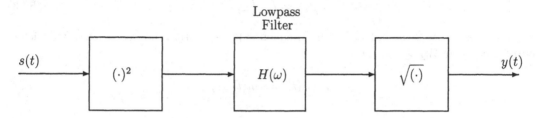

Figure 5.3: Square-Law Envelope Detector

The filter $H(\omega)$ is an ideal lowpass filter with cutoff frequency $2W$ so that its output is $0.5A_c^2[1 + k_a m(t)]^2$. The final box in the detector takes a square-root resulting in an output signal that is proportional to $m(t)$ with a DC offset. In many cases, the baseband message has no spectral components around zero frequency and the DC offset can be removed by a simple highpass filter.

5.2.2 Hilbert Transforms and the Complex Envelope

Another type of envelope detector is based on the Hilbert transform. These transforms are used extensively for analysis and signal processing in passband communication systems. Let $x(t)$ be a signal with Fourier transform $X(\omega)$. The Hilbert transform of $x(t)$ will be denoted by $\hat{x}(t)$ and its Fourier transform by $\hat{X}(\omega)$. In the time-domain, the Hilbert transform is defined by the integral

$$\hat{x}(t) = x(t) * \frac{1}{\pi t} = \frac{1}{\pi} \int_{-\infty}^{\infty} \frac{x(\tau)}{t - \tau}\, d\tau \tag{5.17}$$

where $*$ represents convolution. Thus, the Hilbert transform of a signal is obtained by passing it through a filter with the impulse response

$$h(t) = \frac{1}{\pi t} \tag{5.18}$$

It can be shown that

$$H(\omega) = -j\,\text{sign}\,\omega = \begin{cases} -j & \text{for } \omega > 0 \\ 0 & \text{for } \omega = 0 \\ j & \text{for } \omega < 0 \end{cases} \tag{5.19}$$

The Hilbert transform filter is an ideal 90° phase shifter. Therefore, in the frequency domain

$$\hat{X}(\omega) = H(\omega)X(\omega) = (-j\,\mathrm{sign}\,\omega)\,X(\omega) \tag{5.20}$$

The following Hilbert transform pairs will be useful:

$$\cos\omega_c t \overset{\mathcal{H}}{\Longrightarrow} \sin\omega_c t \tag{5.21}$$

$$\sin\omega_c t \overset{\mathcal{H}}{\Longrightarrow} -\cos\omega_c t \tag{5.22}$$

$$\cos(\omega_c t + \theta) \overset{\mathcal{H}}{\Longrightarrow} \cos\left(\omega_c t + \theta - \frac{\pi}{2}\right) \tag{5.23}$$

Let $m(t)$ be a lowpass signal with cutoff frequency W_1 and $c(t)$ a highpass signal with lower cutoff frequency $W_2 > W_1$. Then

$$m(t)c(t) \overset{\mathcal{H}}{\Longrightarrow} m(t)\hat{c}(t) \tag{5.24}$$

The *analytic signal* or *pre-envelope* associated with $x(t)$ is defined to be

$$x_+(t) = x(t) + j\,\hat{x}(t) \tag{5.25}$$

As an example, the analytic signal associated with $x(t) = \cos\omega_c t$ is

$$x_+(t) = \cos\omega_c t + j\,\sin\omega_c t = e^{j\omega_c t} \tag{5.26}$$

As another example, let $m(t)$ be a lowpass signal with cutoff frequency W less than the carrier frequency ω_c. Using (5.24), the analytic signal associated with $x(t) = m(t)\cos\omega_c t$ can be shown to be

$$x_+(t) = m(t)\cos\omega_c t + j\,m(t)\sin\omega_c t = m(t)e^{j\omega_c t} \tag{5.27}$$

Using (5.20), it can be shown that the Fourier transform of the analytic signal is

$$X_+(\omega) = 2X(\omega)u(\omega) = \begin{cases} 2X(\omega) & \text{for } \omega > 0 \\ X(0) & \text{for } \omega = 0 \\ 0 & \text{for } \omega < 0 \end{cases} \tag{5.28}$$

Thus, the analytic signal has a one-sided spectrum, that is, its Fourier transform is 0 for negative ω.

The *complex envelope* of a signal $x(t)$ with respect to carrier frequency ω_c is defined to be

$$\tilde{x}(t) = x_+(t)\,e^{-j\omega_c t} \tag{5.29}$$

and has the Fourier transform

$$\tilde{X}(\omega) = X_+(\omega + \omega_c) = 2X(\omega + \omega_c)u(\omega + \omega_c) \tag{5.30}$$

When these definitions are used, $x(t)$ is usually a bandpass signal and ω_c is a frequency in the passband. Then $\tilde{x}(t)$ is a lowpass signal. As an example, consider the AM signal $s(t)$ given by (5.1). Its pre-envelope is

$$s_+(t) = A_c[1 + k_a m(t)]e^{j\omega_c t} \tag{5.31}$$

and its complex envelope is

$$\tilde{s}(t) = A_c[1 + k_a m(t)] \tag{5.32}$$

Motivated by the example in the previous paragraph, we will define the *real envelope* of a bandpass signal $x(t)$ to be

$$e(t) = |\tilde{x}(t)| \tag{5.33}$$

Another equivalent formula for the real envelope is

$$e(t) = |x_+(t)| = \left[x^2(t) + \hat{x}^2(t)\right]^{1/2} \tag{5.34}$$

The block diagram of an envelope detector based on (5.34) is shown in Figure 5.4. The required sampling rates will now be investigated when the detector is implemented by digital signal processing techniques. If the input signal $s(t)$ is a standard AM signal and the baseband message $m(t)$ is band limited with cutoff frequency W, then $s(t)$ is band limited with cutoff frequency $W + \omega_c$. Thus, the input to the Hilbert transform filter must be sampled at a rate of at least $2(W + \omega_c)$ to prevent aliasing. The squared envelope is band limited with cutoff frequency $2W$ so the decimation technique described for the square-law envelope detector lowpass filter can be applied to the Hilbert transform filter when it is an FIR filter. Also, the bound relating ω_c and W for the square-law detector is not required for this detector.

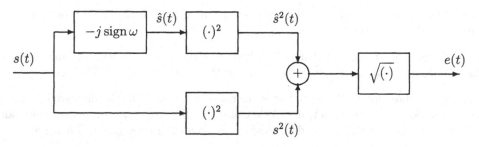

Figure 5.4: Envelope Detector Using the Hilbert Transform

5.3　Laboratory Experiments for AM Modulation and Demodulation

In the following experiments, you will use the TMS320C6713 DSK to modulate and demodulate AM signals. This should reinforce your theoretical knowledge. For each of the exercises, initialize the DSK and the TMS320C6713 DSP as in Chapters 2 and 3.

5.3.1   Experiment 5.1: Making an AM Modulator

1. Initialize McBSP0, McBSP1, and the codec as in Chapters 2 and 3, and use the left channel.

2. Read samples $m(nT)$ from the codec at a 16 kHz rate.

   Convert the samples into 32-bit integers by shifting them arithmetically right by 16 bits. The resulting integers lie in the range $\pm 2^{15}$.

3. AM modulate the input samples to form the sequence

$$s(nT) = A_c[1 + k_a m(nT)] \cos 2\pi f_c nT$$

where the carrier frequency is $f_c = 3$ kHz and A_c is a constant chosen to give a reasonable size output.

   - Convert the input samples to floating point numbers and do the modulation using floating point arithmetic.
   - Since the input samples lie in the range $\pm 2^{15}$, you must choose k_a or scale the 1 in the AM equation so that the signal is not overmodulated.

4. Send $s(nT)$ to the DAC.

   - Convert the modulated samples back into appropriately scaled integers and send them to the left channel of the DAC using polling.

Observing Your Modulator Output

- Attach the signal generator to the DSK left channel line input and the left channel line output to the oscilloscope.

- Set the signal generator to output a 320 Hz sine-wave with an amplitude that creates less than 100% modulation. Calculate and sketch the spectrum of the AM signal.

- Sync the oscilloscope to the signal generator sine wave and sketch the signal you observe on the oscilloscope. Better yet, use Code Composer Studio to capture the output samples and write them to a PC file as described in the following subsection. Then plot the file.

- Increase the amplitude of the input signal until the AM signal is overmodulated and plot the resulting waveform. What is the effect of overmodulation on the spectrum?

- Connect the line output to the speakers of the PC and vary the modulating frequency f_m. You should be able to hear the two sidebands move up and down in frequency as you sweep f_m.

- If you made the spectrum analyzer for Chapter 4, run your AM modulator on one DSK and the spectrum analyzer on another. Connect the modulator output to the analyzer input and observe the spectrum as you change the input message frequency, f_m. You should see the sidebands move as f_m is changed.

5.3.2 How to Capture DSK Output Samples with CCS for Plotting

The File I/O feature of Code Composer Studio can be used to transfer blocks of words from the 'C6713's memory to a text file on the PC. To do this using the polling method:

- Include a loop inside the usual infinite while(1) or for(;;) loop that writes the integer output samples to a *global* array, for example, output[].

- When this inner loop exits, the array is filled and ready to be captured by CCS.

- Immediately after the inner loop, put a dummy statement that involves output[] for a probe point, for example, dummy = output[0];.

The following code segment illustrates this approach.

Code Segment to Allow Capture

```
#define BUFFER_SIZE 256
int output[BUFFER_SIZE]; /* Must be global so CCS can see it */
main()
{
  /* Initialize DSK, etc.   */
  while(1)
  {
    for(i=1; i < BUFFER_SIZE; i++)
    {
    /* 1. Poll XRDY flag and read new input sample.  */
    /* 2, Process the sample as desired.             */
    /* 3. Write output sample to buffer.             */
       output[i] = processed_sample;
    /* 4. Write output sample to codec.              */
    }
    dummy = output[0]; /* Dummy line for probe point  */
  }
}
```

Setting Up CCS for Capture

Connect the probe point to the output file by the following steps:

1. Set the probe point in your code.

2. Choose "File → File I/O" from the CCS main menu.

3. In the dialog box that appears, click the "File Output" tab and then "Add File."

4. Enter the filename of your choice but be sure to choose a file of type "*.dat(Integer)."

5. Return to the File I/O dialog box. Enter output (or whatever you call your output buffer array) and its length in the "address" and "length" boxes respectively.

6. Click "Add Probe Point" and the Break/Probe/Profile Points dialog box will appear.

7. Select "[*your-source=file.c*] line zzz → No Connection" in the Probe Point list.

8. Select "File Out [*your-output-file.out*]" in the "Connect To:" drop-down box. Then click "Replace." The Probe Point list should be updated appropriately.

9. Click "OK" to close the Probe Points dialog box.

10. Click "OK" to close the File I/O box. The "tape player" window should appear.

11. Push the "play button" on the tape player and run your program. Hit the "stop" button when you've collected enough samples. Note that each time you collect more samples using the same output filename, the new samples are appended to the file.

NOTE: The first line in the output file contains irrelevant data. After the first line is removed, the file should be your captured samples, one per line, and can be plotted with your favorite plotting routine.

5.3.3   Experiment 5.2: Making a Square-Law Envelope Detector

Write a program for the TMS320C6713 to implement the square-law envelope detector shown in Figure 5.3. Continue to use a 16 kHz sampling frequency. Use the signal generator as the source of the AM signal $s(t)$. Take the input samples from the ADC, perform the demodulation, and send the demodulated output samples to the DAC. Set the carrier frequency to 3 kHz and assume the baseband message $m(t)$ is band limited with a cutoff frequency of 400 Hz. Use a Butterworth lowpass IIR filter for $H(\omega)$ that has an order sufficient to suppress the unwanted components around $2f_c$ by at least 40 dB.

Assume that $m(t)$ has no spectral components below 50 Hz and remove the DC offset at the output of the square root box by a simple highpass filter of the form

$$G(z) = \frac{1+c}{2} \frac{1 - z^{-1}}{1 - c\,z^{-1}}$$

where c is a constant slightly less than 1 chosen so that the lowest frequency components of $m(t)$ are negligibly distorted. Notice that this filter has an exact null at 0 frequency and is 1 at half the sampling rate. Plot the amplitude response of this filter for various c to select an appropriate value.

Observing the Square-Law Detector Output

Experiment 5.2.1: Demodulating an AM Signal with no Additive Noise

Attach the signal generator to the DSK line input and the oscilloscope to the line output. Set the signal generator to create an AM wave with a sinusoidal modulating signal with a frequency between 100 and 400 Hz. Use a 3 kHz carrier. Sketch or capture and plot the AM input signal and the output of your demodulator.

Experiment 5.2.2: Demodulating an AM Signal Corrupted by Additive Noise

Experiment demodulating signals corrupted by additive, zero mean, Gaussian noise. The lab does not have hardware continuous-time noise generators. You will have to simulate the noise in the DSP using the method described in Appendix A and add the simulated noise sample to the input signal sample. Use the same sinusoidally modulated AM signal as before. Start with a large signal-to-noise power ratio (SNR) and decrease it until the demodulator output is very noisy and barely resembles the message sinusoid. The degradation will increase relatively smoothly over a range of SNR and you will have to use your judgment as to what "barely resembles" means. Estimate the SNR in dB at this point.

Hint: To change the SNR, just turn the amplitude knob for the carrier on the signal generator. Do not change σ in your program and recompile for each new σ! For example, you can let σ be the constant 1.

Attach the line output to the PC's speaker input and listen to the noisy demodulated signal. It may be easier for you to determine when "the demodulator output is very noisy and barely resembles the message" by listening rather than visually.

To calculate the SNR, you will have to measure the average signal power of the samples observed inside the DSP. The average noise power is the value of σ^2 you set in your noise generator code. You can estimate the average power of the received AM signal samples by making the simple power meter shown in Figure 5.5. The constant, a, should be close to but slightly less than 1 to create a narrow band lowpass filter. Try $a = 0.99$, for example. Notice that the gain of the filter at 0 frequency is 1. The lowpass filter generates an estimate of the true statistical expected value $E\{s^2(n)\}$ by time averaging. Make a loop to run the power meter for several thousand samples. Put a break point after the loop. Then you can examine the final value of $p(n)$ with Code Composer Studio when the program halts at the break point.

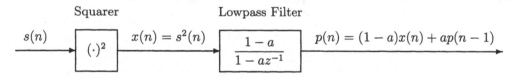

Figure 5.5: A Simple Power Meter

5.3.4   Experiment 5.3: Making an Envelope Detector Using the Hilbert Transform

Write a program to implement the envelope detector shown in Figure 5.4. Again, assume a carrier frequency of 3 kHz and a baseband message band limited to 400 Hz.

You can design the Hilbert transform filter with the program `remez87.exe`. Use an odd number, N, of filter taps. Good results can be achieved by using just one band with a lower cutoff frequency f_1 and upper cutoff frequency f_2 chosen to pass the AM signal. Choosing the band to be centered in the Nyquist band also seems to improve the filter amplitude response generated by `remez87.exe`. To center the band, choose the upper cutoff frequency to be $f_2 = 0.5f_s - f_1$ where f_s is the sampling frequency. Enter 1 for the magnitude of the Hilbert transform in the band and 1 for the weight factor. Select N so that the amplitude response of the filter is quite flat over the signal passband and ripples in the demodulated signal caused by incomplete cancellation of the $2f_c$ components are essentially invisible in the demodulated output.

You can also design the Hilbert transform filter with the program `window.exe`. Try using the Hamming and Kaiser windows. You can make a tradeoff between the transition bandwidth and the out-of-band attenuation with the Kaiser window. Remember that the filter amplitude response is down by a factor of 2 at the band edges for the window function method of design. Be sure to choose the cutoff frequencies far enough outside the signal band so there is little roll-off inside the signal band. Since the impulse response of a Hilbert transform filter has odd symmetry about the center tap, the frequency response will be exactly zero at $\omega = 0$ and $\omega_s/2$.

The resulting FIR filter will have a delay equal to the delay from the input to the center tap of the filter, $T(N-1)/2$. Therefore, $s(nT)$ must be delayed by this amount to match the delay in $\hat{s}(nT)$. This can be easily accomplished by taking $s(nT)$ from the point in the delay-line of the Hilbert transform filter at its center tap.

Test your envelope detector using the same steps as you did for the square-law detector.

5.4 Additional References

There are many good textbooks on communication systems that include discussions of amplitude modulation. For example, see Gibson [II.C.9, Chapter 5], and Haykin [II.D.17, Chapter 3]. See Haykin [II.D.17, Sections 2.10–2.13] for discussions of the Hilbert transform, pre-envelope, complex envelope, and applications to bandpass systems.

Chapter 6

Double-Sideband Suppressed-Carrier Amplitude Modulation and Coherent Detection

The standard AM modulated signal contains a sinusoidal component at the carrier frequency which does not convey any of the baseband message information. This component is included to create a positive envelope which allows demodulation by a simple, inexpensive envelope detector. From an information theory point of view, the power in the sinusoidal carrier component is wasted. In this experiment, you will see that it is not necessary to transmit the carrier component and that the baseband message can be recovered by a *coherent* demodulator. In fact, it can be shown that a coherent demodulator performs better than an envelope detector when the received signal is corrupted by additive noise. The type of modulation that will be studied in this chapter is called *double-sideband suppressed-carrier amplitude modulation* (DSBSC-AM). A close approximation to an ideal coherent demodulator called a *Costas loop* will be implemented.

6.1 Mathematical Form for a DSBSC-AM Signal

As usual, let $m(t)$ be a baseband message signal. The DSBSC-AM signal corresponding to $m(t)$ is

$$s(t) = A_c m(t) \cos \omega_c t \qquad (6.1)$$

This is the same as the AM signal except with the sinusoidal carrier component eliminated. A message $m(t)$ typically has positive and negative values so it can not be recovered from $s(t)$ by an envelope detector. A demodulation method called *coherent demodulation* will be explored in this chapter. The Fourier transform of $s(t)$ is

$$S(\omega) = 0.5 A_c M(\omega - \omega_c) + 0.5 A_c M(\omega + \omega_c) \qquad (6.2)$$

This is the same as the AM spectrum but with the discrete line at the carrier frequency removed. An example is shown in Figure 6.2(b). It will be assumed that $m(t)$ is a lowpass signal with cutoff frequency W. Then, the carrier frequency must satisfy the bound, $\omega_c > W$

so that the two terms on the right-hand side of (6.2) do not overlap as shown in Figure 6.2(b). When they overlap *foldover* is said to have occurred and perfect demodulation cannot be achieved.

When $m(t)$ is a real signal, $M(-\omega) = \overline{M(\omega)}$ and

$$S(\omega_c - \omega) = \overline{S(\omega_c + \omega)} \quad \text{for} \ \ 0 \leq \omega \leq \omega_c \tag{6.3}$$

This equation shows that the component at frequency $\omega_c + \omega$ contains exactly the same information as the component at $\omega_c - \omega$ since one can be uniquely determined from the other by taking the complex conjugate. The portion of the spectrum for $|\omega| > \omega_c$ is called the *upper sideband* and the portion for $|\omega| < \omega_c$ is called the *lower sideband*. The fact that the modulated signal contains both portions of the spectrum explains why the term, double-sideband, is used.

6.2   The Ideal Coherent Receiver

The block diagram for an ideal coherent receiver is shown in Figure 6.1. First, the received signal is passed through a bandpass filter centered at the carrier frequency that passes the DSBSC signal and eliminates out-of-band noise. The output of the receive bandpass filter is then multiplied by a replica of the carrier wave. This replica is generated by a device called the *local oscillator* (LO) in the receiver. Assuming no noise, the product is

$$s_1(t) = 2s(t) \cos \omega_c t = 2A_c m(t) \cos^2 \omega_c t = A_c m(t) + A_c m(t) \cos 2\omega_c t \tag{6.4}$$

The device that performs the product is often called a *product modulator* or *balanced mixer*.

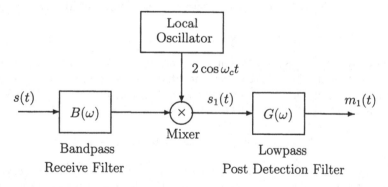

Figure 6.1: Block Diagram of an Ideal Coherent Receiver

The Fourier transform of the product modulator output is

$$S_1(\omega) = A_c M(\omega) + 0.5 A_c M(\omega + 2\omega_c) + 0.5 A_c M(\omega - 2\omega_c) \tag{6.5}$$

and is illustrated in Figure 6.2(c). The first term on the right-hand side of (6.4) is proportional to the desired message. The second term has spectral components centered around

$-2\omega_c$ and $2\omega_c$. The corresponding terms can be seen in $S_1(\omega)$. The undesired high frequency terms are eliminated by the final lowpass filter which has cutoff frequency W. This is often called a *post detection* filter.

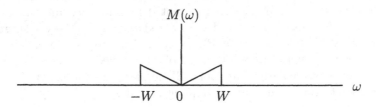

(a) Fourier Transform of Baseband Message

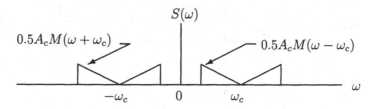

(b) Fourier Transform of DSBSC-AM Signal

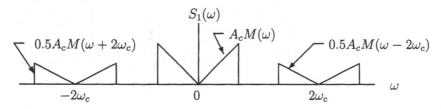

(c) Fourier Transform of Mixer Output

Figure 6.2: Spectra in a DSBSC-AM Communication System

An alternative method of demodulation is to first form the pre-envelope of the received signal. With no additive noise, this is

$$s_+(t) = s(t) + j\hat{s}(t) = A_c m(t) \cos \omega_c t + j A_c m(t) \sin \omega_c t = A_c m(t) e^{j\omega_c t} \qquad (6.6)$$

The baseband message is then recovered to within a scale factor by forming the complex product

$$s_+(t) e^{-j\omega_c t} = A_c m(t) \qquad (6.7)$$

6.3   The Costas Loop as a Practical Approach to Coherent Demodulation

A receiver must have perfect knowledge of the carrier frequency and phase of a received DSBSC-AM signal to perform exact coherent demodulation, and this is almost never the case. However, these parameters can be estimated and tracked very accurately at the receiver by devices called *phase-locked loops* (PLL's) so that nearly optimum coherent demodulation can be achieved. A modification of a type of PLL called a *Costas loop* is shown in Figure 6.3. This form is particularly suited for DSP implementation and the signals shown in the figure are discrete-time signals with sampling period T.

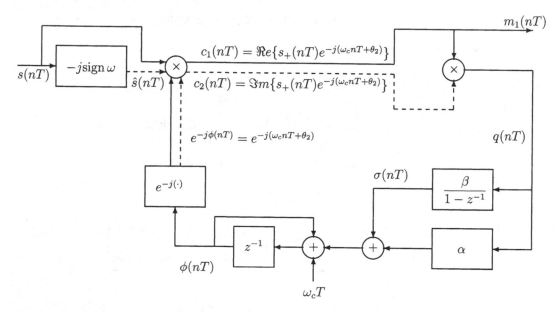

Figure 6.3: Second-Order Costas Loop Demodulator

To add some generality, let the received signal after it has passed through a bandpass receive filter have the form

$$s(nT) = A_c m(nT) \cos(\omega_c nT + \theta_1) \tag{6.8}$$

where ω_c is the nominal carrier frequency and θ_1 is a constant or slowly changing phase angle. When there is a frequency offset between the nominal and actual carrier frequencies due to Doppler shifts or misalignment of the transmitter and receiver local oscillators, θ_1 takes the form

$$\theta_1 = \Delta\omega\, nT + \gamma \tag{6.9}$$

where $\Delta\omega$ is the frequency offset and γ is a constant phase offset.

The first step in the system is to form the complex envelope

$$s_+(nT) = s(nT) + j\hat{s}(nT) = A_c m(nT) e^{j(\omega_c nT + \theta_1)} \tag{6.10}$$

The parallel solid and dotted lines in the figure represent complex signals with the solid line corresponding to the real part and dotted line to the imaginary part.

The system generates an estimate $\phi(nT)$ of the angle of the received signal that can expressed as

$$\phi(nT) = \omega_c nT + \theta_2(nT) \tag{6.11}$$

The method for generating this angle will be explained shortly. It is passed through the complex exponential box to give the local oscillator signal $e^{-j\phi(nT)}$.

The local oscillator signal is multiplied by the complex envelope resulting in the signal

$$c(nT) = s_+(nT)e^{-j\phi(nT)} = A_c m(nT)e^{j[\theta_1 - \theta_2(nT)]} \tag{6.12}$$

which is separated into its real part

$$\begin{aligned}
c_1(nT) &= s(nT)\cos\phi(nT) + \hat{s}(nT)\sin\phi(nT) \\
&= A_c m(nT)\cos[\theta_1 - \theta_2(nT)]
\end{aligned} \tag{6.13}$$

and imaginary part

$$\begin{aligned}
c_2(nT) &= \hat{s}(nT)\cos\phi(nT) - s(nT)\sin\phi(nT) \\
&= A_c m(nT)\sin[\theta_1 - \theta_2(nT)]
\end{aligned} \tag{6.14}$$

The loop is said to be *in lock* when the phase error $\theta_1 - \theta_2$ remains small. When the phase error is exactly zero, the demodulated message appears at the point labeled $c_1(nT) = m_1(nT)$ and $c_2(nT) = 0$. A lock detection strategy is to lowpass filter $c_2^2(nT)$ and declare that the loop is in lock when this signal falls below a threshold for a period of time.

The real and imaginary parts are multiplied, resulting in the signal

$$\begin{aligned}
q(nT) &= c_1(nT)c_2(nT) = A_c^2 m^2(nT)\cos[\theta_1 - \theta_2(nT)]\sin[\theta_1 - \theta_2(nT)] \\
&= 0.5 A_c^2 m^2(nT)\sin\{2[\theta_1 - \theta_2(nT)]\}
\end{aligned} \tag{6.15}$$

Notice that when θ_1 and θ_2 differ by less than 90 degrees, $q(nT)$ has the same sign as the phase error $\theta_1 - \theta_2$, so it indicates in which direction the local phase estimate θ_2 should be changed to reduce the phase error to zero. When the loop is in lock, the small angle approximation, $\sin x \simeq x$, can be used to accurately approximate $q(nT)$ by

$$q(nT) \simeq A_c^2 m^2(nT)[\theta_1 - \theta_2(nT)] \quad \text{for } |\theta_1 - \theta_2(nT)| \ll 1 \tag{6.16}$$

The lower half of the block diagram generates the loop's estimate of the phase of the received signal by computing

$$\phi((n+1)T) = \phi(nT) + \omega_c T + \alpha q(nT) + \sigma(nT) \tag{6.17}$$

where

$$\sigma(nT) = \beta q(nT) + \sigma((n-1)T) \tag{6.18}$$

and α and β are small positive constants with $\beta < \alpha/50$, typically. The basic philosophy behind these equations is that at each new sampling instant the loop's phase estimate is incremented by the nominal change in carrier phase between samples, $\omega_c T$, plus a small correction term $\alpha q(nT)$ roughly proportional to the phase error. Notice that when $q(nT) = 0$ for all n, $\phi(nT)$ is the linear ramp

$$\phi(nT) = \omega_c nT + \phi(0) \tag{6.19}$$

which has a slope equal to the nominal carrier frequency.

The accumulator block $\beta/(1 - z^{-1})$ is included to allow the loop to track a carrier input phase $\theta_1(nT)$ that is a linear ramp with zero steady-state error. The input phase has this form when there is a frequency offset between the received and local carrier frequencies. This block along with the rest of the lower branch introduces a second-order pole at $z = 1$ in the open loop gain which is equivalent to a double accumulation in the time-domain. It is well known in automatic control theory that a loop with the double accumulation can track a first-order input polynomial (linear ramp) with zero steady-state error while with a single accumulation it cannot. The output $\sigma(nT)$ of the accumulator reaches the steady-state value $\Delta\omega\, T$ which is the phase change between samples caused by the frequency offset $\Delta\omega$.

The Costas loop is a nonlinear and time-varying system because of the sin() and $m^2(nT)$ terms in $q(nT)$. Therefore, it cannot be characterized by a transfer function. However, when $m(nT)$ is a stationary process and the loop is in lock, it can be accurately approximated by a linear, time-invariant system by using the small angle approximation (6.16) and replacing $m^2(nT)$ by its expected value. Replacing $m^2(nT)$ by its expected value can be justified by the fact that the loop filters act as lowpass filters on $q(nT)$ resulting in a time-averaged estimate of its statistical mean. Let

$$k_1 = A_c^2 \, E\{m^2(nT)\} \tag{6.20}$$

and further approximate $q(nT)$ by

$$q(nT) \simeq k_1[\theta_1 - \theta_2(nT)] \tag{6.21}$$

These approximate equations can be represented by the linearized loop shown in Figure 6.4. The transfer function for the linearized loop is

$$H(z) = \frac{\Theta_2(z)}{\Theta_1(z)} = \frac{k_1(\alpha + \beta)\left(1 - \frac{\alpha}{\alpha+\beta}z^{-1}\right)z^{-1}}{1 - [2 - k_1(\alpha + \beta)]z^{-1} + (1 - k_1\alpha)z^{-2}} \tag{6.22}$$

The frequency response is obtained by letting $z = e^{j\omega T}$ and has the shape of a narrowband lowpass filter for small α and β. The closed loop gain at zero frequency is $H(1) = 1$.

6.4   Exercises and Experiments for the Costas Loop

Now it is time for you to design, implement, and test a Costas loop coherent receiver. The continuous-time modulated input signal for your receiver can be generated by the signal generator. The signal generator should be set to generate an output voltage of the form

$$s(t) = A_c m(t) \cos 2\pi f_c t \tag{6.23}$$

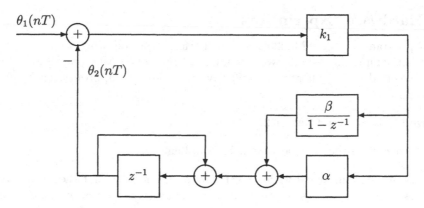

Figure 6.4: Linearized Costas Loop

where

$A_c = 1$
$m(t) = 1 + 0.4 \cos 2\pi f_m t$
$f_c = 4000$ Hz
$f_m = 400$ Hz

Actually, $s(t)$ is an AM signal with modulation index $\mu = 0.4$. However, it can also be considered to be a DSBSC-AM signal with $m(t)$ containing a DC value and all the theory for the Costas loop still holds.

6.4.1   Theoretical Design Exercises

In these exercises you will do theoretical computations to select the Costas loop parameters for a reasonable design. Do the following steps:

1. Compute k_1 by (6.20).

2. Choose some small values for the loop filter constants, for example, $\alpha = 0.01$ and $\beta = 0.0002$. You will find that β should be small relative to α, perhaps, less than $\alpha/50$, to get a transient response without excess ripple. Recursively compute the response of the linearized loop to a unit step in $\theta_1(nT)$ using the following formula which is based on (6.22):

$$\theta_2(nT) = k_1[(\alpha + \beta)\theta_1((n-1)T) - \alpha\theta_1((n-2)T)] + [2 - k_1(\alpha + \beta)]\theta_2((n-1)T)$$
$$- (1 - k_1\alpha)\theta_2((n-2)T)$$

Continue the computations until $\theta_2(nT)$ gets close to its final value and plot the result. Repeat this step for different values of α and β until you find a pair for which the step response settles to its final value in about 0.2 seconds.

3. Compute and plot the closed loop amplitude response $A(f) = 20 \log_{10} |H(e^{j2\pi f/f_s})|$ for the values of α and β finally selected in the previous step.

6.4.2   Hardware Experiments

Write a C program for the TMS320C6713 to perform the following steps. Keep in mind that the input samples are 16-bit two's complement integers in the range $\pm 2^{15}$ so that once you have converted them to floats, you will have to scale them appropriately to match your theoretical design.

1. Initialize the DSK as in Chapter 2.

2. Read samples from the ADC at a 16 kHz sampling rate.

3. Demodulate the input signal with a Costas loop. Remember that the VCO phase angle, $\phi(nT)$, essentially increases linearly but could, possibly, become negative. Make sure to add or subtract multiples of 2π to keep $\phi(nT)$ in the range $[0, 2\pi)$ to eliminate any overflow problems.

4. Send the demodulated signal samples to the DAC.

To test your Costas loop demodulator, do the following:

1. Connect the signal generator to the line input and set it to generate the AM signal $s(t)$ defined above. Connect the DAC output to the oscilloscope and debug your DSP program if the output is not $m(t)$.

2. Once your coherent demodulator is working properly, investigate its performance in the presence of a frequency offset. First set the signal generator carrier frequency to the nominal 4 kHz value and let your loop achieve lock. Then slowly change the carrier frequency to a slightly different value and see if the loop remains locked. You may need to increase α and β to make the loop bandwidth large enough so that the loop can track at the speed you turn the frequency knob. The loop should be able to track frequency offsets up to the point where the input signal spectrum falls off the edges of the Hilbert transform filter.

   Use Code Composer Studio to watch $\sigma(nT)$ in your DSP program and check that it has the correct steady-state value for the frequency offset you are using.

3. The linearized equations describe the loop behavior accurately when it is in lock. When there is a large initial frequency offset, the behavior is quite different. Experimentally investigate this behavior by setting the signal generator carrier frequency to a value that differs by 30 Hz or more from the nominal 4 kHz value and starting the Costas loop. The loop may take a much longer time than you expected to achieve lock. This is called its *pull in* behavior.

4. Experiment demodulating signals corrupted by additive, zero mean, Gaussian noise. Use the approach described on page 131 of Chapter 5. In particular, observe the behavior of the coherent demodulator as the signal-to-noise ratio is decreased and make a rough estimate of the SNR at which the demodulator no longer works. If your loop was designed properly, it should work at lower SNR's than the envelope detectors.

Connect the line output to the PC's speakers and listen to the demodulated signal as the SNR is changed. Compare the sound with the envelope detector noisy outputs.

5. If you are interested in doing more and time permits, set up an array and write the first few hundred samples of $q(nT)$ to the array. Do not add noise to the input samples. Then send the array to the PC and plot the resulting signal to get a clearer picture of the loop transient response. You can also do this with other signals in the loop.

6.5 Additional References

For more complete presentations of DSBSC-AM, see Gibson [II.D.9, Chapter 5] and Haykin [II.D.17, Chapter 3]. For Costas loop discussions see Gibson [II.D.9, Section 8.8], Gitlin, Hayes, and Weinstein [II.D.11, Section 6.3], and Stiffler [II.D.34, Section 8.5]. To my knowledge, the modified Costas loop structure shown in this experiment cannot be found in other textbooks.

Chapter 7

Single-Sideband Modulation and Frequency Translation

AM and DSBSC-AM modulation do not use the frequency spectrum efficiently. Their spectral components equal distances above and below the carrier frequency contain identical information because they are complex conjugates of each other. The portion above the carrier frequency is called the *upper sideband* and the portion below the *lower sideband*. In this experiment you will see how a baseband message can be transmitted by using only one of the sidebands and, consequently, half the bandwidth of AM or DSBSC-AM. This type of modulation is called *single-sideband* (SSB) modulation. It has been extensively used in many radio transmission systems and in the telephone network.

Translating the frequency spectrum of a signal is closely related to SSB modulation and is commonly used at various points in communication systems. A technique for frequency translation particularly suited to DSP implementations will be described in this experiment. No frequency translation experiments will be performed since it is actually the same as upper sideband SSB modulation.

7.1 Single-Sideband Modulators

An obvious type of SSB modulator is shown in Figure 7.1. As usual, we will assume that the baseband message signal $m(t)$ is band limited with a cutoff frequency W which is less than the carrier frequency ω_c. The first stage of this modulator generates the DSBSC-AM signal

$$a(t) = A_c m(t) \cos \omega_c t \tag{7.1}$$

which has the Fourier transform

$$A(\omega) = 0.5 A_c M(\omega - \omega_c) + 0.5 A_c M(\omega + \omega_c) \tag{7.2}$$

and is centered around the carrier frequency ω_c.

The DSBSC-AM signal is then passed through the filter $H(\omega)$ to select the desired sideband. Upper sideband SSB modulation is created with the ideal highpass filter

$$H_u(\omega) = \left\{ \begin{array}{ll} 1 & \text{for } |\omega| > \omega_c \\ 0 & \text{elsewhere} \end{array} \right. \tag{7.3}$$

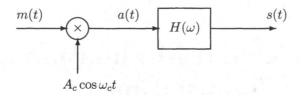

Figure 7.1: SSB Modulation by Combining DSBSC-AM and Filtering

and the lower sideband SSB modulation by the ideal lowpass filter

$$H_\ell(\omega) = \begin{cases} 1 & \text{for } |\omega| < \omega_c \\ 0 & \text{elsewhere} \end{cases} \tag{7.4}$$

It will now be shown that an SSB signal can be expressed in terms of the baseband message $m(t)$ and its Hilbert transform $\hat{m}(t)$. The pre-envelope of the SSB signal has the transform

$$S_+(\omega) = 2S(\omega)u(\omega) = 2A(\omega)H(\omega)u(\omega) = A_c M(\omega - \omega_c)H(\omega) \tag{7.5}$$

and the transform of its complex envelope is

$$\tilde{S}(\omega) = S_+(\omega + \omega_c) = A_c M(\omega)H(\omega + \omega_c) \tag{7.6}$$

Now consider the upper sideband case. On substituting $H_u(\omega)$ for $H(\omega)$ in (7.6), it can be seen after a little thought that

$$\begin{aligned} \tilde{S}(\omega) &= A_c M(\omega)u(\omega) = 0.5A_c M(\omega)(1 + \text{sign}\,\omega) \\ &= 0.5A_c M(\omega)[1 + j(-j\text{sign}\,\omega)] = 0.5A_c M(\omega) + j0.5A_c \hat{M}(\omega) \end{aligned} \tag{7.7}$$

So, the complex envelope is

$$\tilde{s}(t) = 0.5A_c[m(t) + j\hat{m}(t)] \tag{7.8}$$

Therefore, the SSB signal can be expressed as

$$s(t) = \Re\{\tilde{s}(t)e^{j\omega_c t}\} = 0.5A_c m(t)\cos\omega_c t - 0.5A_c \hat{m}(t)\sin\omega_c t \tag{7.9}$$

Similarly, in the lower sideband case, it follows that the transform of the complex envelope is

$$\begin{aligned} \tilde{S}(\omega) &= A_c M(\omega)u(-\omega) = 0.5A_c M(\omega)(1 - \text{sign}\,\omega) \\ &= 0.5A_c M(\omega)[1 - j(-j\text{sign}\,\omega)] = 0.5A_c M(\omega) - j0.5A_c \hat{M}(\omega) \end{aligned} \tag{7.10}$$

Therefore, the complex envelope is

$$\tilde{s}(t) = 0.5A_c[m(t) - j\hat{m}(t)] \tag{7.11}$$

The corresponding SSB signal is

$$s(t) = \Re\{\tilde{s}(t)e^{j\omega_c t}\} = 0.5A_c m(t)\cos\omega_c t + 0.5A_c \hat{m}(t)\sin\omega_c t \tag{7.12}$$

Equations (7.9) and (7.12) suggest the SSB modulator structure shown in Figure 7.2. Upper or lower sideband selection is accomplished by simply changing the sign of the input to the lower side of the output adder.

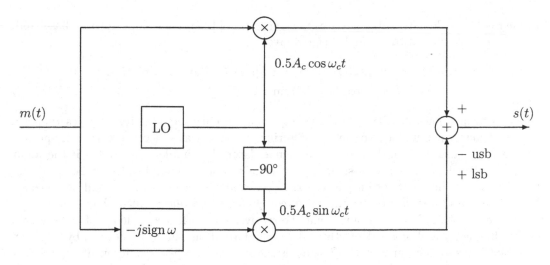

Figure 7.2: A Single-Sideband Modulator Using a Hilbert Transform

7.2 Coherent Demodulation of SSB Signals

One approach to demodulating SSB signals is to first multiply the received signal by a locally generated replica of the carrier signal. Multiplying (7.9) or (7.12) by $2\cos\omega_c t$ yields

$$
\begin{aligned}
b(t) &= A_c m(t) \cos^2 \omega_c t \mp A_c \hat{m}(t) \sin \omega_c t \cos \omega_c t \\
&= 0.5 A_c m(t) + 0.5 A_c m(t) \cos 2\omega_c t \mp 0.5 A_c \hat{m}(t) \sin 2\omega_c t \quad (7.13)
\end{aligned}
$$

The first term on the right-hand side of (7.13) is proportional to the desired message. The second and third terms have spectra centered about $2\omega_c$ and can be removed by passing $b(t)$ through a lowpass filter with cutoff frequency W. The effect in the frequency domain of multiplying by $\cos\omega_c t$ in the time domain is to shift $S(\omega)$ to the right and left by ω_c and take the sum. This translates the sidebands around $\pm\omega_c$ down to baseband and forms $M(\omega)$ which is the desired term and also translates them up to $\pm 2\omega_c$ which are the terms removed by the lowpass filter. A block diagram for this demodulator has the same form as Figure 7.1 except with the input $m(t)$ replaced by the received signal $s(t)$ and the filter $H(\omega)$ by a lowpass filter. In practice, this system should be preceded by a bandpass filter that passes $s(t)$ and eliminates out-of-band noise.

Another SSB demodulator that uses a Hilbert transform and is well suited to DSP implementation will now be described. The first step is to take the Hilbert transform of the received signal $s(t)$ and form the pre-envelope. Using (7.8) and (7.11), the pre-envelope can be expressed as

$$
s_+(t) = s(t) + j\hat{s}(t) = \tilde{s}(t)e^{j\omega_c t} = 0.5 A_c [m(t) \pm j\hat{m}(t)]e^{j\omega_c t} \quad (7.14)
$$

where the plus sign is for upper sideband and the minus sign is for lower sideband modulation. Multiplying the pre-envelope by $e^{-j\omega_c t}$ generates the complex envelope

$$
\tilde{s}(t) = s_+(t)e^{-j\omega_c t} = 0.5 A_c [m(t) \pm j\hat{m}(t)] \quad (7.15)
$$

In the frequency domain, this translates the transform of the pre-envelope down to baseband. Taking the real part of this complex signal gives

$$0.5 A_c m(t) = \Re\{s_+(t)e^{-j\omega_c t}\} = \Re\{[s(t) + j\hat{s}(t)][\cos\omega_c t - j\sin\omega_c t]\}$$
$$= s(t)\cos\omega_c t + \hat{s}(t)\sin\omega_c t \qquad (7.16)$$

which is proportional to the desired message signal. This demodulator requires taking a Hilbert transform but does not require filtering out terms at twice the carrier frequency. Figure 7.2 is also a block diagram for a demodulator that implements (7.16) if the input $m(t)$ is replaced by the received signal $s(t)$, the cosine and sine amplitudes are set to 1, and the plus sign is chosen at the output adder. In practice, the demodulator would be preceded by an bandpass filter that passes the signal components and rejects out-of-band noise.

These two demodulators assume that the receiver has perfect knowledge of the received carrier frequency and phase. Unfortunately, this information cannot be derived by a system like the Costas loop because the SSB signal is the sum of an *inphase* component $m(t)\cos\omega_c t$ and a *quadrature* component $\hat{m}(t)\sin\omega_c t$. It can be shown that when the demodulator's frequency is correct but the phase is in error, the demodulated output is a phase shifted version of the transmitted message. Since the human ear is relatively insensitive to phase, this does not degrade voice or music. However, the phase offset can cause unacceptable degradation when the shape of the message must be maintained like in digital data communication systems. A frequency error results in a demodulated signal which has all its spectral components shifted by this error. A standard approach to solving this problem is to add a small sinusoidal component called a *pilot tone* whose frequency is not in the SSB signal band and has a known relationship to the carrier frequency. The pilot tone frequency is often chosen to be the carrier frequency when the baseband message signal has no DC components. The receiver can then generate a local carrier reference by using a narrow bandwidth bandpass filter to select the pilot tone and possibly following this filter by a phase-locked loop.

7.3   Frequency Translation

The spectrum of a bandpass signal must often be translated from one center frequency to another in communication systems. One reason is to place the signal spectrum in an allocated channel. Several messages can be multiplexed together by shifting them to non-overlapping adjacent spectral bands and transmitting the sum of the resulting signals. This is called *frequency division multiplexing* (FDM). Another reason is to correct for carrier frequency offsets caused by oscillator inaccuracies or Doppler shifts. Of course, AM, DSBSC-AM, and SSB modulators translate signal spectra from baseband to passband and the coherent demodulators do the reverse.

A method for frequency translation that is well suited to DSP applications will now be described. Let $s(t)$ be a bandpass signal with the frequency ω_0 somewhere in its passband. The problem is to translate the spectrum so that ω_0 is moved to $\omega_1 = \omega_0 + \Delta\omega$. The first step is to form the pre-envelope

$$s_+(t) = s(t) + j\hat{s}(t) \qquad (7.17)$$

The corresponding Fourier transform is

$$S_+(\omega) = 2S(\omega)u(\omega) \tag{7.18}$$

The next step is to multiply by a complex exponential with frequency $\Delta\omega$ to get

$$r_+(t) = s_+(t)e^{j\Delta\omega t} = [s(t) + j\hat{s}(t)][\cos\Delta\omega t + j\sin\Delta\omega t] \tag{7.19}$$

which has the transform
$$R_+(\omega) = S_+(\omega - \Delta\omega) \tag{7.20}$$

This translates the original spectrum to the right by $\Delta\omega$ and moves the value at ω_0 to the frequency ω_1. Taking the real part of $r_+(t)$ gives the desired translated signal

$$r(t) = s(t)\cos\Delta\omega t - \hat{s}(t)\sin\Delta\omega t \tag{7.21}$$

The real part of $r_+(t)$ can also be expressed as

$$r(t) = [r_+(t) + \bar{r}_+(t)]/2 \tag{7.22}$$

so its Fourier transform is

$$
\begin{aligned}
R(\omega) &= [R_+(\omega) + \bar{R}_+(-\omega)]/2 \\
&= S(\omega - \Delta\omega)u(\omega - \Delta\omega) + \bar{S}(-\omega - \Delta\omega)u(-\omega - \Delta\omega) \tag{7.23}
\end{aligned}
$$

Figure 7.2 is also the block diagram for a frequency translator if the input $m(t)$ is replaced by the bandpass signal $s(t)$, the frequency ω_c is replaced by $\Delta\omega$, $0.5A_c$ is replaced by 1, and the negative sign is used at the output adder. Because frequency translation is functionally the same as upper sideband SSB modulation with the appropriate carrier frequency, no translation experiments will be performed.

Notice that (7.21) can be used even when the passband of the translated signal overlaps that of the original signal. To do this using real signals would require a double conversion process where the signal is first shifted to a non-overlapping band by multiplying by $\cos\omega_3 t$ and selecting the upper sideband with a highpass filter and then repeating the process to translate the spectrum back to the desired frequency. This is generally not as convenient as (7.21) for DSP applications.

7.4   Laboratory Experiments

Initialize the DSP and codec as in Chapter 2 for the following SSB experiments. Once again, you will be using a sampling rate of $f_s = 1/T = 16$ kHz. The DSP will be programmed to both generate and demodulate SSB signals since the lab does not have a hardware SSB signal generator.

7.4.1 Experiment 7.1: Making an SSB Modulator

Perform the following tasks to make and test an SSB modulator:

- Initialize the DSP and codec for a 16 kHz sampling rate.

- Write a program to implement the SSB modulator shown in Figure 7.2. Implement both the upper and lower sideband modulators. Take the message samples $m(nT)$ from the ADC. Send the modulated signal samples to the DAC. Use the carrier frequency $f_c = 4$ kHz and an amplitude A_c that scales the output samples appropriately for the codec. Be sure to match the delay introduced by your Hilbert transform filter.

- Attach the signal generator to the line input and set it to generate a 1200 Hz sine wave.

- Determine the theoretical formulas for the transmitted SSB signals for both the upper and lower sideband cases with the input $m(t) = A_m \cos 2\pi 1200\, t$.

- Observe the signals generated by your modulator on the scope for both cases and compare them with the theoretical ones.

- Vary the frequency of $m(t)$ from 0 to 4 kHz and observe $s(t)$ on the scope. Report what happens to the frequency of $s(t)$ for both the upper and lower sideband modulators.

- Next add a pilot tone $p(t) = A_p \cos 2\pi f_c t$ to the SSB output signal to provide a carrier reference for the demodulator you will make next. Use your judgment in choosing the value for A_p.

7.4.2 Experiment 7.2: Coherent Demodulation of an SSB Signal

In this exercise you will make a demodulator for the SSB signal with the added pilot tone. A block diagram of one possible demodulator structure is shown in Figure 7.3. The input signal $s(nT)$ is passed through a highpass filter $G(\omega)$ that rejects the pilot tone but passes the SSB component in the upper sideband case. $G(\omega)$ should be replaced by a lowpass filter that passes the SSB signal and rejects the pilot tone in the lower sideband case. A notch filter could also be used to eliminate the pilot tone. The resulting signal is passed through a Hilbert transform filter to form the pre-envelope.

The portion enclosed by dotted lines is a pair of bandpass filters that extract replicas of the pilot tone and its $-90°$ phase shift. The transfer functions of these two filters are

$$B_1(z) = \frac{(1 - r)(1 - rz^{-1}\cos\omega_c T)}{1 - 2rz^{-1}\cos\omega_c T + r^2 z^{-2}} \tag{7.24}$$

and

$$B_2(z) = \frac{(1 - r)rz^{-1}\sin\omega_c T}{1 - 2rz^{-1}\cos\omega_c T + r^2 z^{-2}} \tag{7.25}$$

The denominators of these filters have the factorization

$$1 - 2rz^{-1}\cos\omega_c T + r^2 z^{-2} = (1 - re^{j\omega_c T}z^{-1})(1 - re^{-j\omega_c T}z^{-1}) \tag{7.26}$$

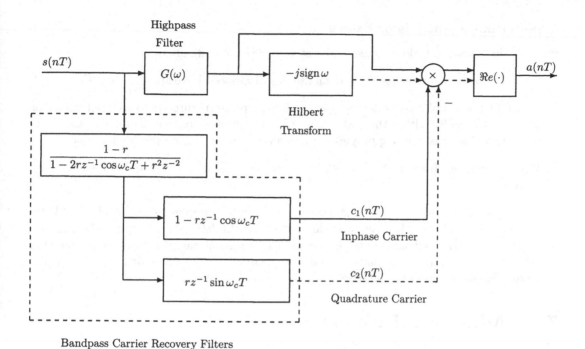

Bandpass Carrier Recovery Filters

Figure 7.3: Using a Pilot Tone in SSB Demodulation

Thus the filter poles are at $z = re^{\pm j\omega_c T}$. The quantity r is a number slightly less than 1 and controls the bandwidth of the filters. The closer it is to 1, the narrower the bandwidth.

The pre-envelope is then demodulated by the recovered complex carrier reference and the real part is taken to give the output signal $a(nT)$. Let the signal input to the multiplier be $v(nT) = v_1(nT) + jv_2(nT)$. Then

$$a(nT) = \Re\{v(nT)[c_1(nT) - jc_2(nT)]\} = v_1(nT)c_1(nT) + v_2(nT)c_2(nT) \qquad (7.27)$$

Theoretical Exercise

Prove that at the carrier frequency ω_c and when r is very close to 1, the transfer functions of the pilot tone extraction filters are approximately

$$B_1(e^{j\omega_c T}) \simeq 0.5 \qquad (7.28)$$

and

$$B_2(e^{j\omega_c T}) \simeq -0.5j \qquad (7.29)$$

By trial and error, choose a value of r that gives roughly a 50 Hz 3 dB bandwidth. A value too close to 1 can cause significant computational problems because of the high internal gain of the filter which results in numerical overflows and, possibly, instability. A too small value will make the bandwidth too large and the filter will not sufficiently attenuate the unwanted signal components.

Experimental Demodulator Exercises

Perform the following tasks to make and test your SSB demodulator:

1. Write a program to implement the demodulator discussed above and shown in Figure 7.3.

2. Now pipe the samples generated by your modulator program directly to your demodulator program internally within the DSP. Write the demodulator output samples to the DAC and check that it is working properly by observing the output on the oscilloscope.

3. Vary the message frequency and check that your modulator and demodulator are working correctly.

4. When your demodulator is working, send the modulator samples to the left channel DAC output. Connect the left channel analog output to the left channel input. Demodulate the left channel input and write the demodulated output samples to the right channel output. Observe the modulated signal (left channel output) and demodulated signal (right channel output) on the scope.

7.5   Additional References

For more complete discussions of SSB modulation and a generalization known as vestigial sideband (VSB) modulation see Gibson [II.D.9, Sections 5.4 and 5.5] and Haykin [II.D.17, Sections 3.5–3.9].

Chapter 8

Frequency Modulation

Frequency modulation (FM) was invented and commercialized after amplitude modulation. Its main advantage is that it is more resistant to additive noise than AM. In addition to commercial radio, it is used as a component of television signals, for satellite and microwave communications, and for digital data transmission. In this chapter the basic theory of FM modulation and demodulation will be presented and you will implement two types of demodulators, the frequency discriminator and the phase-locked loop.

8.1 The FM Signal and Some of its Properties

8.1.1 Definition of Instantaneous Frequency and the FM Signal

An FM signal is generated by using the baseband message signal to change the instantaneous frequency of a carrier sinusoid rather than its amplitude. The *instantaneous frequency* of a sinusoid $\cos \theta(t)$ is defined to be

$$\omega(t) = \frac{d}{dt}\theta(t) \tag{8.1}$$

This definition can be justified by observing that when $\theta(t) = \omega_c t$, its derivative is ω_c which is the frequency of $\cos \omega_c t$. The instantaneous frequency of an FM wave with carrier frequency ω_c is related to the baseband message $m(t)$ by the equation

$$\omega(t) = \omega_c + k_\omega m(t) \tag{8.2}$$

where k_ω is a positive constant called the *frequency sensitivity*. An oscillator whose frequency is controlled by its input $m(t)$ in this manner is called a *voltage controlled oscillator*. The angle of the FM signal, assuming the value is 0 at $t = 0$, is

$$\theta(t) = \int_0^t \omega(\tau)\, d\tau = \omega_c t + \theta_m(t) \tag{8.3}$$

where

$$\theta_m(t) = k_\omega \int_0^t m(\tau)\, d\tau \tag{8.4}$$

151

is the carrier phase deviation caused by $m(t)$. The FM signal generated by $m(t)$ is

$$s(t) = A_c \cos[\omega_c t + \theta_m(t)] \tag{8.5}$$

A discrete-time approximation to the FM wave can be obtained by replacing the integral by a sum. The approximate phase angle is

$$\theta(nT) = \sum_{k=0}^{n-1} \omega(kT)T = \omega_c nT + \theta_m(nT) \tag{8.6}$$

where

$$\theta_m(nT) = k_\omega T \sum_{k=0}^{n-1} m(kT) \tag{8.7}$$

The total carrier angle can be computed recursively by the formula

$$\theta(nT) = \theta((n-1)T) + \omega_c T + k_\omega T m((n-1)T) \tag{8.8}$$

The resulting FM signal sample is

$$s(nT) = A_c \cos \theta(nT) \tag{8.9}$$

8.1.2   Single Tone FM Modulation

A simple formula for the Fourier transform of the FM wave in terms of the transform of the baseband message like that for the AM wave does not exist. However, in the special case of the sinusoidal message $m(t) = A_m \cos \omega_m t$ interesting results can be derived. This is called *single tone* FM modulation. The FM wave generated by this message is

$$s(t) = A_c \cos \left(\omega_c t + \frac{k_\omega A_m}{\omega_m} \sin \omega_m t \right) \tag{8.10}$$

The *modulation* index for this FM signal is defined as

$$\beta = \frac{k_\omega A_m}{\omega_m} = \frac{\text{peak frequency deviation}}{\text{modulating frequency}} \tag{8.11}$$

An example for $\beta = 5$, $f_m = 100$ Hz, and $f_c = 1$ kHz is shown in Figure 8.1. Observe how the oscillations are fastest when $m(t)$ is at its positive peak and slowest at its negative peak.

It can be shown [II.D.17, p. 163] that $s(t)$ has the series expansion

$$s(t) = A_c \sum_{n=-\infty}^{\infty} J_n(\beta) \cos[(\omega_c + n\omega_m)t] \tag{8.12}$$

where $J_n(x)$ is the n-th order Bessel function of the first kind and is the value of the following integral:

$$J_n(x) = \frac{1}{2\pi} \int_{-\pi}^{\pi} e^{-j(n\tau - x \sin \tau)} \, d\tau \tag{8.13}$$

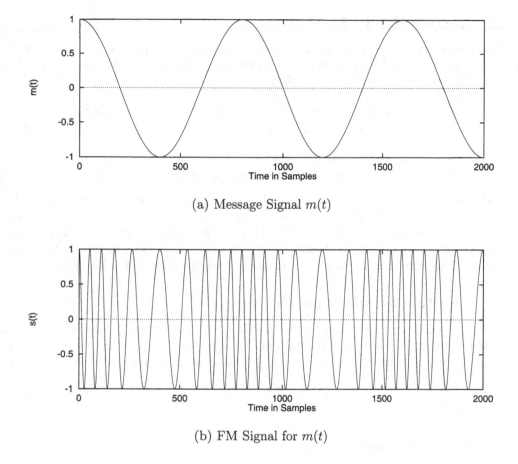

(a) Message Signal $m(t)$

(b) FM Signal for $m(t)$

Figure 8.1: Example for $f_c = 1$ kHz, $f_m = 100$ Hz, $f_s = 80$ kHz, $\beta = 5$

These functions can be computed by the series

$$J_n(x) = \sum_{m=0}^{\infty} (-1)^m \frac{\left(\frac{1}{2}x\right)^{n+2m}}{m!(n+m)!} \tag{8.14}$$

More properties of the Bessel functions can be found in [II.D.17, Appendix 4] and plots of the first few functions are shown on page 164 of the same reference.

Clearly, the spectrum of the FM signal is much more complex than that of the AM signal. It has components at the infinite set of frequencies $\{\omega_c + n\omega_m; \; n = -\infty, \cdots, \infty\}$. Plots of the spectra for various β can be found in [II.D.17] as well as most other undergraduate communication systems textbooks. The sinusoidal component at the carrier frequency has amplitude $J_0(\beta)$ and can actually become zero for some β.

8.1.3   Narrow Band FM Modulation

Another case where interesting results can be derived occurs when $|\theta_m(t)| \ll 1$ for all t and is called *narrow band* FM. Using the approximations that $\cos x \simeq 1$ and $\sin x \simeq x$ for $|x| \ll 1$, the FM signal can be approximated as follows:

$$
\begin{aligned}
s(t) &= A_c \cos[\omega_c t + \theta_m(t)] = A_c \cos \omega_c t \cos \theta_m(t) - A_c \sin \omega_c t \sin \theta_m(t) \\
&\simeq A_c \cos \omega_c t - A_c \theta_m(t) \sin \omega_c t
\end{aligned} \tag{8.15}
$$

or in complex notation

$$
s(t) \simeq A_c \Re e \left\{ e^{j\omega_c t} [1 + j\theta_m(t)] \right\} \tag{8.16}
$$

This is similar to the AM signal except that the discrete carrier component $A_c \cos \omega_c t$ is $90°$ out of phase with the sinusoid $A_c \sin \omega_c t$ multiplying the phase angle $\theta_m(t)$. The spectrum of narrow band FM is similar to that of AM. This narrow band approximation is sometimes used in analog FM modulators along with a frequency multiplier. However, with DSP implementations there is little reason not to use (8.9) to generate a true FM signal.

8.1.4   The Bandwidth of an FM Signal

In general, an exact simple formula for the bandwidth of an FM signal does not exist. The bandwidth depends on the form of the baseband message and the peak frequency deviation. The following formula, known as Carson's rule, is often used as an estimate of the bandwidth:

$$
B_T = 2(\Delta f + f_m) \quad \text{Hz} \tag{8.17}
$$

where

   Δf is the peak frequency deviation

and

   f_m is the maximum frequency at which the baseband message has a component.

For example, commercial FM signals use a peak frequency deviation of $\Delta f = 75$ kHz and a maximum baseband message frequency of $f_m = 15$ kHz. Carson's rule estimates the FM signal bandwidth as $B_T = 2(75 + 15) = 180$ kHz which is six times the 30 kHz bandwidth that would be required for AM modulation.

8.2   FM Demodulation by a Frequency Discriminator

A frequency discriminator is a device that converts a received FM signal into a voltage that is an estimate of the instantaneous frequency of its input without using a local oscillator and, consequently, in a noncoherent manner. Typically, the conversion is performed in analog discriminators by applying the FM signal to a bandpass filter with a relatively wide bandwidth and a center frequency that is shifted somewhat from the FM carrier frequency so that the instantaneous frequency of the input signal falls in a band on one side of the filter's amplitude peak where the response is monotonically increasing or decreasing. When the instantaneous frequency changes slowly relative to the time-constants of the filter, its output is approximately an FM signal with the same instantaneous frequency but with an envelope that varies according to the amplitude response of the filter at the instantaneous frequency. This approximation is called *quasi-static* analysis. The amplitude variations are then detected with an envelope detector like the ones used for AM demodulation. The block diagram of a very elementary discriminator is shown in Figure 8.2. In a more advanced discriminator, the input FM signal $s(t)$ is applied to a second bandpass filter with a center frequency of $f_1 = f_c + \Delta$ so the instantaneous frequency of $s(t)$ varies to the left of f_1 and the amplitude variations of the filter output are in the opposite direction of those of the first filter. The envelope of the output of the first filter is subtracted from that of the second filter to get the discriminator output. If the amplitude responses have the same shapes relative to their center frequencies and are symmetric about their center frequencies, the discriminator output will be zero when the input frequency is at the carrier frequency, positive when it is above, and negative when below. The filters can be designed to result in a discriminator output that is almost linearly related to the deviation of the input frequency from the carrier frequency over an adequately wide range.

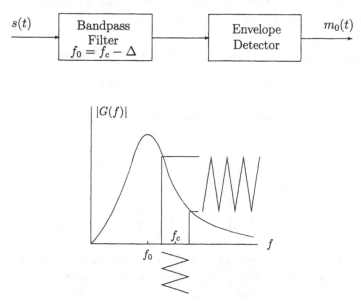

Figure 8.2: An Elementary Discriminator

8.2.1 An FM Discriminator Using the Pre-Envelope

Rather than implementing a filter and envelope detector as shown in Figure 8.2, a better approach for DSP applications is make a discriminator based on using the pre-envelope. When $\theta_m(t)$ is sufficiently small and band-limited so that $\cos\theta_m(t)$ and $\sin\theta_m(t)$ are essentially band-limited signals with cutoff frequencies less than ω_c, it can be shown that the pre-envelope of the FM signal is

$$s_+(t) = s(t) + j\hat{s}(t) = A_c e^{j[\omega_c t + \theta_m(t)]} \tag{8.18}$$

The angle of the pre-envelope is

$$\varphi(t) = \arctan[\hat{s}(t)/s(t)] = \omega_c t + \theta_m(t) \tag{8.19}$$

The derivative of the phase is

$$\frac{d}{dt}\varphi(t) = \frac{s(t)\dfrac{d}{dt}\hat{s}(t) - \hat{s}(t)\dfrac{d}{dt}s(t)}{s^2(t) + \hat{s}^2(t)} = \omega_c + k_\omega m(t) \tag{8.20}$$

which is exactly the instantaneous frequency. This equation can be approximated for a discrete-time implementation by using FIR filters to form the derivatives and Hilbert transform. Notice that the denominator is the squared envelope of the FM signal.

The bandwidth of this and other FM discriminators must be at least as great as that of the received FM signal. Thus, the required bandwidth is usually significantly greater than that of the baseband message. This limits the degree of noise reduction that can be achieved by preceding the discriminator by a bandpass receive filter.

8.2.2   A Discriminator Using the Complex Envelope

A discriminator using the complex envelope also can be derived. The complex envelope for the FM signal is

$$\tilde{s}(t) = s_+(t)e^{-j\omega_c t} = s_I(t) + j\,s_Q(t) = A_c e^{j\theta_m(t)}$$

The angle of the complex envelope is

$$\tilde{\varphi}(t) = \arctan[s_Q(t)/s_I(t)] = \theta_m(t)$$

and the derivative of the phase is

$$\frac{d}{dt}\tilde{\varphi}(t) = \frac{s_I(t)\dfrac{d}{dt}s_Q(t) - s_Q(t)\dfrac{d}{dt}s_I(t)}{s_I^2(t) + s_Q^2(t)} = k_\omega m(t)$$

which is proportional to the message signal and has no added constant term.

A block diagram for implementing this discriminator is shown in Figure 8.3. First the pre-envelope is formed and demodulated to get the complex envelope whose real part is the inphase component and imaginary part is the quadrature component. The inphase and quadrature components are both lowpass signals. Therefore, the frequency response of the differentiators must approximate $j\omega$ over a band centered around $\omega = 0$ out to the cut-off frequency for the I and Q components. If the differentiators are implemented as FIR filters, their amplitude responses will automatically pass through 0 at the origin and excellent designs can be achieved. Notice how the delays through the Hilbert transform filter and differentiation filter are matched by taking signals out of the center taps. In the discriminator using the pre-envelope, the differentiators must be approximate $j\omega$ over a passband centered around the carrier frequency and it is harder to make these bandpass differentiators.

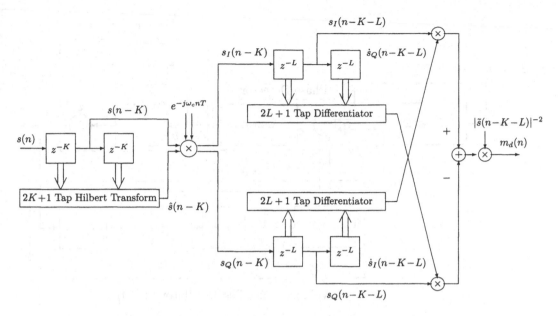

Figure 8.3: Discrete-Time Discriminator Realization Using the Complex Envelope

8.3   Using a Phase-Locked Loop for FM Demodulation

A device called a *phase-locked loop* (PLL) can be used to demodulate an FM signal with better performance in a noisy environment than a frequency discriminator. The block diagram of a discrete-time version of a PLL is shown in Figure 8.4. It is similar to the Costas loop.

The PLL input shown in the figure is the noiseless FM signal

$$s(nT) = A_c \cos[\omega_c nT + \theta_m(nT)] \tag{8.21}$$

as described by (8.6) through (8.9). This input is passed through a Hilbert transform filter to form the pre-envelope

$$s_+(nT) = s(nT) + j\hat{s}(nT) = A_c e^{j[\omega_c nT + \theta_m(nT)]} \tag{8.22}$$

The pre-envelope is multiplied by the output of the voltage controlled oscillator (VCO) block. The phase of the VCO one sample into the future is the input to the z^{-1} block which is described by the equation

$$\phi((n+1)T) = \phi(nT) + \omega_c T + k_v T y(nT) \tag{8.23}$$

Starting at $n = 0$ and iterating the equation, it follows that

$$\phi(nT) = \omega_c nT + \theta_1(nT) \tag{8.24}$$

where

$$\theta_1(nT) = \theta(0) + k_v T \sum_{k=0}^{n-1} y(kT) \tag{8.25}$$

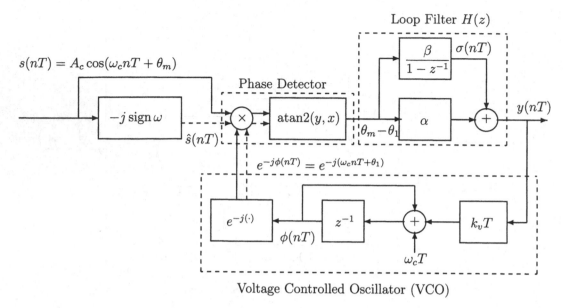

Figure 8.4: A Discrete-Time Phase-Locked Loop

The VCO output is

$$v(nT) = e^{-j\phi(nT)} = e^{-j[\omega_c nT + \theta_1(nT)]} \tag{8.26}$$

The multiplier output is

$$p(nT) = A_c e^{j[\theta_m(nT) - \theta_1(nT)]} \tag{8.27}$$

The phase error between the angle of the FM input signal and the VCO output is the angle of the multiplier output $p(nT)$ and can be computed as

$$\theta_m(nT) - \theta_1(nT) = \arctan\left[\frac{\Im m\{p(nT)\}}{\Re e\{p(nT)\}}\right] \tag{8.28}$$

This is shown in the figure as being computed by the C library function atan2(y,x) which is a four quadrant arctangent and gives angles between $-\pi$ and π. The block consisting of the multiplier and arctan function is called a *phase detector*.

A less accurate, but computationally simpler, estimate of the phase error when the error is small is

$$
\begin{aligned}
\Im m\{p(nT)\} &= \hat{s}(nT)\cos[\omega_c nT + \theta_1(nT)] - s(nT)\sin[\omega_c nT + \theta_1(nT)] \\
&= A_c \sin[\theta_m(nT) - \theta_1(nT)] \simeq A_c[\theta_m(nT) - \theta_1(nT)]
\end{aligned} \tag{8.29}
$$

The phase detector output is applied to the loop filter which has a transfer function of the form

$$H(z) = \alpha + \frac{\beta}{1 - z^{-1}} = (\alpha + \beta)\frac{1 - \frac{\alpha}{\alpha+\beta}z^{-1}}{1 - z^{-1}} \tag{8.30}$$

The accumulator portion of the loop filter which has the output $\sigma(nT)$ enables the loop to track carrier frequency offsets with zero error. It will be shown shortly that the output $y(nT)$ of the loop filter is an estimate of the transmitted message $m(nT)$.

The PLL is a nonlinear system because of the characteristics of the phase detector. If the discontinuities in the arctangent are ignored, the PLL can be represented by the linearized model shown in Figure 8.5. The transfer function for the linearized PLL is

$$
\begin{aligned}
L(z) &= \frac{Y(z)}{\Theta_m(z)} = \frac{H(z)}{1 + H(z)\dfrac{k_v T z^{-1}}{1 - z^{-1}}} \\
&= (1 - z^{-1})\frac{\alpha + \beta - \alpha z^{-1}}{1 - [2 - (\alpha + \beta)k_v T]z^{-1} + (1 - \alpha k_v T)z^{-2}}
\end{aligned}
\tag{8.31}
$$

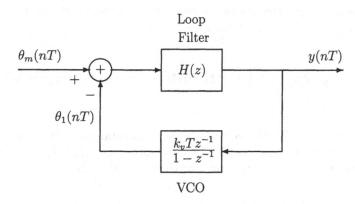

Figure 8.5: Linearized Model for the Phase-Locked Loop

At low frequencies, which corresponds to $z \simeq 1$, $L(z)$ can be approximated by

$$
L(z) \simeq \frac{z - 1}{k_v T}
\tag{8.32}
$$

Thus

$$
Y(z) \simeq \Theta_m(z)\frac{z - 1}{k_v T}
\tag{8.33}
$$

and in the time-domain

$$
y(nT) \simeq \frac{\theta_m((n + 1)T) - \theta_m(nT)}{k_v T}
\tag{8.34}
$$

Using (8.7) for θ_m gives

$$
y(nT) \simeq \frac{k_\omega}{k_v}m(nT)
\tag{8.35}
$$

This last equation demonstrates that the PLL is an FM demodulator under the appropriate conditions.

The frequency response of the linearized loop has the characteristics of a band-limited differentiator. The loop parameters must be chosen to provide a loop bandwidth that is sufficient to pass the desired baseband message signal but the bandwidth should also be kept as small as possible to suppress out-of-band noise. The PLL performs better than a frequency discriminator when the FM signal is corrupted by additive noise. The reason is that the bandwidth of the frequency discriminator must be large enough to pass the modulated FM signal while the PLL bandwidth only has to be large enough to pass the baseband message. With wideband FM, the bandwidth of the modulated signal can be significantly larger than that of the baseband message. Therefore, the PLL can reject more out-of-band noise because of its narrower bandwidth.

The PLL described in this experiment is very similar to the Costas loop presented in Chapter 6 for coherent demodulation of DSBSC-AM. However, it should be noted that there is a significant difference in the loop bandwidths required for these two systems. The bandwidth of the PLL used for FM demodulation must be large enough to pass the baseband message signal. On the other hand, the Costas loop is used to generate a stable carrier reference signal so its bandwidth should be very small and just wide enough to track carrier drifts and allow a reasonable acquisition time.

8.4   Laboratory Experiments for Frequency Modulation

The following experiments should give you a deeper understanding of the theoretical concepts presented above for FM modulation and demodulation. As usual, initialize the DSK as in Chapter 2 and use a 16 kHz sampling rate for these experiments. .

8.4.1   Experiment 8.1: Measuring the Spectrum of an FM Signal

The experiments described in this section assume that a spectrum analyzer is not available. If your lab has a spectrum analyzer or you made the spectrum analyzer for Chapter 4, you can use your imagination and expand on the suggestions of this subsection and observe a wider band of spectral lines. To investigate the spectral properties of FM, perform the following tasks:

1. Set the signal generator to FM modulate an $f_c = 4$ kHz sinusoidal carrier with an $f_m = 150$ Hz sine wave. Look at the signal generator manual to see how to get an FM signal with the desired carrier frequency, modulating frequency and waveform, and frequency deviation.

2. Connect the FM output signal to the oscilloscope and observe the resulting waveforms as you vary the frequency deviation.

3. Write a program for the DSP to implement an IIR bandpass filter that passes the 4 kHz carrier component and strongly attenuates the other FM spectral components. Send the filter output to the DAC so it can be observed on the oscilloscope.

4. Watch the amplitude of the 4 kHz carrier component on the scope as the modulation index is increased from 0. Remember that this component should be proportional to $J_0(\beta)$.

5. Increase the modulation index slowly until the carrier component becomes zero. Compare this experimentally determined value of β with the theoretical value for the the first zero of $J_0(\beta)$. You can generate values of the Bessel function by using the series expansion given by (8.14) on page 153 or with MATLAB.

6. Plot the theoretical power spectrum of a sinusoidally modulated FM signal with $\beta = 2$, 5, and 10.

8.4.2 Experiment 8.2: FM Demodulation Using a Frequency Discriminator

Write a C program that implements the frequency discriminator shown in Figure 8.3. Assume that the carrier frequency is 4 kHz, the baseband message is band-limited with a cutoff frequency of 500 Hz, and use a sampling rate of 16 kHz. Synchronize the sample processing loop with the transmit ready flag (XRDY) of McBSP1. Read samples from the ADC, apply them to your discriminator, and write the output samples to the DAC.

Use remez87.exe, window.exe, or MATLAB to design the FIR differentiation and Hilbert transform filters. Use enough taps to approximate the desired Hilbert transform frequency response well from 1200 to 6800 Hz. Try a differentiator bandwidth extending from 0 to 4000 Hz. (**Be sure to match the delays of your filters in your implementation.**)

Experimentally test your discriminator by doing the following:

1. Use the signal generator to create a sinusoidally modulated FM signal as you did for the FM spectrum measurement experiments. Attach the signal generator to the line input and observe your demodulated signal on the oscilloscope to check that the program is working.

2. Modify your program to add Gaussian noise to the input samples and observe the discriminator output as you increase the noise variance. Does the performance degrade gracefully as the noise gets large?

3. Attach the line output to the PC's speakers and listen to the noisy demodulator output as you vary the SNR.

8.4.3 Experiment 8.3: Using a Phase-Locked Loop for FM Demodulation

Design and implement a PLL like the one shown in Figure 8.4 to demodulate a sinusoidally modulated FM signal with the same parameters used previously in the discriminator experiments. Let $\alpha = 1$ and choose β to be a factor of 100 or more smaller than α. Perform the following theoretical exercises to select your loop parameters:

1. Compute and plot the amplitude response of the linearized loop using (8.31) for different loop parameters until you find a set that gives a reasonable response.

2. Theoretically compute and plot the time response of the linearized loop to a unit step input for your selected set of parameters by iterating a difference equation corresponding to the transfer function.

Write a C program for the DSP to implement the PLL. Use a 16 kHz sampling rate. Remember to take into account that the ADC input samples are in the range $\pm 2^{15}$. Test your PLL demodulator by the following steps:

1. Connect an FM signal from the signal generator to the DSK line input and observe the DAC output on the oscilloscope.

2. See if your PLL will track carrier frequency offsets by changing the carrier frequency on the signal generator slightly and observing the output. Experiment and see how large an offset your loop will track. Also observe any differences in behavior when you change the carrier frequency smoothly and slowly or make step changes.

3. Modify your program to add Gaussian noise to the input samples and observe the demodulated output as the noise variance increases. How does the quality of the demodulated output signal compare with that of the frequency discriminator at the same signal-to-noise ratio, particularly when this ratio gets small? You should find that the PLL works at a lower SNR than the discriminator.

4. Connect the line output to the PC's speaker and listen to the demodulated output as you vary the SNR. How does it sound compared to the discriminator output?

8.5   Additional References

All the senior level textbooks on communication systems contain sections on frequency modulation. For example, see Gibson [II.D.9, Chapter 6] and Haykin [II.D.17, Sections 3.10–3.14]. See the references just cited, Gitlin [II.D.11, Section 6.2], Gardner [II.D.8], and Lee and Messerschmitt [II.D.26, Chapter 13] for further discussions of phase-locked loops. A brief discussion of discrete-time PLL's can be found in Lee and Messerschmitt [II.D.26, Section 13.2]. To my knowledge, the PLL structure described in this experiment using the pre-envelope cannot be found in other textbooks.

Chapter 9

Pseudo-Random Binary Sequences and Data Scramblers

This chapter begins a series on digital communications. DSP chips have made a dramatic impact on this field, initially in narrow band systems like voice-band telephone line modems and cellular telephones. In 1970, a plain 9600 bps telephone line modem was the size of a big microwave oven; contained many analog chips for filters, delay lines, and adaptive equalizer coefficient scalars; required a fan because of significant power consumption; and cost at least $15,000. It was basically just a data pump with no extra features. A few years later, medium scale integrated (MSI) digital chips were used to make a micro-coded digital signal processing unit to replace the analog functions, but the modems were still large and costly. The MSI chips included cascadable 4-bit wide ALU slices and an AMD multiplier chip. As soon as DSP chips were introduced in the early 1980's they were used to further reduce the size and cost of telephone line modems. Typically, several DSP's were required to implement a transmitter and receiver. VLSI technology rapidly improved and now a state-of-the-art V.92 56 kbps modem can be bought for less than $100 and fits in a small box or on a small card. In addition, this modem has many features like data compression, error detection and correction, trellis coded modulation, fax modes, automatic dialing, network management functions, a secondary channel, and the ability to fall back to most of the past popular telephone line modem standards ranging from speeds of 300 bps up to 33,600 bps. It is now possible to concurrently run at least 12 full duplex V.92 modems in a single state-of-the-art DSP core and chips with multiple cores are currently being sold commercially. These high-end chips are used in remote access servers (RAS) by Internet service providers for voice over IP (VOIP) and modem pools. Because of the flexibility of the software approach to implementing signal processing algorithms with DSP's, new theoretical developments were almost instantaneously included in commercial telephone line modems. These techniques later found their way into higher speed systems that use greater channel bandwidths like high speed digital subscriber lines (DSL), microwave systems, satellite communications, and HDTV.

Broadband data transmission via DSL, cable, fiber optic lines, and wireless systems is rapidly making voiceband telephone line modems obsolete for dial-up access. However, they are still used in the large FAX machine market. New generations of DSP's like TI's

TMS320C6000 series are being used in the broadband systems. DSP's with special accelerator units for FFT's, turbo and low density parity check codes, and encryption are being produced. The DSP manufacturers are working hard to be competitive with FPGA's for high speed signal processing tasks.

In order to simulate and test digital communication systems, sequences that approximate ideal binary random sequences are required. In this chapter, you will see how to generate pseudo-random binary sequences using linear feedback shift registers. Then you will see how a simple variation of these circuits can be used to make a self synchronizing digital data scrambler and descrambler. These scramblers are required to break up long strings of 1's or 0's to allow tracking loops in the receiver to maintain lock, rather than for secrecy.

At very high data rates, these scramblers and descramblers can be implemented by very simple VLSI circuits. At moderate data rates like found in telephone line modems, they would be implemented by a few lines of simple DSP code. Incorporating this function and as many others as possible into the DSP code eliminates extra hardware. This improves reliability and reduces manufacturing cost. The ability to manufacture products at the lowest possible cost is extremely important to companies operating in the highly competitive commercial market.

9.1   Using Linear Feedback Shift Registers to Generate Pseudo-Random Binary Sequences

An ideal binary random sequence is an infinite sequence of independent, identically distributed, random variables, each taking on the values 0 or 1 with probability 0.5. These sequences are often used as a models for the data streams generated by binary sources. Excellent approximations to binary random sequences can be generated by linear feedback shift registers. The resulting sequences are called pseudo-random, pseudo-noise (PN), maximal length, or m sequences. Suggestions for additional references are included at the end of this chapter.

9.1.1   The Linear Feedback Shift Register Sequence Generator

The block diagram of a linear feedback shift register sequence generator is shown in Figure 9.1. It is common in digital sequential circuit analysis to represent a delay element by the symbol D rather than z^{-1}. The D represents a single stage of a shift register when the circuit is implemented in hardware or a memory location when it is implemented by software. All the adders perform modulo 2 addition which is equivalent to the exclusive-or logical function. Addition and subtraction are identical in modulo 2 arithmetic. The h's can be 0 or 1, with 0 indicating no connection to the adder and 1 indicating a connection. The input $x(n)$ is a binary sequence which will be assumed to be identically 0 in this section.

The z-transform of a binary sequence with z^{-1} replaced by D is called its Huffman transform. The coefficients of D in the transform are interpreted using modulo 2 arithmetic. The Huffman transform has the same basic properties as the z-transform so binary sequences

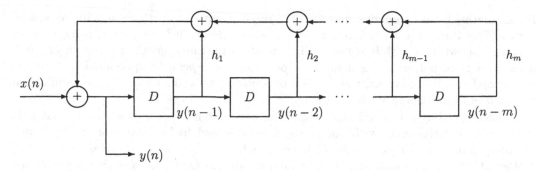

Figure 9.1: Linear Feedback Shift Register Sequence Generator

and linear sequential circuits can be analyzed and represented by these transforms in the same way as continuous-amplitude discrete-time systems.

Assuming that the input $x(n)$ is identically 0, the output of the feedback shift register sequence generator is

$$y(n) = \sum_{k=1}^{m} h_k y(n-k) \tag{9.1}$$

where the summation represents modulo 2 additions. Equivalently, any output sequence must satisfy the homogeneous difference equation

$$y(n) + \sum_{k=1}^{m} h_k y(n-k) = 0 \tag{9.2}$$

In this chapter, addition of binary quantities will always be understood to be performed by modulo 2 arithmetic. Let the *state* of the shift register generator be defined to be the m-tuple

$$\mathbf{s}(n) = [y(n-1), y(n-2), \cdots, y(n-m)] = [s_1(n), s_2(n), \cdots, s_m(n)] \tag{9.3}$$

Given the state at any time, the present and all future outputs can be uniquely computed from (9.1).

9.1.2   The Connection Polynomial and Sequence Period

If the initial shift register state is **0**, then the present and all future outputs must be **0** and all future states remain at **0**. This is the trivial solution to (9.2). It can be shown that if the initial state is not **0**, the future states can never become 0. However, the m-component state vector can take at most $2^m - 1$ nonzero values so it must repeat at some time. Then the output and state sequence will repeat, so the shift register will generate a periodic output sequence with period no greater than $2^m - 1$.

The properties of a shift register sequence are determined by its *connection polynomial*

$$h(D) = 1 + \sum_{k=1}^{m} h_k D^k \tag{9.4}$$

For simulating binary random sequences, shift register sequences with the maximum possible period $2^m - 1$ are of primary interest and these are often called *maximal length* sequences. It can be shown that a shift register will generate a maximal length sequence if and only if its connection polynomial is a special type known as a *primitive* polynomial. Primitive polynomials of all degrees exist and tables of primitive polynomials of degree up to 34 can be found in Peterson and Weldon [II.E.12].

A necessary but not sufficient condition for a polynomial to be primitive is that it be *irreducible*. A polynomial with binary coefficients is said to be irreducible over the field of binary numbers if it cannot be factored into the product of polynomials with binary coefficients and degrees at least 1. Otherwise it is called *reducible*. A very simple partial test of whether a polynomial is irreducible is to check whether 0 or 1 are roots. If 0 is a root, D must be a factor, and if 1 is a root, $D + 1$ must be a factor. Notice that if 1 is a root of $h(D)$, then $h(1) = 0$ and $h(D)$ must have an even number of 1 coefficients. Thus, irreducible polynomials over the binary field must have an odd number of 1 coefficients. As an example, consider the polynomial $h_1(D) = 1 + D + D^2$. Clearly, 0 and 1 are not roots so D and $D + 1$ are not factors. Since there are no other first degree factors, $h(D)$ must be irreducible. On the other hand, $h_2(D) = h_1^2(D)$ does not have 0 or 1 as roots but it is reducible.

The period of a shift register sequence with an irreducible connection polynomial $h(D)$ of degree m can be shown to be the smallest nonzero integer N such that $D^N - 1$ is divisible by $h(D)$ using modulo 2 arithmetic for the coefficients. The period N is called the *exponent* of $h(D)$. It can be shown that $h(D)$ always must divide $D^{2^m-1} - 1$. However, N may be smaller than $2^m - 1$ but must divide it. The polynomial is primitive when $N = 2^m - 1$. For primitive connection polynomials, the shift register state goes through all $2^m - 1$ nonzero values before repeating.

The situation is more complicated when the connection polynomial is reducible. Suppose

$$h(D) = \prod_{k=1}^{L} f_k(D) \tag{9.5}$$

where each factor is irreducible and has exponent n_k. In this case, the shift register will generate sequences with different periods depending on its initial state. Each period will be the product of a subset of the exponents.

9.1.3   Properties of Maximal Length Sequences

Maximal length sequences have several properties that make them good approximations to ideal binary random sequences when $N = 2^m - 1$ is large. Derivations of these properties can be found in Golomb [II.E.8, pp. 43-45].

Frequency of Occurrence of 1's and 0's

The number of 1's in one period of a maximal length sequence is 2^{m-1} and the number of 0's is $2^{m-1} - 1$. Thus, each period contains one more 1 than 0. For large N, 1's and 0's essentially appear with equal likelihood.

Frequency of Runs of 1's and 0's

Just because 1's and 0's are essentially equally likely for large N does not mean they are randomly arranged in time. All the 1's could be clumped together. However, this is not the case. A run of k 1's is defined to be a string starting with a 0, followed by k 1's, and ending with a 0. The probability of this string occurring in an ideal binary random sequence is $2^{-(k+2)}$. Similarly, a run of k 0's is a string starting with a 1, followed by k 0's, and ending with a 1 and has the same probability.

In one period of a maximal length sequence, there is one run of m 1's. There is no run of $m - 1$ 1's. For $1 \le k \le m - 2$, there are 2^{m-k-2} runs of k 1's. There is no run of m 0's, one run of $m - 1$ 0's, and 2^{m-k-2} runs of k 0's for $1 \le k \le m - 2$.

Correlation Property

In discussing correlation properties, it will be convenient to transform sequences of 0's and 1's into sequences of +1's and −1's. Let $y(n)$ be a sequence with period N that can have the value 0 or 1. The transformed sequence is

$$\breve{y}(n) = \begin{cases} +1 & \text{if } y(n) = 0 \\ -1 & \text{if } y(n) = 1 \end{cases} \tag{9.6}$$

The periodic autocorrelation function is defined to be

$$R(n) = \frac{1}{N} \sum_{k=0}^{N-1} \breve{y}(k)\breve{y}(n+k) \tag{9.7}$$

where the sum is performed using ordinary addition.

For maximal length sequences, $N = 2^m - 1$ and the periodic autocorrelation function is

$$R(n) = \begin{cases} -\frac{1}{N} & \text{for } n \text{ not a multiple of } N \\ 1 & \text{for } n \text{ a multiple of } N \end{cases} \tag{9.8}$$

For ideal binary random sequences, the statistical autocorrelation function is 1 for $n = 0$ and 0 otherwise.

9.2 Self Synchronizing Data Scramblers

9.2.1 The Scrambler

Long strings of 1's or 0's can appear at the output of devices like terminals and computer serial ports when they are idle. These signals must be randomized before being applied to the transmitter in a modem or else the symbol clock tracker and adaptive equalizer in the remote receiver will not work properly. The randomization is frequently accomplished by a self synchronizing scrambler. Figure 9.1 is also the block diagram for this kind of scrambler where $x(n)$ is the scrambler input and $y(n)$ is its output. A primitive shift register connection polynomial is usually used. Two primitive connection polynomials used in all recent ITU-T standard modems are

$$h_1(D) = 1 + D^{18} + D^{23} \tag{9.9}$$

and

$$h_2(D) = 1 + D^5 + D^{23} = D^{23}h_1(D^{-1}) \tag{9.10}$$

Notice that one of these polynomials is obtained from the other by turning its coefficients around backwards. It can be shown that this always turns one primitive polynomial into another. With 0 input, the scrambler generates a maximal length sequence of period $2^{23} - 1$ if the initial state is nonzero.

The scrambler input and output are related by the equation

$$y(n) = x(n) + \sum_{k=1}^{m} h_k y(n - k) \tag{9.11}$$

In terms of the transform notation

$$Y(D) = \frac{X(D)}{h(D)} \tag{9.12}$$

when the initial state is **0**. In ITU-T standards the scrambler is defined by saying that the output is generated by dividing the input by $h(D)$.

In some ITU-T modem standards, a pseudo-random binary sequence is specified to be generated by making the scrambler input 1 for an initial training period. Let $y_h(n)$ be a solution to the homogeneous equation formed by letting the input $x(n)$ be identically 0 in (9.11). This is a maximal length sequence. A particular solution to (9.11) for $x(n) \equiv 1$ is $y_p(n) = 1$ for all n. This can be seen as follows. The connection polynomial has an odd number of nonzero coefficients. Thus the feedback sum on the right-hand side of (9.11) has an even number of terms which are all 1 and add to 0 modulo 2. Since $x(n) = 1$, the total sum on the right-hand side becomes 1 which matches the $y_p(n) = 1$ on the left-hand side. The total scrambler output is

$$y(n) = y_h(n) + y_p(n) = y_h(n) + 1 = \bar{y}_h(n) \tag{9.13}$$

where the over-bar denotes logical complement. Thus, the scrambler output is the complement of a maximal length sequence when the input is 1.

From the previous paragraph, it can be seen that when the initial state of the scrambler is all 1's and the input is 1, the output is always 1 and the state remains all 1's. This is called a *lock-up* condition of the scrambler and can easily be detected and corrected. It is the only initial state that causes lock-up with an all 1's input. This situation is the complement of having an identically 0 input and starting in the all 0's state. A detailed analysis of the lock-up phenomenon with periodic inputs is presented in Gitlin, Hayes, and Weinstein [II.D.11, pp. 454–460].

9.2.2   The Descrambler

At the receiver, the input is recovered from the received sequence $y(n)$ by inverting (9.11). The resulting descrambling equation is

$$x(n) = y(n) + \sum_{k=1}^{m} h_k y(n-k) \tag{9.14}$$

or in transform notation

$$X(D) = Y(D)h(D) \tag{9.15}$$

The descrambler is just an FIR filter with $m+1$ taps that uses modulo 2 arithmetic. A block diagram of the descrambler is shown in Figure 9.2. If $y(n)$ is corrupted by channel errors, the scrambler output will also contain errors. If $h(D)$ has K nonzero coefficients, a single isolated error in $y(n)$ will cause K output errors as it propagates by the nonzero coefficients. Therefore, a connection polynomial with the least number of nonzero coefficients should be chosen.

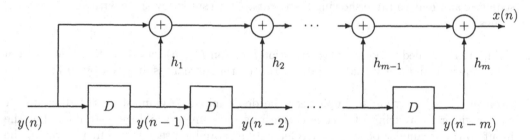

Figure 9.2: Self Synchronizing Descrambler

9.3   Theoretical and Simulation Exercises for Shift Register Sequence Generators and Scramblers

The goal of this chapter is to introduce the shift register sequence method for generating pseudo-random binary sequences so they can be used to simulate binary data sources in the remaining chapters on digital communications. The "experiments" in this section are really theoretical and computer simulation exercises to reinforce your understanding of shift register sequence generators and self synchronizing scramblers. The programs you will write can serve as the basis for ones you may need in future experiments.

9.3.1   Exercises for a Shift Register Sequence Generator with a Primitive Connection Polynomial

It can be shown that $h(D) = 1 + D^2 + D^5$ is a primitive polynomial. Perform the following exercises:

1. Assume the input to the scrambler in Figure 9.1 is 0 and that the initial state is not **0**. Find the period N of the sequence the shift register generates.

2. Write a C program to implement the scrambler including an input $x(n)$ which is set to 0. Store the shift register state as 5 consecutive bits in a single integer variable. Update the state by or-ing each new output into the appropriate bit in this integer variable and then shifting it with one of the C shift operators. You can use the C compiler for the PC or DSP.

3. Set the state of your shift register to a nonzero value. Generate and record enough outputs to verify your calculation of the period.

4. Count the number of 1's and 0's in one period of your sequence and check that the results agree with the theory.

5. Count the number of runs of 1's and 0's of each possible length in one period of your sequence and make a table showing the results. Make sure they agree with the theoretical values.

6. Compute the scaled periodic autocorrelation function $NR(n)$ for $n = 0, 1, \ldots, N-1$ from your sequence and check that it agrees with the theoretical result given by (9.7).

7. Now write a C program to implement the descrambler. Again, store the descrambler shift register as a string of 5 consecutive bits in a single C integer variable. Let the input to the scrambler be $x(n) = 1$. Generate a scrambled sequence, put it through the descrambler, and check that the descrambler output is all 1's. Notice that the initial states of the scrambler and descrambler do not have to be identical if an initial burst of errors is acceptable as the descrambler shift register fills up with received bits.

9.3.2   Exercises for a Shift Register Sequence Generator with an Irreducible but not Primitive Connection Polynomial

Let $h(D) = 1 + D + D^2 + D^3 + D^4$ be the connection polynomial for a shift register sequence generator. It can be shown that $h(D)$ is irreducible but not primitive. Perform the following exercises:

1. Find the period N for this sequence generator.

2. The four stage shift register can have $2^4 - 1 = 15$ nonzero values. Let $y(n)$ be the sequence generated by a particular nonzero state. Consider $y(n)$ and the sequences obtained by delaying $y(n)$ by $1, 2, \ldots, N-1$ samples to be an *equivalence class* of N sequences. It can be shown that each member of the equivalence class corresponds to a unique shift register initial state. Thus, there must be $(2^4 - 1)/N$ equivalence classes. Find one member of each equivalence class and its corresponding initial shift register state.

9.3.3 Exercises for a Shift Register Sequence Generator with a Reducible Connection Polynomial

Now let the connection polynomial for the shift register sequence generator be

$$h(D) = (1 + D + D^2)(1 + D + D^3) = 1 + D^4 + D^5 \tag{9.16}$$

Perform the following exercises:

1. Verify that the product is correct.

2. It can be shown that both factors are irreducible. Find the exponents for the two factors.

3. Find initial states that result in sequences with periods 3, 7, and 21. Record the states and corresponding sequences.

9.4 Additional References

The most complete book on binary pseudo-random sequences is Golomb [II.E.8]. Other references that discuss them are Gallager [II.E.7, Chapter 6], Gibson [II.D.9, Appendix G], and Gitlin, Hayes, and Weinstein [II.D.11, Section 6.7]. A detailed analysis of self synchronizing scramblers is contained in this last reference.

Chapter 10

Introduction to the RS-232C Protocol and a Bit-Error Rate Tester

In this chapter you will learn about a commercial instrument called a *bit-error rate tester* that is commonly used to evaluate the performance of digital communication systems. First, you will be introduced to the EIA RS-232C interface protocol which is a very common method for serially transmitting digital data between nearby devices. Then you will connect a commercial bit-error rate tester to the TMS320C6713 DSK, use the DSP to add noise to the serial bit stream, run a bit-error rate test, and compare measured and theoretical results. See the last section of this chapter for additional references on the theory of optimum signal detection and bit-error probability.

10.1   The EIA RS-232C Serial Interface Protocol

One of the most common methods for serially transmitting digital data between devices has been the EIA RS-232C interface protocol. EIA stands for *Electronics Industries Association* and RS for *recommended standard*. It is typically used at data rates below 38.4 kbps and between devices that are less than 15 meters apart. It was not designed for data transmission over long distances. A full implementation of the standard uses a 25 pin D connector with the pin connections shown in Table 10.1. Often, only a small subset of the signals is actually used and a 9 pin connector is employed. RS-232 connections have been replaced by USB and IEEE 1394 Firewire serial ports on many PC's and other devices. However, many devices with RS-232C connectors still exist and you can buy inexpensive USB to RS-232C adapter cables.

This chapter introduces the concept of binary antipodal data transmission, and the theory and measurement of bit errors caused by additive Gaussian noise. These concepts do not depend on the exact type of serial transmission used. The RS-232C method provides a simple example for exploring the theory.

In the data communications jargon, a data terminal is called a *data terminal equipment* which is abbreviated by the letters DTE. Examples of a DTE are a dumb terminal or the serial port in a PC. To transmit data over long distance channels like a voice-band telephone channel, the DTE is connected to a modem with an RS-232C cable. The modem is called

173

a *data communications equipment* which is abbreviated by DCE. Other types of devices can also be configured to act as a DCE.

The RS-232C signals nominally have the values 12 and −12 volts. A logical 0 is called a *space* and is represented by the 12 volt level. A logical 1 is called a *mark* and is represented by the −12 volt level. A voltage above 3 volts is often quantized to a space and a voltage below −3 volts to a mark.

The most important connector signals are described in the following paragraphs. Pin 6 is called Data Set Ready (DSR) and is controlled by the DCE (modem). A high voltage (12 v) indicates that the DCE has been turned on and is ready to make a connection with a remote modem. The term "remote" is used to mean "at the far end" of the communications channel. Pin 20 is named Data Terminal Ready (DTR) and is controlled by the DTE (data terminal). A high voltage on pin 20 indicates to the DCE that the DTE is turned on and ready to accept data. Pin 8 is named Data Carrier Detect (DCD) and is controlled by the DCE. A high voltage on pin 8 indicates to the DTE that the local modem has made a connection with the remote modem and is ready to begin transmitting data. The process of making a connection between two modems is often called *handshaking*. These basic control signals must usually be high before data transmission can proceed.

The actual data is sent and received over pins 2 and 3. Pin 2 is the transmitted data (TD). This is the serial binary data stream sent from the DTE to the DCE for transmission to the remote DCE. Pin 3 is the received data (RD) sent to the local DTE from the local DCE which has been transmitted by the remote DCE to the local DCE. Data is typically transmitted and received independently and simultaneously and this is called *full duplex* operation. When data flows in one direction at a time over the same channel, it is called *half duplex* operation.

Pin 15 is named Transmitter Clock (TC) and is generated by the DCE. In many cases, the modem (DCE) controls the data transmission by clocking bits out of the DTE with TC. In a few cases, the DTE can control the data transmission by supplying a signal called Serial Clock Transmit External (SCTE) to the modem on pin 24. Pin 17 is named Receiver Clock (RC) and is generated by the modem. It clocks received data from the modem into the DTE. These clocks are phased so that the data is clocked into the DTE or DCE in the middle of a bit where the voltage level is stable.

Once DTR, DSR, and DCD are high, the DTE asks to begin data transmission by raising pin 4 which is named Request to Send (RTS). When the DCE is ready to receive data from the DTE and send it to the the remote DCE, the DCE replys by raising pin 5 which is called Clear to Send (CTS). The DTE then begins sending the data timed by the transmitter clock.

The signals just described and pin 7 which is Signal Ground (SG) are used in most RS-232C cables. The Ring Indicator (RI) signal on pin 22 is often included when the DCE is a telephone line modem. This signal is generated by the modem and indicates that the modem has detected a ringing signal from a remote site in the dial network that is trying to make a connection. The communications software in the DTE can then send a command to the modem instructing it to connect to the telephone line and answer the call. The remaining signals are used only in special situations.

PIN	NAME	FUNCTION	SOURCE
1	FG	Frame Ground	–
2	TD	Transmitted Data	DTE
3	RD	Received Data	DCE
4	RTS	Request to Send	DTE
5	CTS	Clear to Send	DCE
6	DSR	Data Set Ready	DCE
7	SG	Signal Ground	–
8	DCD	Data Carrier Detect	DCE
9		Positive Test Voltage	DCE
10		Negative Test Voltage	DCE
11	QM	Equalizer Mode	DCE
12	SDCD	Secondary Data Carrier Detect	DCE
13	SCTS	Secondary Clear to Send	DCE
14	STD	Secondary Transmitted Data	DTE
	NS	New Sync	DTE
15	TC	Transmitter Clock	DCE
16	SRD	Secondary Received Data	DCE
	DCT	Divided Clock, Transmitter	DCE
17	RC	Receiver Clock	DCE
18	DCR	Divided Clock, Receiver	DCE
19	SRTS	Secondary Request to Send	DTE
20	DTR	Data Terminal Ready	DTE
21	SQ	Signal Quality Detect	DCE
22	RI	Ring Indicator	DCE
23	DRS	Data Rate Selector	DCE
		Data Rate Selector	DTE
24	SCTE	Serial Clock Transmit External	DTE
25	BUSY	Busy	DCE

Table 10.1: RS-232C Interface Table

10.2   Bit-Error Probability for Binary Signaling on the Additive, White, Gaussian Noise Channel

The performance of a digital communication system is often evaluated by measuring its bit-error probability as a function of the channel signal-to-noise ratio (SNR) and comparing the results with theoretical values. The measured bit-error probability is often called the *bit-error rate*. The theoretical bit-error probability depends on the modulation and demodulation schemes used to transmit the digital data over an analog link as well as the type of noise and distortion the channel introduces. In high speed digital communication systems like satellite systems, an excellent model for the channel is that it simply adds signal independent, white, Gaussian noise to the transmitted signal. The noise and distortion introduced by other kinds of channels like voice-band telephone line links can be significantly more complicated but the additive Gaussian noise model is often used as a first-order approximation.

As a simple example, we will analyze the case of binary transmission over an additive Gaussian noise channel. Let the transmitted signal $s(t)$ be a binary waveform that can have the value A or $-A$ over each bit period $nT \leq t < (n+1)T$ where T is the bit duration and $f_b = 1/T$ is the bit or data rate. In each bit period, the values A and $-A$ are equally likely and the values in different bit periods are independent random variables. Assume the channel adds white Gaussian noise $v(t)$ with two-sided power spectral density $N_0/2$ to the signal $s(t)$, so the received signal is

$$r(t) = s(t) + v(t) \tag{10.1}$$

It can be shown that a receiver that is optimum in the sense of minimizing the bit-error probability first computes the statistic

$$r_n = \frac{1}{T}\int_{nT}^{(n+1)T} r(t)\, dt = \pm A + \frac{1}{T}\int_{nT}^{(n+1)T} v(t)\, dt = \pm A + v_n \tag{10.2}$$

where

$$v_n = \frac{1}{T}\int_{nT}^{(n+1)T} v(t)\, dt \tag{10.3}$$

The integrator is often called a *matched filter* or *integrate and dump circuit*. It can be shown that v_n is a Gaussian random variable with zero mean and variance $\sigma^2 = N_0/(2T)$. The receiver then decides that A was transmitted in the n-th bit period if $v_n > 0$ and $-A$ was transmitted if $v_n \leq 0$.

The probability a decision error is made given that $-A$ was transmitted is

$$P(\text{error} \,|\, s(nT) = -A) = P(-A + v_n > 0) = P\left(\frac{v_n}{\sigma} > \frac{A}{\sigma}\right) \tag{10.4}$$

The random variable v_n/σ is a Gaussian random variable with zero mean and variance 1. Therefore,

$$P(\text{error} \,|\, s(nT) = -A) = \int_{A/\sigma}^{\infty} \frac{1}{\sqrt{2\pi}}\, e^{-\frac{v^2}{2}}\, dv = Q(A/\sigma) \tag{10.5}$$

where

$$Q(x) = \int_{x}^{\infty} \frac{1}{\sqrt{2\pi}}\, e^{-\frac{v^2}{2}}\, dv \tag{10.6}$$

Similarly, it can be shown that the probability of an error given that A was transmitted is the same thing. Therefore, the average probability of error is

$$P(\text{error}) = P(A)P(\text{error}\,|\,s(nT) = A) + P(-A)P(\text{error}\,|\,s(nT) = -A) = Q(A/\sigma) \quad (10.7)$$

The signal power in the integrator output is A^2 and the noise power is σ^2, so the output signal-to-noise ratio is $\rho = A^2/\sigma^2$. Therefore, the bit-error probability in terms of the output signal-to-noise ratio is

$$P(\text{error}) = Q(\sqrt{\rho}) \quad (10.8)$$

Different digital communication schemes can be compared and ranked by evaluating their bit-error probabilities for a given data rate, channel bandwidth, and channel signal-to-noise ratio.

In mathematics texts and MATLAB, the *complementary error function* is defined as

$$\text{erfc}(x) = \frac{2}{\sqrt{\pi}} \int_x^\infty e^{-t^2}\, dt \quad (10.9)$$

The tail probability for a standard normal random variable (Gaussian with 0 mean and variance 1) is

$$Q(x) = \frac{1}{\sqrt{2\pi}} \int_x^\infty e^{-u^2/2}\, du \quad (10.10)$$

Making the substitution $u/\sqrt{2} = t$ in the integral for $Q(x)$ gives

$$Q(x) = \frac{1}{2} \text{erfc}\left(\frac{x}{\sqrt{2}}\right) \quad (10.11)$$

so another formula for the error probability is

$$P(\text{error}) = \frac{1}{2} \text{erfc}\left(\sqrt{\frac{\rho}{2}}\right) \quad (10.12)$$

You can use this result in MATLAB to compute the theoretical bit-error probability.

Another approach to computing the error probability is to use the fact that the Gaussian tail probability $Q(x)$ can be accurately approximated for x greater than 2 by

$$Q(x) = \int_x^\infty \frac{1}{\sqrt{2\pi}} e^{-\frac{v^2}{2}}\, dv \simeq \frac{1}{x\sqrt{2\pi}} e^{-\frac{x^2}{2}} \quad (10.13)$$

This is actually an upper bound for $Q(x)$ and becomes more accurate as x increases.

10.3   The Navtel Datatest 3 Bit Error Rate Tester

There are many commercial test instruments for measuring bit-error rates. These measurements are often called BERT tests in the communications jargon. Each station in this lab has a Navtel bit-error rate tester. These instruments are designed to operate in the full duplex mode at data rates commonly used with RS-232C connections. These rates extend from 50 to 19200 bps in the normal async/sync mode and include the rates 56, 57.6, and 64 kbps in the high-speed mode. The Navtel tester can act as a DCE or DTE.

The Navtel tester performs a BERT test by transmitting a specified pattern and assumes the same pattern is transmitted from the remote end. It synchronizes to the received pattern and then counts errors. The test pattern selection includes several ASCII text sequences as well as pseudo-random shift register sequences of length 63, 511, 2047, and 4095. The test duration can be set to be a variety of fixed times, fixed of number bits, or continuous. During the test, the Navtel counts the number of bits received and the number of errors and continually computes the bit-error rate as the ratio of the current cumulative number of errors and the number of bits received. This value can be observed in the tester's display while the test is in progress. Some other variables that can be displayed are the number of bits received, the number of bit errors, the number of blocks received, the number of block errors, the number of synchronization losses, and the elapsed time. See the Navtel manual for more details.

The Navtel tester also has a full RS-232C breakout box. LED indicators show the status of the key RS-232C interface leads. Each key lead has a red and green monitor. An illuminated red LED indicates an ON, space, or 12 volt signal while an illuminated green LED indicates an OFF, mark, or -12 volt signal. Neither LED is illuminated if the level is between -3 and 3 volts. The tester can be connected between a DTE and DCE to monitor the leads.

Since the Navtel tester uses RS-232C level signals, it can not be directly connected to the DSP which uses the 0 and 5 v TTL levels. Each station has a home made TTL to RS-232C converter daughter card for connecting the tester to McBSP0. Details of the converter card can be found in Appendix B. McBSP0 is connected to the control port and McBSP1 to the data port of the AIC23 codec by default. However, they can independently be routed to the peripheral expansion connector on the DSK instead of to the McBSP's by writing the appropriate word to the MISC register of the CPLD. McBSP0 was chosen for the converter card so McBSP1 can be connected to the codec's data port. This allows an external device like the Navtel tester to be used as a data terminal and the DSP to be programmed to act as a modem, sending and receiving analog channel signals through the codec.

10.4   Bit-Error Rate Test Experiment

In this experiment, you will learn how to use the Navtel bit-error rate tester. Since the lab does not have Gaussian noise generators, the data pattern from the tester will be sent to the DSP which will introduce bit errors according to the additive Gaussian noise channel model. The corrupted bit pattern will then be sent back to the Navtel tester. You will generate a graph of the bit-error rate vs. SNR and compare it with theoretical results.

Perform the following exercises:

1. Connect one end of the RS-232C cable to the DCE socket on the Navtel tester and the other end to the DB25 RS-232C connector for the daughter card on the rear of the PC.

2. Turn on the Navtel tester and press the right and left arrows until BERT blinks. Then press SETUP/CLEAR. Now set the Navtel parameters as follows:

(a) Press the right arrow to make the entry under MESSAGE blink and then press SETUP. Use the arrows to select 4095 and press SETUP again. This selects the shift register sequence of length 4095.

(b) Similarly, set the message length to Cont (continuous).

(c) Set CLOCK to Ext (external). The clocks will be generated by the DSP's serial port McBSP0 and converted to RS-232C levels by the daughter card.

(d) Set the MODE to DTE so the Navtel looks like a terminal (DTE).

(e) Set the LEVEL to 8 and PARITY to No. This means that in the async mode each character will consist of 8 data bits with no parity bit.

(f) Set SY/ASY to Syn (synchronous). In this mode, a continuous bit stream with no start and stop bits is transmitted.

(g) Set the SPEED to 19200.

3. Write a program for the DSP to take bits from the Navtel tester and simply loop them back to the Navtel through McBSP0. Later you will be asked to introduce errors into the bit stream. The sample rate generator (SRG) in McBSP0 should be used to generate the transmit frame sync (FSX), receive frame sync (FSR), transmit bit clock (CLKX), and receive bit clock (CLKR). A block diagram of the SRG is shown in Figure 10.1. FSR and FSX are not connected to the Navtel tester but are used internally in McBSP0. However, CLKR and CLKX are connected to the Navtel tester and used as its external data clocks, so McBSP0 must be configured to make them outputs. The clock for the sample rate generator should be an external clock supplied by Timer 0. The timer output, TOUT0, is looped back to the SRG0 clock input pin, CLKS0, by the converter box. You should look at Chapter 12 of the *TMS320C6000 Peripherals Reference Guide* [I.10] for complete details on the serial ports and their sample rate generator. You will notice there that the SRG can use an internal clock whose frequency is CPU clock/2 for the 'C6713 and that it has dividers to generate the bit clocks and frame syncs. The reason for choosing TOUT0 as the clock source is that the dividers in the SRG cannot divide the internal clock by a large enough factor to achieve the desired 19200 bps rate.

The timers were discussed in Section 2.3 and the following formula for the frequency of TOUT for the timer in clock mode was given:

$$f_{\text{TOUT}} = \frac{\text{CPU clock frequency}}{8 \times (\text{Period Register value})} \qquad (10.14)$$

where the "Period Register value" is an unsigned 32-bit integer. The CPU clock frequency is 225 MHz for the 'C6713 DSK. We will configure the McBSP0 sample rate generator to use TOUT0 as its clock which can be inverted in the SRG according to the value of CLKSP and the result, CLKSRG, is selected rather than the internal clock source and applied to a pair of dividers if the clock select mode bit (CLKSM) is 0. The first divider uses the value of the 8-bit unsigned integer, CLKGDV, as the divide-down number to generate the signal, CLKG, which is possibly inverted to form the bit clocks CLKX and

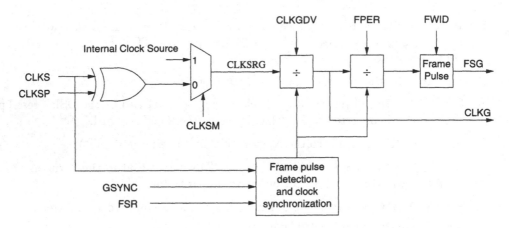

Figure 10.1: Structure of the Sample Rate Generator (SRG)

CLKR. The frequency of CLKG when the timer is in clock mode is

$$\text{bit clock frequency} \quad = \quad \text{CLKG frequency} = \frac{\text{CLKSRG frequency}}{\text{CLKGDV} + 1}$$

$$= \frac{\text{CPU clock frequency}}{8 \times (\text{Period Register})(\text{CLKGDV} + 1)} \quad (10.15)$$

When CLKGDV is odd or zero, CLKG has a 50% duty cycle. When CLKGDV is an even number, $2p$, the high state lasts $p + 1$ cycles and the low state p cycles.

The signal CLKG is then applied to a divider with the unsigned integer 12-bit divide-down number, FPER, to generate the signal FSG which is at the frame sync frequency. The frequency of FSG is

$$\text{frame sync frequency} = \text{FSG frequency} = \frac{\text{CLKG frequency}}{\text{FPER} + 1} \quad (10.16)$$

The sample rate generator includes one additional counter to generate the width of the frame sync pulse. The unsigned 8-bit integer, FWID, determines the frame sync pulse width. The FSG pulse width is FWID + 1 pulses of CLKG.

The following code segment will help you write your program. First, a structure is initialized with the values required to configure McBSP0 for the desired mode of operation. McBSP0 is configured to transmit and receive 32-bit words and the divide-down number, FPER, for the frame syncs is set to 31. According to the peripherals manual, the actual frame sync period will be FPER + 1 = 32 serial bit clocks. Therefore, data bits will be transmitted and received with no gaps between words. For details on the mnemonics see the *TMS320C6000 Chip Support Library API Reference Guide*, [I.6], Appendix B.10. Next a configuration structure for Timer0 is constructed. In order to use the CSL functions, `MCBSP_open()` is called which returns the handle hMcBSP0 and `TIMER_open()` is called which returns the handle `hTimer0`. The structure values for McBSP0 are then loaded into the McBSP0 registers by the function `MCBSP_config()` but McBSP0 is not started yet. Then Timer0 is initialized by the function `TIMER_config()` and started by the function `Timer_start()`. After the timer is running, McBSP0 is started by the function `McBSP_start()`. Finally a dummy word of 0 is written to DXR0 to get the transmitter actually running.

Program 10.1 Code Segment for Configuring and Starting McBSP0

```
#include <csl.h>
#include <csl_mcbsp.h>
#include <csl_timer.h>

MCBSP_Handle hMcBSP0;
TIMER_Handle hTimer0;
    ...

main()
{
    ...

MCBSP_Config mcbspCfgData = {
        MCBSP_FMKS(SPCR, FREE, NO)              |
        MCBSP_FMKS(SPCR, SOFT, NO)              |
        MCBSP_FMKS(SPCR, FRST, NO)              |
        MCBSP_FMKS(SPCR, GRST, YES)             |
        MCBSP_FMKS(SPCR, XINTM, XRDY)           |
        MCBSP_FMKS(SPCR, XSYNCERR, NO)          |
        MCBSP_FMKS(SPCR, XRST, YES)             |
        MCBSP_FMKS(SPCR, DLB, OFF)              |
        MCBSP_FMKS(SPCR, RJUST, RZF)            |
        MCBSP_FMKS(SPCR, CLKSTP, DISABLE)       |
        MCBSP_FMKS(SPCR, DXENA, OFF)            |
        MCBSP_FMKS(SPCR, RINTM, RRDY)           |
        MCBSP_FMKS(SPCR, RSYNCERR, NO)          |
        MCBSP_FMKS(SPCR, RRST, YES),

        MCBSP_FMKS(RCR, RPHASE, SINGLE)         |
        MCBSP_FMKS(RCR, RFRLEN2, OF(0))         |
        MCBSP_FMKS(RCR, RWDLEN2, DEFAULT)       |
        MCBSP_FMKS(RCR, RCOMPAND, MSB)          |
        MCBSP_FMKS(RCR, RFIG, NO)               |
        MCBSP_FMKS(RCR, RDATDLY, 0BIT)          |
        MCBSP_FMKS(RCR, RFRLEN1, OF(0))         |
        MCBSP_FMKS(RCR, RWDLEN1, 32BIT)         |
        MCBSP_FMKS(RCR, RWDREVRS, DISABLE),
```

```
        MCBSP_FMKS(XCR, XPHASE, SINGLE)          |
        MCBSP_FMKS(XCR, XFRLEN2, DEFAULT)        |
        MCBSP_FMKS(XCR, XWDLEN2, DEFAULT)        |
        MCBSP_FMKS(XCR, XCOMPAND, MSB)           |
        MCBSP_FMKS(XCR, XFIG, NO)                |
        MCBSP_FMKS(XCR, XDATDLY, OBIT)           |
        MCBSP_FMKS(XCR, XFRLEN1, OF(0))          |
        MCBSP_FMKS(XCR, XWDLEN1, 32BIT)          |
        MCBSP_FMKS(XCR, XWDREVRS, DISABLE),

        MCBSP_FMKS(SRGR, GSYNC, FREE)            |
        MCBSP_FMKS(SRGR, CLKSP, RISING)          |
        MCBSP_FMKS(SRGR, CLKSM, CLKS)            |
        MCBSP_FMKS(SRGR, FSGM, FSG)              |
        MCBSP_FMKS(SRGR, FPER, OF(31))           |
        MCBSP_FMKS(SRGR, FWID, OF(1))            |
        MCBSP_FMKS(SRGR, CLKGDV, OF(0)),

        MCBSP_MCR_DEFAULT,
        MCBSP_RCER_DEFAULT,
        MCBSP_XCER_DEFAULT,

        MCBSP_FMKS(PCR, XIOEN, SP)               |
        MCBSP_FMKS(PCR, RIOEN, SP)               |
        MCBSP_FMKS(PCR, FSXM, INTERNAL)          |
        MCBSP_FMKS(PCR, FSRM, INTERNAL)          |
        MCBSP_FMKS(PCR, CLKXM, OUTPUT)           |
        MCBSP_FMKS(PCR, CLKRM, OUTPUT)           |
        MCBSP_FMKS(PCR, CLKSSTAT, DEFAULT)       |
        MCBSP_FMKS(PCR, DXSTAT, DEFAULT)         |
        MCBSP_FMKS(PCR, FSXP, ACTIVEHIGH)        |
        MCBSP_FMKS(PCR, FSRP, ACTIVEHIGH)        |
        MCBSP_FMKS(PCR, CLKXP, RISING)           |
        MCBSP_FMKS(PCR, CLKRP, FALLING)
    };

TIMER_Config timer0CfgData ={
        TIMER_FMKS(CTL, INVINP, NO)        |
        TIMER_FMKS(CTL, CLKSRC, CPUOVR4)   |
        TIMER_FMKS(CTL, CP, CLOCK)         |
        TIMER_FMKS(CTL, HLD, YES)          |
        TIMER_FMKS(CTL, INVOUT, NO)        |
        TIMER_FMKS(CTL, FUNC, TOUT),
```

```
           put your PDR value here, /* Period register value */

           0x00000000      /* Initial counter register value */
    };

    /* Open McBSP0 and get handle */
    hMcBSP0 = MCBSP_open(MCBSP_DEV0, MCBSP_OPEN_RESET);

    /* Open Timer0 and get handle */
    hTimer0 = TIMER_open(TIMER_DEV0, 0);

    /* Configure McBSP0.. */
    MCBSP_config(hMcBSP0, &mcbspCfgData);

    /* Configure Timer 0 */
    TIMER_config(hTimer0, &timer0CfgData);

    /* Start Timer0 */
    TIMER_start(hTimer0);

    /* Start McBSP0 */
    MCBSP_start(hMcBSP0, MCBSP_XMIT_START | MCBSP_RCV_START |
            MCBSP_SRGR_START | MCBSP_SRGR_FRAMESYNC, 220);

    /* Write a dummy word to DXR0 to get transmitter started */
    MCBSP_write(hMcBSP0, 0);

       . . .
```

4. Include a line in your program to connect McBSP0 to the Peripheral Expansion Connector rather than the AIC23 control port. You can use a function in the BSL to do this. For documentation on the BSL functions, start Code Composer Studio, click on **Help**, **Contents**, **TMS320C6713 DSK**, **Software**, and, finally, **Board Support Library**. Bit 0 in the MISC register of the CPLD controls the McBSP0 connection. You must change it from 0 to 1 to connect McBSP0 to the Peripheral Expansion Connector. You can do this by including the following line in your program:

   ```
   DSK6713_rset(DSK6713_MISC, 0x01)
   ```

5. Select the Period Register value for Timer0 with McBSP0 sample rate generator divider CLKGDV = 0 to give a serial bit rate as close to 19200 bps as possible. Check your initial program by starting a bit-error rate test. To do this, press RUN on the Navtel tester. Press the up or down arrows until BERT is displayed. If you are looping the data back

correctly, BERT should remain 0. You can introduce a single error by pressing INSERT ERROR and observe the effect on the BERT display.

6. Now modify your program to add errors to the bit stream. For each of the 16 pairs of bits in a received 32-bit word, generate a pair of zero mean, uncorrelated, Gaussian noise samples by the method described in Appendix A. Assume that a logical 0 is represented by A volts on the channel and logical 1 by $-A$ volts as discussed in Section 10.2. Let the integrator output noise variance be $\sigma^2 = 1$ and adjust A to get the desired output signal-to-noise ratio. Then, the SNR in dB is

$$S = 10 \log_{10} \frac{A^2}{\sigma^2} = 20 \log_{10} A \tag{10.17}$$

and the required value for A is

$$A = \sigma \times 10^{S/20} = 10^{S/20} \tag{10.18}$$

Let a particular pair of Gaussian noise samples be denoted by (X, Y). According to the theory, a transmitted 1 is changed to a 0 if $X > A$ and a transmitted 0 is changed to a 1 if $X \leq -A$. The probabilities of these two events are identical because the probability density function for the zero mean, Gaussian random variables is even. Thus, to determine when to introduce a bit error, check to see if $X > A$. If this is true, an error should be introduced in the first bit of a pair. Similarly, when $Y > A$ an error should be introduced in the second bit of a pair. The errors can be introduced by XOR-ing 1's into the error locations in the received 32-bit serial word to complement the correct bits.

7. Now experimentally generate a bit-error rate *vs.* SNR plot. Start with a 13 dB SNR and work down to 7 dB in one dB increments. Make sure to run your test long enough at each SNR to obtain a statistically reliable estimate of the error rate. Plot the BER as the ordinate on a logarithmic scale. Plot the SNR in dB on a linear dB scale.

The Gaussian noise generator algorithm causes an upper limit on the SNR value that can cause errors. It is shown in Appendix A, Equation (A.9) that with the 'C6x compiler the maximum magnitude for X or Y is

$$|X_{\max}| = |Y_{\max}| = \sigma \sqrt{30 \log_e 2} = 4.56009\sigma \tag{10.19}$$

No errors can occur if $X_{\max} < A$. Using (10.18) and (10.19), it follows that this bound is equivalent to

$$\sigma \sqrt{30 \log_e 2} < \sigma \times 10^{S/20}$$

or

$$S > 20 \log_{10} \sqrt{30 \log_e 2} = 13.1795 \ \text{dB} \tag{10.20}$$

8. Theoretically compute the BER *vs.* SNR plot and compare it with your experimentally measured curve.

10.5   Additional References

There are many books that discuss digital communication systems and have sections on the optimum detection of signals corrupted by additive Gaussian noise and how to compute error probabilities for various signal sets. For example, see Gibson [II.D.9, Chapter 10], Gitlin, Hayes and Weinstein [II.D.11, Section 2.2], and Haykin [II.D.17, Sections 7.2–7.3].

Chapter 11

Digital Data Transmission by Baseband Pulse Amplitude Modulation

In this chapter you will be introduced to a common method for digital data transmission known as *baseband pulse amplitude modulation* (PAM). The presentation is slanted towards transmission over band limited channels and DSP implementation. The concepts learned here will be generalized to passband digital communication systems in Chapters 13–16. Some of the concepts and terms you will be introduced to are: baseband shaping filters and raised cosine shaping, intersymbol interference and the Nyquist criterion, eye diagrams, symbol error probability formulas, interpolation filter banks, and a symbol clock recovery method.

11.1 General Description of a Baseband Pulse Amplitude Modulation System

In pulse amplitude modulation, information symbols are transmitted at discrete, uniformly spaced time intervals. The carrier signal is a train of pulses uniformly spaced at the same interval as the information symbols. The amplitude of each pulse is a one-to-one function of the corresponding information symbol. The information is recovered at the receiver by measuring the amplitude of each pulse and mapping it back to the information symbol. In wideband systems, non-overlapping, rectangular, full period pulses are often used. In band limited systems, the pulses overlap but are selected so that the information symbols can be measured by sampling the received signal at the symbol rate as will be explained below.

The block diagram of a typical baseband PAM system is shown in Figure 11.1. The transmitter input d_i is a serial binary data sequence with a bit rate of R_d bits/sec. Input bits are blocked into J-bit words by the serial-to-parallel converter and mapped into the sequence of symbols a_n which are selected from an alphabet of $M = 2^J$ distinct voltage levels. These symbols are generated at the rate of $f_s = R_d/J$ symbols/sec and we will designate the interval between symbols as $T = 1/f_s$. The term, *baud*, is commonly used for the symbol rate f_s in honor of Baudot who invented a binary code for representing alphanumeric characters. Baud is also frequently used (or misused) to mean one symbol period. The levels are often selected to be equally spaced and have an arithmetic average of zero. In this chapter, we will use levels spaced by $2d$ with the M possible values

$$\ell_i = d(2i - 1) \quad \text{for} \quad i = -\frac{M}{2} + 1, \ldots, 0, \ldots, \frac{M}{2} \tag{11.1}$$

Thus, the minimum level is $-(M-1)d$ and the maximum level is $(M-1)d$.

It is mathematically convenient to represent the symbol sequence by the Dirac impulse train

$$s^*(t) = \sum_{k=-\infty}^{\infty} a_k \delta(t - kT) \tag{11.2}$$

The Impulse Modulator block forms this function. This impulse train is applied to a Transmit Filter with impulse response $g_T(t)$ which band limits the signal to the channel bandwidth. The resulting transmitted signal is

$$s(t) = \sum_{k=-\infty}^{\infty} a_k g_T(t - kT) \tag{11.3}$$

which is a superposition of amplitude modulated pulses. The combination of the impulse modulator and transmit filter is a mathematical model for a DAC followed by a lowpass filter.

In this chapter, the channel will be modeled as a linear, time-invariant filter with the frequency response $C(\omega)$ followed by an additive noise source.

At the receiver, the channel output $r(t)$ is first passed through a receive filter which eliminates out-of-band noise and, in conjunction with the transmit filter, forms a properly shaped pulse. The combined transmit filter, channel, and receive filter frequency response is

$$G(\omega) = G_T(\omega)C(\omega)G_R(\omega) \tag{11.4}$$

and the corresponding impulse response is

$$g(t) = g_T(t) * c(t) * g_R(t) = \mathcal{F}^{-1}\{G(\omega)\} \tag{11.5}$$

where $*$ represents convolution. The combined filter represented by $G(\omega)$ is called the *baseband shaping filter*. The output of the receive filter is

$$x(t) = \sum_{k=-\infty}^{\infty} a_k g(t - kT) + v(t) * g_R(t) \tag{11.6}$$

Now assume that the noise is zero and the combined impulse response is zero at the time instants nT except for $n = 0$ where it is 1, that is,

$$g(nT) = \delta_{n,0} = \begin{cases} 1 & \text{for} \quad n = 0 \\ 0 & \text{otherwise} \end{cases} \tag{11.7}$$

An impulse response with this property is said to have *no intersymbol interference* (ISI). This property is examined in more detail in Section 11.2. With these assumptions, it can be

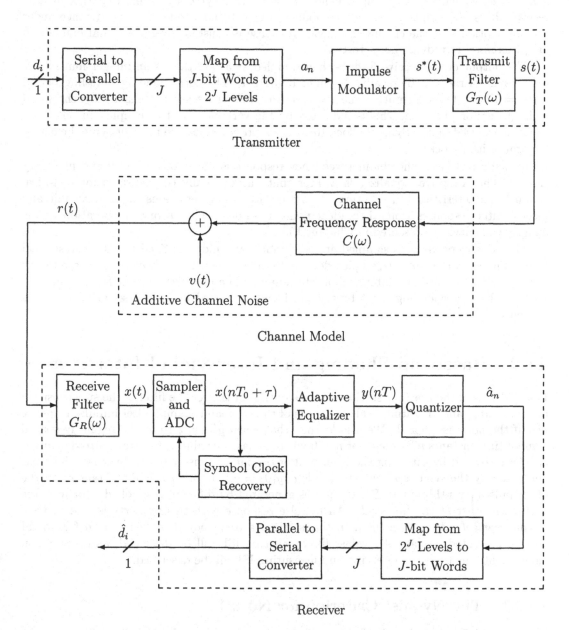

Figure 11.1: Block Diagram of a Baseband PAM System

seen from (11.6) that the samples of the receive filter output taken at times nT reduce to $x(nT) = a_n$ which is exactly the sequence of transmitted symbols. Often, the transmit and receive filters are designed to give a combined filter with no intersymbol interference under the assumption that the channel frequency response is constant over the signal bandwidth and, therefore, introduces no distortion.

The output of the receive filter is then sampled at a rate that is an integer multiple N of the symbol rate f_s. Typically, N might be 3 or 4. The corresponding sampling period is $T_0 = T/N$. The transmitter and receiver time references are not exactly synchronized in frequency or phase and this is indicated by the variable τ in the sampler output. The samples are used by the Symbol Clock Recovery system to lock the receiver symbol clock to the transmitter clock.

In many instances, the channel frequency response is not known exactly and may vary slowly. The Adaptive Equalizer is a filter that automatically compensates for non-ideal channel characteristics. In addition, it corrects for small deviations in the transmit and receive filter responses from their ideal nominal values. The theory of operation of the adaptive equalizer is presented in Chapter 15.

The equalizer output is sampled at the symbol rate and quantized to the nearest ideal level. The equalizer output samples deviate from the ideal levels because of the additive channel noise and residual intersymbol interference. The quantizer output level is mapped back to the corresponding J-bit binary word and converted back to a serial output data sequence.

11.2   Baseband Shaping and Intersymbol Interference

It was shown in the previous section that the output of the receive filter is the superposition of time shifted and amplitude scaled versions of the baseband shaping filter impulse response $g(t)$ if the noise is ignored. We also learned that when $g(t)$ is zero at the regularly spaced symbol time instants nT except for $n = 0$ where it has the value 1, the transmitted symbols can be recovered by sampling the receive filter output at times nT. In this case, the pulse generated by the symbol a_n has amplitude $a_n g(0) = a_n$ at time nT and the tails of all the other pulses pass through 0. So, the pulses generated by different symbols do not interfere with each other at the times nT. An impulse response with this property is said to have no *intersymbol interference* (ISI). In this section, a frequency domain criterion for no ISI will be presented, a class of lowpass filters with no ISI will be discussed, and a common experimental method for observing and measuring ISI will be described.

11.2.1   The Nyquist Criterion for No ISI

Equation (11.7) states the criterion for no ISI in terms of the baseband shaping filter impulse response samples $g(nT)$. These samples can be computed from the shaping filter frequency response by the formula

$$g(nT) = \frac{1}{2\pi} \int_{-\infty}^{\infty} G(\omega) e^{j\omega nT} \, d\omega \tag{11.8}$$

Let $\omega_s = 2\pi f_s = 2\pi/T$. The integral can be computed by partitioning the ω axis into the infinite set of intervals

$$\left(-\frac{\omega_s}{2} - k\omega_s, \frac{\omega_s}{2} - k\omega_s\right] \quad \text{for} \quad k = -\infty, \ldots, \infty$$

and taking the sum of integrals over the intervals to get

$$g(nT) = \sum_{k=-\infty}^{\infty} \frac{1}{\omega_s} \int_{-\frac{\omega_s}{2}-k\omega_s}^{\frac{\omega_s}{2}-k\omega_s} \frac{1}{T} G(\omega) e^{j\omega nT} \, d\omega = \sum_{k=-\infty}^{\infty} \frac{1}{\omega_s} \int_{-\frac{\omega_s}{2}}^{\frac{\omega_s}{2}} \frac{1}{T} G(\omega - k\omega_s) e^{j(\omega - k\omega_s)nT} \, d\omega \quad (11.9)$$

Recognizing that $e^{-jkn\omega_s T} = e^{-kn2\pi} = 1$ and taking the sum inside the integral gives

$$g(nT) = \frac{1}{\omega_s} \int_{-\frac{\omega_s}{2}}^{\frac{\omega_s}{2}} G^*(\omega) e^{j\omega nT} \, d\omega \quad (11.10)$$

where

$$G^*(\omega) = \frac{1}{T} \sum_{k=-\infty}^{\infty} G(\omega - k\omega_s) \quad (11.11)$$

The function $G^*(\omega)$ is called the *aliased* or *folded* spectrum and is well known in digital signal processing theory.

The criterion (11.7) for no ISI is satisfied if and only if $G^*(\omega) = 1$. This can be seen by evaluating (11.10) for this special case. The constraint that the aliased spectrum $G^*(\omega)$ must be a constant for no ISI is known as the *Nyquist criterion*.

11.2.2 Raised Cosine Baseband Shaping Filters

The requirement for no ISI only makes constraints on the symbol rate samples of the baseband shaping filter impulse response. There are an infinite number of impulse responses that meet the constraints. One class of filters that is often specified for use with lowpass channels has *raised cosine* frequency domain shaping. The frequency response of a raised cosine filter is

$$G(\omega) = \begin{cases} T & \text{for } |\omega| \leq (1-\alpha)\frac{\omega_s}{2} \\ \frac{T}{2}\left\{1 - \sin\left[\frac{T}{2\alpha}\left(|\omega| - \frac{\omega_s}{2}\right)\right]\right\} & \text{for } (1-\alpha)\frac{\omega_s}{2} \leq |\omega| \leq (1+\alpha)\frac{\omega_s}{2} \\ 0 & \text{elsewhere} \end{cases} \quad (11.12)$$

where α is a constant in the interval $[0, 1]$ and is called the *excess bandwidth factor*. The raised cosine frequency response is flat over the central portion of the passband and rolls off sinusoidally to zero at the band edge. It is down from the 0 frequency value by a factor of 2, which is equivalent to 6 dB, at the frequency $\omega_s/2$. The frequency $\omega_s/2$ is called the *Nyquist frequency*.

It can be shown that the corresponding impulse response is

$$g(t) = \frac{\sin\left(\frac{\omega_s}{2}t\right)}{\frac{\omega_s}{2}t} \frac{\cos\left(\alpha\frac{\omega_s}{2}t\right)}{1 - 4(\alpha t/T)^2} \quad (11.13)$$

The program `C:\digfil\rascos.exe` computes samples of the raised cosine filter impulse response modified by the Hamming window.

Notice that when $\alpha = 0$, the raised cosine filter becomes an ideal flat lowpass filter with cutoff frequency $\omega_s/2$. When $\alpha = 1$, the frequency response has no flat region and is one cycle of a cosine function raised up so it becomes 0 at the cutoff frequency of ω_s. As the bandwidth is increased by making α closer to 1, the impulse response decays more rapidly.

11.2.3   Splitting the Shaping Between the Transmit and Receive Filters

When the channel amplitude response is flat across the signal passband and the noise is white, it can be shown [II.D.29, p. 54] that the amplitude response of the combined baseband shaping filter should be equally split between the transmit and receive filters to maximize the output signal-to-noise ratio, that is,

$$|G_T(\omega)| = |G_R(\omega)| = |G(\omega)|^{1/2} \tag{11.14}$$

Their phases can be arbitrary as long as the combined phase is linear.

When raised cosine shaping is used, the resulting optimum transmit and receive filters are called *square-root of raised cosine* filters. The program `C:\digfil\sqrtraco.exe` can be used to compute the impulse response of a square-root of raised cosine filter.

11.2.4   Eye Diagrams

The *eye diagram* is a useful diagnostic tool for qualitatively evaluating the optimality of a PAM system. An eye diagram is formed by superimposing oscilloscope traces of the receive filter output. Each trace is triggered at the same phase within a symbol interval and lasts for a few symbols. A sketch of some traces in an eye diagram for a combined channel with no ISI and a two-level input symbol alphabet is shown in Figure 11.2. Notice that all the traces pass through the ideal symbol values of $\pm d = \pm 1$ at the sampling instants $\{nT\}$. The empty area inside the traces around nT is called an *eye opening*. With an M level input alphabet, there are $M - 1$ openings stacked vertically at each symbol instant. The quantizer slicing or decision levels are usually chosen to be the values $k2d$ half way between the ideal symbol levels. Somewhat surprisingly, the optimum time at which to sample the receive filter output is not at the peaks of the eye diagram with band limited channels. It is somewhat down in a valley. The more band limited the channel, the more rapidly the eye closes away from the symbol instants and the higher the peaks between symbols become.

When intersymbol interference is present, the traces do not all pass through the ideal levels at the sampling instants and these points become dispersed. As the ISI increases, the dispersion grows and the eye begins to close. As long as the eye is open, no decision errors are made when the additive channel noise is zero. However, when noise is present, the error rate increases as the eye closes.

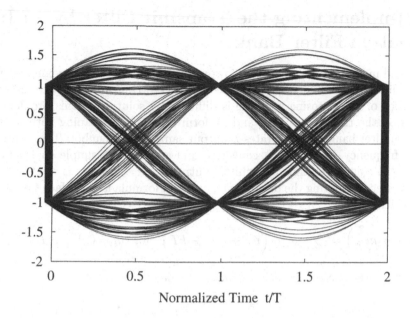

Figure 11.2: Eye Diagram for a Channel with no ISI and $M = 2$. Raised Cosine Shaping with $\alpha = 0.12$.

With zero additive noise, the symbol rate samples of the receive filter output are

$$x(nT) = \sum_{k=-\infty}^{\infty} a_k g(nT - kT) = g(0) \left[a_n + \sum_{\substack{k=-\infty \\ k \neq n}}^{\infty} a_k \frac{g(nT - kT)}{g(0)} \right] \tag{11.15}$$

The right-hand summation in (11.15) is the ISI for the current received symbol. Decision errors are made when the magnitude of this ISI exceeds d except when the ISI carries an outer level outside the eye diagram. We will assume that $g(0) > 0$. The ISI is a random variable and depends on the symbol sequence. The worst case or peak ISI occurs when the symbols a_k have their maximum magnitude $(M - 1)d$ and the same sign as $g(nT - kT)$. Then the sum for the ISI becomes

$$D = (M - 1)d \sum_{\substack{k=-\infty \\ k \neq n}}^{\infty} \left| \frac{g(nT - kT)}{g(0)} \right| = (M - 1)d \sum_{\substack{k=-\infty \\ k \neq 0}}^{\infty} \left| \frac{g(kT)}{g(0)} \right| \tag{11.16}$$

The peak fractional eye closure is defined to be

$$\eta = \frac{D}{d} = (M - 1) \sum_{\substack{k=-\infty \\ k \neq 0}}^{\infty} \left| \frac{g(kT)}{g(0)} \right| \tag{11.17}$$

When η is less than 1, the eyes are open.

11.3　Implementing the Transmit Filter by an Interpolation Filter Bank

Designing and manufacturing an analog lowpass filter that closely approximates the impulse response or linear phase frequency response of a desired transmit filter like the raised cosine or square-root of raised cosine response is difficult. A solution to this problem is to place the burden on the transmitter DSP and perform most of the shaping with a discrete-time interpolation filter bank that generates L output samples per symbol. These samples, which occur with frequency Lf_s, are D/A converted and applied to a simple analog lowpass filter to generate the continuous-time transmitter output.

The first step in deriving the desired interpolation formula is to replace t by $nT + m(T/L)$ in (11.3) which gives

$$s\left(nT + m\frac{T}{L}\right) = \sum_{k=-\infty}^{\infty} a_k g_T\left(nT + m\frac{T}{L} - kT\right) \quad \text{for} \quad m = 0, 1, \ldots, L-1 \qquad (11.18)$$

Now let L discrete-time *interpolation subfilters* be defined as

$$g_{T,m}(n) = g_T\left(nT + m\frac{T}{L}\right) \quad \text{for} \quad m = 0, 1, \ldots, L-1 \qquad (11.19)$$

Then, the L output samples required during the symbol period starting at time nT can be expressed as

$$s\left(nT + m\frac{T}{L}\right) = \sum_{k=-\infty}^{\infty} a_k g_{T,m}(n-k) \quad \text{for} \quad m = 0, 1, \ldots, L-1 \qquad (11.20)$$

Notice that for each m, (11.20) is equivalent to passing the T spaced input symbol sequence $\{a_n\}$ through a T spaced digital filter with impulse response $g_{T,m}(n)$. The resulting DSP output words are then multiplexed to a DAC at the rate of Lf_s samples/second. Finally, the DAC output is passed through a simple analog lowpass filter to eliminate the unwanted high frequency spectral components around multiples of Lf_s. This process is illustrated in Figure 11.3. In practice, the transmit filter impulse response is truncated to a finite duration by a window function like the Hamming window so each subfilter becomes a finite tap FIR digital filter.

11.4　Symbol Error Probability for a Channel with a Perfect Frequency Response and Additive Gaussian Noise

An important measure of the performance of a digital communication system is its error probability as a function of the channel SNR. A formula for the symbol error probability of PAM will be derived in this section under the following assumptions:

1. The frequency response of the channel is a constant over the signal bandwidth.

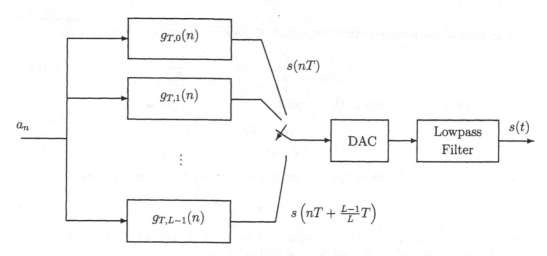

Figure 11.3: Implementing the Transmit Filter by an Interpolation Filter Bank

2. Symbols from the M level alphabet are used with equal probability.

3. Symbols selected at different times are uncorrelated random variables.

4. The additive noise is white and Gaussian with two-sided power spectral density $N_0/2$.

5. The combined baseband shaping filter has a raised cosine response with excess bandwidth factor α and the shaping is split equally between the transmit and receive filters. Therefore, the transmit and receive filters both have square-root of raised cosine responses.

It can be shown [II.D.29, pp. 52-53] that the average transmitted power is

$$P_s = \frac{E\{a_n^2\}}{T} \frac{1}{2\pi} \int_{-\infty}^{\infty} |G_T(\omega)|^2 \, d\omega \qquad (11.21)$$

With the square-root of raised cosine transmit filter, this reduces to

$$P_s = \frac{E\{a_n^2\}}{T} \qquad (11.22)$$

Using (11.1) and the fact that the levels are equally likely, the expected squared symbol value is found to be

$$a^2 = E\{a_n^2\} = \frac{2}{M} \sum_{k=1}^{M/2} [d(2k-1)]^2 = (M^2 - 1)\frac{d^2}{3} \qquad (11.23)$$

Therefore, the average transmitted power is

$$P_s = (M^2 - 1)\frac{d^2}{3T} \qquad (11.24)$$

The noise at the output of the square-root of raised cosine receive filter has the variance

$$\sigma^2 = \frac{1}{2\pi} \int_{-\infty}^{\infty} \frac{N_0}{2} |G_R(\omega)|^2 \, d\omega = \frac{N_0}{2} \tag{11.25}$$

Also, the channel noise power in the Nyquist band $(-\omega_s/2, \omega_s/2)$ is

$$P_N = \frac{1}{2\pi} \int_{-\omega_s/2}^{\omega_s/2} \frac{N_0}{2} \, d\omega = \frac{N_0}{2T} \tag{11.26}$$

Since there is no ISI, the samples of the receive filter output have the form

$$x(nT) = a_n + v_R(nT) \tag{11.27}$$

where $v_R(nT)$ is a sample of the channel noise filtered by the receive filter. When a_n is one of the $M-2$ inner levels, the symbol error probability is

$$P_I = P(|v_R(nT)| > d) = 2Q(d/\sigma) \tag{11.28}$$

where $Q(x)$ is the Gaussian tail probability defined in (10.13). For the outer level $(M-1)d$ the error probability is

$$P_{O+} = P(v_R(nT) < -d) = Q(d/\sigma) \tag{11.29}$$

and for the outer level $-(M-1)d$ the error probability is

$$P_{O-} = P(v_R(nT)) > d) = Q(d/\sigma) = P_{O+} \tag{11.30}$$

The total symbol error probability is

$$P_e = \frac{M-2}{M} P_I + \frac{1}{M} P_{O+} + \frac{1}{M} P_{O-} = 2\frac{M-1}{M} Q(d/\sigma) \tag{11.31}$$

Solving (11.24) for d and using (11.25) and (11.26), the error probability can be expressed in terms of the channel signal-to-noise ratio P_s/P_N as

$$P_e = 2\frac{M-1}{M} Q\left[\left(\frac{3}{M^2-1} \frac{P_s}{P_N} \right)^{1/2} \right] \tag{11.32}$$

11.5 Symbol Clock Recovery

In typical PAM systems, the receiver has a reasonably good, but not perfect, knowledge of the transmitter's symbol clock frequency. It must lock its local symbol clock frequency and phase to those of the received signal to maintain the proper sampling instants. A method for deriving the symbol clock from the received PAM signal is presented in this section. It is particularly suited to band limited systems. Wideband systems in which the signals have sharp transitions often use other clock recovery methods.

The block diagram of a clock recovery system is shown in Figure 11.4. The receive filter output $x(t)$ is first passed through a prefilter with frequency response $B(\omega)$. The prefilter is typically a bandpass filter centered at $f_s/2$, half the symbol frequency. Let the combined baseband shaping filter and prefilter frequency and impulse responses be

$$G_1(\omega) = G(\omega)B(\omega) \quad \text{and} \quad g_1(t) = g(t) * b(t) \tag{11.33}$$

Then, the prefilter output is

$$q(t) = \sum_{k=-\infty}^{\infty} a_k g_1(t - kT) \tag{11.34}$$

The prefilter output is passed through a squarer whose output is

$$p(t) = q^2(t) = \sum_{k=-\infty}^{\infty} \sum_{m=-\infty}^{\infty} a_k a_m g_1(t - kT) g_1(t - mT) \tag{11.35}$$

The squarer output is passed through a narrowband bandpass filter $H(\omega)$ whose center frequency is the symbol rate f_s. The output $z(t)$ looks like a sinusoid at the symbol clock frequency with a slowly varying amplitude and phase. Its zero crossings tend to cluster together. This signal can then be applied to a narrow band phase-locked loop to generate a stable symbol clock.

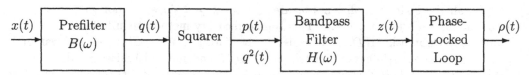

Figure 11.4: Block Diagram of a Clock Recovery System

As before, it will be assumed that the symbols are a sequence of zero-mean uncorrelated random variables. Therefore,

$$E\{a_k a_m\} = a^2 \delta_{k,m} \tag{11.36}$$

where a^2 is given by (11.23). The expected value of the squarer output is

$$E\{p(t)\} = \sum_{k=-\infty}^{\infty} \sum_{m=-\infty}^{\infty} E\{a_k a_m\} g_1(t - kT) g_1(t - mT) \tag{11.37}$$

On using (11.36), this reduces to

$$E\{p(t)\} = a^2 \sum_{k=-\infty}^{\infty} g_1^2(t - kT) \tag{11.38}$$

The expected value of the squarer output is periodic with period equal to the symbol period T. Therefore, it can be expressed as a Fourier series of the form

$$E\{p(t)\} = \sum_{k=-\infty}^{\infty} p_k e^{jk\omega_s t} \tag{11.39}$$

where

$$p_k = \frac{1}{T} \int_0^T E\{p(t)\} e^{-jk\omega_s t}\, dt \tag{11.40}$$

After several lines of manipulations, it can be shown that

$$p_k = \frac{a^2}{T} \int_{-\infty}^{\infty} g_1^2(t) e^{-jk\omega_s t}\, dt = \frac{a^2}{T 2\pi} \int_{-\infty}^{\infty} G_1(\omega) G_1(k\omega_s - \omega)\, d\omega \tag{11.41}$$

The expected value of the output bandpass filter is also periodic with the Fourier series expansion

$$E\{z(t)\} = \sum_{k=-\infty}^{\infty} z_k e^{jk\omega_s t} \tag{11.42}$$

where

$$z_k = p_k H(k\omega_s) = H(k\omega_s) \frac{a^2}{T 2\pi} \int_{-\infty}^{\infty} G_1(\omega) G_1(k\omega_s - \omega)\, d\omega \tag{11.43}$$

By selecting the prefilter $B(\omega)$ to be a narrow band filter that passes components only near $\pm \omega_s/2$, it can be seen from (11.41) that $p_k = 0$ except for $k = -1, 0$, or 1. By selecting the output bandpass filter $H(\omega)$ so that it only passes spectral components near $\pm \omega_s$, the $k = 0$ term is removed and only the symbol frequency components for $k = \pm 1$ remain.

When the baseband shaping filter has zero excess bandwidth, that is, when $G(\omega) = 0$ for $|\omega| \geq \omega_s/2$, all the Fourier coefficients p_k are zero for $k \neq 0$ since the nonzero portions of $G(\omega)$ and $G(k\omega_s - \omega)$ do not overlap. This timing recovery method then fails.

Franks and Bubrouski [II.D.7] have also derived formulas for $E\{z^2(t)\}$ and var $z(t)$. They show that when $G_1(\omega)$ is symmetric about $\omega_s/2$ and is band limited to the interval $\omega_s/4 < |\omega| < 3\omega_s/4$ and $H(\omega)$ is symmetric about ω_s, the variance of $z(t)$ is zero and perfect timing recovery is possible. When these symmetry conditions are nearly met, the variations in the zero crossings of the timing wave $z(t)$ are very small and the receiver can track the symbol clock frequency by locking to the zero crossings. The filters in the timing recovery system can introduce a phase shift in the timing wave which must be taken into account. The fractionally spaced equalizer can automatically correct for this phase shift as long as the recovered symbol clock frequency is correct.

11.6 Simulation and Theoretical Exercises for PAM

These exercises are designed to improve your understanding of PAM by generating signals with C programs before doing real-time implementations in C or assembly language and having to deal with the hardware as well as the software. You will generate four level PAM signals using raised cosine and square-root of raised cosine shaping and create eye diagrams. You will also be asked to make plots of the symbol error probability for several cases.

11.6.1 Generating Four-Level Pseudo-Random PAM Symbols

Write a C function to generate pseudo-random four-level symbols. The function should use a 23-stage self synchronizing shift register sequence generator with the connection polynomial

(9.9) or (9.10), as discussed in Chapter 9, to generate the binary sequence d_n. Generate the four-level sequence a_n from pairs of binary symbols (d_{2n}, d_{2n+1}) according to the rule

$$a_n = (-1)^{d_{2n}}(1 + 2d_{2n+1})d \qquad (11.44)$$

where d is a desired scale factor.

Draw a vertical axis and show the 4 levels. Label each level with its value and also the corresponding pair of binary digits. Observe that the binary labels for adjacent levels differ in only one bit. This is called *Gray coding*. With additive Gaussian noise, the most likely symbol decision error in the receiver is to an adjacent level. Gray coding minimizes the bit-error probability at the receiver.

11.6.2 Eye Diagram for a PAM Signal Using a Raised Cosine Shaping Filter

Use the program `rascos.exe` to design an interpolation filter bank for a raised cosine shaping filter with an excess bandwidth factor of $\alpha = 1.0$. Use a symbol rate of $f_s = 1/T = 4$ kHz. Truncate the shaping filter impulse response to the interval $[-4T, 4T]$ with a Hamming window. Generate L=16 samples of the PAM signal per symbol interval, that is, generate the sequence $s(kT/16)$.

Use the filter bank and the four-level random symbol sequence to generate data for an eye diagram that extends over two symbol intervals. Do this by writing enough pairs $(\mathrm{mod}(k, 32), s(kT/16))$ to a file to form a reasonably filled out 4-level eye diagram. The function, $\mathrm{mod}(k, 32)$, is the remainder when k is divided by 32 and ranges from 0 to 31. As k increases, $\mathrm{mod}(k, 32)$ cycles through the values $0, 1, \ldots, 31$. This performs the function of resetting the trace to the left-hand side every two symbols. When k reaches a multiple of 32 and $\mathrm{mod}(k, 32) = 0$, you should write the three extra points $(32, s(kT/16))$, $(32, 0)$ and $(0, 0)$ to the file before writing $(0, s(kT/16))$. The first point continues the trace to the right edge of the plot, the second point moves the trace vertically to 0, and the third point moves the trace horizontally at 0 from the right edge back to the origin. Writing these three extra points is necessary for some plotting functions that draw lines through the eye diagram from the last point on the right-hand side to the first point on the left-hand side. These lines are blanked out on an oscilloscope when it retraces. Some plotting programs behave nicely, blank the retrace like an oscilloscope, and the extra points are not needed. Print the eye diagram using your favorite plotting program.

Now change α to 0.125 and generate a new eye diagram. Print the new diagram. Discuss differences in the two eye diagrams and comment on how the excess bandwidth factor affects the required symbol sampling time accuracy in the receiver.

11.6.3 Eye Diagram for a PAM Signal Using a Square-Root of Raised Cosine Shaping Filter

Repeat the exercises of the previous section for a square-root of raised cosine baseband shaping filter, but only for $\alpha = 0.125$. Also, compute the peak fractional eye closure defined by (11.17) from the shaping filter impulse response.

11.6.4   Theoretical Error Probability for a PAM System

Plot the symbol error probability P_e given by (11.32) for $M = 2, 4$, and 8 as a function of the channel signal-to-noise ratio P_s/P_N. Plot P_e on a logarithmic scale and $10 \log_{10}(P_s/P_N)$ dB on a linear scale.

11.7   Hardware Exercises for PAM

Now use the DSP to generate a PAM signal and synthesize the symbol clock recovery system. Use a symbol rate of $f_s = 4$ kHz. Generate $L = 4$ PAM signal samples per symbol with an interpolation filter bank, so the DAC sampling rate should be set to $4 \times 4 = 16$ kHz.

11.7.1   Generating a PAM Signal and Eye Diagram

Generate a four-level PAM signal using a raised cosine baseband shaping filter. Generate the four-level pseudo-random input symbol sequence by the same method used in Section 11.6.1. Choose $\alpha = 0.125$ for the excess bandwidth factor of your shaping filter and truncate the impulse response to the interval $[-4T, 4T]$ by a Hamming window. Generate $L = 4$ output samples per symbol period by the interpolation method. Write the output samples to the left channel. Make sure to use level o3 optimization when compiling your program.

Let the impulse response of the baseband shaping filter, viewed as an FIR filter with $T/4$ tap spacing, be

$$g_n = \begin{cases} g(nT/4) & \text{for } n = -16, -15, \ldots, 16 \\ 0 & \text{elsewhere} \end{cases} \qquad (11.45)$$

The frequency response of this filter is

$$G(\omega) = \sum_{n=-16}^{16} g_n e^{-j\omega nT/4} = g_0 + 2 \sum_{n=1}^{16} g_n \cos(\omega nT/4) \qquad (11.46)$$

The right-hand expression is a result of the fact that the impulse response has even symmetry. Compute and plot the amplitude response of the filter in dB over the frequency range of 0 to 8 kHz.

There are a variety of ways to structure a program to generate the real-time output signal. Here is one approach to try. Write output samples to the McBSP1 data transmit register (DXR) with an interrupt routine that is triggered by the serial port transmit interrupts (XINT) that occur when the data transmit register is loaded into the serial port transmit shift register (XSR). Determine the symbol timing by counting interrupts modulo 4. Set up an 8-word circular buffer as a "mail box." One half of the buffer (4 words) will be used to hold the output samples for the current symbol period, and the remaining half will be used to store the four samples for the next symbol period. Each symbol period, the input and output halves will be swapped. These are sometimes called *ping-pong buffers*. The mail box structure is illustrated in Figure 11.5. Initialize an output pointer to the address of the first word, word 0, in the buffer and an input pointer to the fifth word, word 4. Before starting data transmission, set the interrupt count to 0 to indicate the start of a symbol. At the start of each symbol period, generate four output samples, and write them to the mailbox. Do this in the main routine. The input pointer should be incremented circularly modulo 8 after each sample is written to the mailbox. After the four samples are written to the mailbox, the main routine should wait for the interrupt count to become 0. When a transmit interrupt occurs, write the sample addressed by the output pointer to the DXR, increment the output pointer circularly modulo 8, and increment the interrupt count modulo 4.

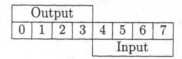

Figure 11.5: The Output Mailbox Structure

Test your transmitter program by observing the eye diagram on the oscilloscope. Use DC coupling for the scope input.

Generating a Baud Sync Signal

You will need a signal to synchronize the sweeps with the symbol period to get a display like your theoretical plot. One way to generate a sync signal is to create a 4000 Hz square-wave on the right channel codec output. You can do this by putting an integer like $A = 16000$ in the lower half of the word sent to the codec for first two samples in a baud and $-A$ in the second two samples. Of course, the codec filters will round off the corners and make it look more like a sine wave.

11.7.2 Testing the Square-Law Symbol Clock Frequency Generator

Write a program for the DSP to implement the symbol clock recovery system discussed in Section 11.5 up to the point labeled $z(t)$ in Figure 11.4. Do not implement the phase-locked loop. Use the same raised cosine baseband shaping filter you designed in Section 11.7.1. The sampling rate for all operations in the symbol clock generator should be $4f_s = 16$ kHz. Write the PAM output samples to the DAC left channel and also pipe them to your clock recovery system. For the prefilter, $B(\omega)$, design a second-order IIR filter with a center frequency of $f_s/2 = 2$ kHz and roughly a 100 Hz 3 dB bandwidth. For the postfilter, $H(\omega)$, design a second-order bandpass IIR filter with a center frequency of 4 kHz and a 3 dB bandwidth of roughly 25 Hz. You should experiment with these bandwidths and observe how they affect the system performance. Write the clock recovery system output samples $z(nT/4)$ to the right channel output.

To test your program, first drive your baseband shaping filter with the alternating two-level symbol sequence $a_n = (-1)^n d$. This alternating sequence is called a *dotting* sequence in the modem jargon. Send the shaping filter output samples to the left channel output, pipe them to the clock tone generator, and send the clock tone generator output samples to the right channel output. Observe the left and right channel outputs simultaneously on the oscilloscope.

Notice that

$$(-1)^n = \cos\left(\frac{\omega_s}{2}nT\right) \tag{11.47}$$

which are symbol rate samples of a cosine wave that has a frequency of half the symbol rate. According to DSP theory, the sampled signal has spectral components at the set of frequencies $\{\omega_s/2 + k\omega_s; k = -\infty, \ldots, \infty\}$. The shaping filter will pass the 2 kHz component and heavily attenuate the other components in the $[0, 8)$ kHz band. In other words, the PAM output signal should be very close to a 2 kHz sine wave. Check that the tone generator output is a 4 kHz sine wave locked to the PAM signal.

Next, use a two-level pseudo-random symbol sequence having values $\pm d$. Use the shift register generator to select the levels. Observe the PAM signal and clock tone generator output simultaneously on the oscilloscope. They should still be locked together. Comment on how the tone generator output looks compared to the output with the dotting sequence.

Finally, use the full four-level pseudo-random input symbol sequence. Observe the output of the clock recovery system on the oscilloscope and compare it with the previous cases.

11.7.3 Optional Team Exercise

If you are interested in doing more with PAM, team up with an adjacent group. Make one setup a PAM transmitter and the other a PAM receiver. This exercise is nontrivial and should be considered to be equivalent to a complete lab experiment.

For simplicity put all of the raised cosine shaping in the transmitter. Transmit a two-level PAM signal. The transmitted levels should be selected by the output of the 23-stage scrambler described in Section 9.2.1 with an input of 0. Make sure the initial scrambler state is non-zero. Again, use a 4 kHz symbol rate. Connect the line output of the transmitter to the line input of the receiver. Sample the received signal at 16 kHz.

Even though the transmitter and receiver both use a sampling frequency of 16 kHz, there will be slight differences due to small physical and temperature differences in the oscillator crystals and circuit components. You will have to devise a method for synchronizing the symbol clock in the receiver to the symbol clock in the transmitter. The sampling phase of the codec cannot be altered, so you will have to pass the received samples through a variable phase interpolator that compensates for the phase difference between the transmit and receive clocks. Variable phase interpolators are discussed in Chapter 12. You can lock the phase of the receiver symbol clock to the positive zero crossings of the symbol clock tone generator. In addition you will have to compensate for any delays in the system so that samples are taken at the symbol instants, that is, at the point where the eye has its maximum opening. You could also add a baseband version of a $T/2$ spaced adaptive equalizer. See Chapter 15 for a discussion of passband equalizers. The fractionally spaced equalizer will automatically compensate for a fixed symbol phase offset but not for a frequency offset.

Quantize the selected symbol rate samples to a binary sequence. Descramble this sequence and check that the output is all 0's.

Add Gaussian noise to the received samples in the DSP and make a plot of the bit-error rate vs. SNR.

11.8   Additional References

For a very complete discussion of baseband digital data transmission by PAM, see the classic book by Lucky, Salz, and Weldon [II.D.29, Chapter 4]. Also see Gibson [II.D.9, Chapter 8], Gitlin, Hayes, and Weinstein [II.D.11, Chapter 4], Lee and Messerschmitt [II.D.26, Chapter 6], and Proakis [II.D.32, Chapter 6]. All these references discuss the basic idea of PAM, baseband shaping filters, intersymbol interference, eye diagrams, Nyquist's criterion for no ISI, raised cosine pulses, and symbol error probability formulas.

Discussions of interpolation filter banks can be found in Crochiere and Rabiner [II.C.4, Section 3.3] and Proakis and Manolakis [II.C.15, Chapter 10].

A very thorough analysis of the symbol clock recovery scheme described in this chapter is presented in Franks and Bubrouski [II.D.7]. Further discussions of this scheme and others can be found in Gitlin, Hayes, and Weinstein [II.D.11, Section 6.5] and Lee and Messerschmitt [II.D.26, Chapter 15].

Chapter 12

Variable Phase Interpolation

The receiver in a digital communication system usually knows the nominal symbol rate used by the transmitter. Since the receiver is at a distance from the transmitter, has slightly different components, and is at a different temperature, the locally generated symbol clock in the receiver will differ in phase and slightly in frequency from the transmitter's clock. Therefore, the receiver must synchronize its symbol clock to the clock in the signal received from the transmitter. It must do this just using information derived from the received signal. Some codecs designed for modem front ends have built-in hardware capability for changing their sampling phase by small increments as directed by commands from the DSP they are connected to. The clock tone generator discussed in Chapter 11 can be used to determine the needed phase increments. The codec for the TMS320C6713 DSK does not have this capability and runs with a fixed phase. In this chapter, we will see how to implement the phase shifting in the DSP by a variable phase interpolator. First, a continuously variable phase shifter will be presented. Then a phase shifter using fine quantized steps will be discussed.

Another problem arises when the modem output samples are connected directly to a digital link like a T1 channel where the sampling rate is fixed at 8 kHz and is different from the rate needed for the modem input or output samples which is a multiple of the symbol rate. We will see how to solve this problem by using an interpolation filter bank to convert between two sampling rates that are rationally related.

12.1 Continuously Variable Phase Interpolation

Let $x(t)$ be a band limited signal with cutoff frequency ω_c, that is, $X(\omega) = 0$ for $|\omega| \geq \omega_c$. Let the sampling rate be $\omega_s \geq 2\omega_c$ and sampling period be $T = 2\pi/\omega_s$. The sampling frequency in Hertz is $f_s = \omega_s/(2\pi) = 1/T$. The ideal impulse sampled signal is

$$x^*(t) = \sum_{n=-\infty}^{\infty} x(nT)\delta(t - nT) \tag{12.1}$$

and it can be shown the Fourier transform of $x^*(t)$ is given by the *aliasing formula*

$$X^*(\omega) = f_s \sum_{n=-\infty}^{\infty} X(\omega - n\omega_s) \tag{12.2}$$

According to the sampling theorem, $x(t)$ can be exactly reconstructed for any t from its samples $\{x(nT)\}$ by applying $x^*(t)$ to an ideal lowpass filter with the frequency response

$$H_0(\omega) = \begin{cases} 1/f_s = T & \text{for } |\omega| < f_s/2 \\ 0 & \text{elsewhere} \end{cases} \tag{12.3}$$

and impulse response

$$h_0(t) = \frac{\sin \frac{\omega_s}{2} t}{\frac{\omega_s}{2} t} \tag{12.4}$$

The reconstruction formula or sampling theorem is

$$x(t) = \sum_{k=-\infty}^{\infty} x(kT) h_0(t - kT) = \sum_{k=-\infty}^{\infty} x(kT) \frac{\sin \frac{\omega_s}{2}(t - kT)}{\frac{\omega_s}{2}(t - kT)} \tag{12.5}$$

If the signal is over-sampled so that ω_s is strictly greater than $2\omega_c$, the following narrower band ideal lowpass filter can be used for signal reconstruction to eliminate out-of-band noise:

$$H(\omega) = \begin{cases} 1/f_s & \text{for } -\omega_c < \omega < \omega_c \\ 0 & \text{elsewhere} \end{cases} \tag{12.6}$$

The impulse response of this filter is

$$h(t) = 2\frac{\omega_c}{\omega_s} \frac{\sin \omega_c t}{\omega_c t} \tag{12.7}$$

Then, $x(t)$ can be reconstructed from its samples by the formula

$$x(t) = \sum_{k=-\infty}^{\infty} x(kT) h(t - kT) = \sum_{k=-\infty}^{\infty} x(kT) 2\frac{\omega_c}{\omega_s} \frac{\sin \omega_c(t - kT)}{\omega_c(t - kT)} \tag{12.8}$$

Letting $t = nT + dT$ gives the following formula for interpolating between samples of $x(t)$:

$$\begin{aligned} x(nT + dT) &= \sum_{k=-\infty}^{\infty} x(kT) h(nT - kT + dT) \\ &= \sum_{k=-\infty}^{\infty} h(kT + dT) x(nT - kT) \end{aligned} \tag{12.9}$$

The variable, d, is the time advance normalized by the sampling period. Values for $x(\cdot)$ around the time nT can be computed by varying d by the desired fraction of the symbol period T.

For actual computation, the sum in (12.9) must be truncated. This can be done by truncating $h(t)$ with a Hanning window. Suppose the impulse response is to be truncated to the time interval $-(L+0.5) < t/T < L+0.5$ where L is an integer, and that the normalized advance is limited to plus or minus half a symbol, that is, $-0.5 \le d < 0.5$. The required Hanning window is

$$w(t) = \begin{cases} 0.5 + 0.5 \cos \dfrac{\pi t}{(L+0.5)T} & \text{for } -(L+0.5) < t/T < L+0.5 \\ 0 & \text{elsewhere} \end{cases}$$

Other windows like the Hamming window, for instance, can be used. Let the windowed impulse response be

$$
\begin{aligned}
g(t) &= h(t)w(t) \\
&= \begin{cases} 2\dfrac{\omega_c}{\omega_s}\dfrac{\sin\omega_c t}{\omega_c t}\left(\dfrac{1}{2}+\dfrac{1}{2}\cos\dfrac{\pi t}{(L+\frac{1}{2})T}\right) & \text{for } -(L+\tfrac{1}{2}) < \tfrac{t}{T} < L+\tfrac{1}{2} \\ 0 & \text{elsewhere} \end{cases}
\end{aligned}
\tag{12.10}
$$

Then, the approximation to (12.9) becomes

$$
\hat{x}(nT + dT) = \sum_{k=-L}^{L} g(kT + dT)x(nT - kT)
\tag{12.11}
$$

Of course, a delay of L input samples must be added to make the filter physically realizable, so the actual formula that would be computed is

$$
\hat{x}(nT - LT + dT) = \sum_{k=-L}^{L} g(kT + dT)x(nT - LT - kT)
\tag{12.12}
$$

With fixed d, this interpolation formula represents a $2L+1$ tap FIR filter with tap coefficients $g(nT + dT)$ and input sequence $x(nT)$. Inside the window the desired interpolation filter impulse response with normalized advance d is

$$
\begin{aligned}
g(nT + dT) &= h(nT + dT)w(nT + dT) \\
&= 2\frac{\omega_c}{\omega_s}\frac{\sin\omega_c(n+d)T}{\omega_c(n+d)T}\left(0.5 + 0.5\cos\frac{\pi(n+d)}{L+0.5}\right) \\
&\quad \text{for } n = -L, -L+1, \ldots, L \text{ and } -0.5 \le d < 0.5
\end{aligned}
\tag{12.13}
$$

Notice that (12.13) represents $g(t)$ over the interval $[-(L + 0.5)T, (L + 0.5)T)$ by $2L + 1$ sections over the sub-intervals

$$
[(n - 0.5)T, (n + 0.5)T) \quad \text{for } n = -L, \ldots, L
$$

as d varies between -0.5 and 0.5 for each section.

Direct computation of the tap coefficients for each new value of d requires evaluation of trigonometric functions and division which takes significantly more time than addition or multiplication in DSP's. A solution to this problem is to approximate the $2L + 1$ sections of $g(t)$ by low degree polynomials. We will approximate the sections by least-squares cubic polynomial fits of the form

$$
g_k(d) = c_{0,k} + c_{1,k}d + c_{2,k}d^2 + c_{3,k}d^3 \quad \text{for } k = -L, \ldots, L \text{ and } -0.5 \le d < 0.5
\tag{12.14}
$$

It is easy to store the $4(2L + 1)$ polynomial coefficients.

The resulting approximate interpolator can be implemented for a given d by first using (12.14) to compute the tap values and then performing the convolution

$$
\tilde{x}(n; d) = \sum_{k=-L}^{L} g_k(d)x(nT - kT)
\tag{12.15}
$$

An alternative realization can be obtained by substituting (12.14) into (12.15) to give

$$\tilde{x}(n; d) = \sum_{i=0}^{3} \left[\sum_{k=-L}^{L} c_{i,k} x(nT - kT) \right] d^i \qquad (12.16)$$

For each i, the sum inside the square brackets in (12.16) is a $2L + 1$ tap FIR filter with tap coefficients $c_{i,k}$ and input sequence $x(nT)$. The outputs of these filters are multiplied by powers of d and summed. This structure was also suggested in [II.C.6]. For a single fixed d, computation by (12.16) is not more efficient than by (12.15). However, when interpolated values for several values of d are required, (12.16) is more efficient.

12.1.1   Computing the Least-Squares Fits

The program `interp.exe` in the directory `C:\digfil\interpol` can be used to compute the least-squares fit polynomial coefficients for the interpolator sections. The source code, `interp.for` is also included in the directory. In the first cut at the interpolator program, the least-squares cubic fit for a section was computed using finely spaced samples of $g(t)$ confined to the section. It was found that the resulting impulse response had small discontinuities at the section boundaries. Then cubic splines were used to eliminate these discontinuities but this forced larger errors in the interiors of the sections. Finally, some samples from adjoining sections were used in computing the section polynomial approximation and the discontinuities were significantly reduced. More precisely, to approximate $g(t)$ over the interval $[(n-0.5)T, (n+0.5)T]$ a least-squares fit was performed by using samples of $g(t)$ taken uniformly over the interval $[(n - 0.5 - \alpha)T, (n + 0.5 + \alpha)T]$ with $0 \le \alpha \le 1$. Experimentally, it was found that $\alpha = 0.04$ gave the best results. In the program, 101 uniformly spaced samples of $g(t)$ over the extended intervals are used to compute the least-squares fit cubic polynomials. The user is given the option of choosing the cutoff frequency f_c, sampling rate f_s, number of sections, and overlap factor α which is called G in the program.

Another lesson was learned in developing the program. The ideal lowpass impulse response was initially truncated with a rectangular window. The resulting amplitude responses and envelope delays had large ripples. Using the Hanning window nicely solved this problem.

The program `response.exe` in the `interpol` directory computes the amplitude response and envelope delay relative to the center tap for the filters designed by `interp.exe`. The program asks you to "ENTER ALPHA" which is the desired normalized advance, d.

12.2   Quantized Variable Phase Interpolation

Another approach to variable phase interpolation is to divide the symbol period into relatively finely spaced points and design a fixed interpolation filter to achieve the phase shift corresponding to each separate point. The symbol clock recovery and tracking system then selects the filter with the phase shift closest to the desired value. The design of an interpolation filter bank for implementing a PAM transmit shaping filter was discussed in Section 11.3. Exactly the same technique can be used to design a multi-step phase shifting filter bank. It is only necessary to replace the transmit filter impulse response, $g_T(t)$, by the reconstruction filter impulse response, $g(t)$, given by (12.10) and the data symbol sequence, a_k, by the T-spaced signal samples, $x(kT)$. In this section, T is the period between samples, not the symbol period. The symbol period is typically an integer like 3 or 4 times T. Then, the interpolation formula (11.20) for $t = nT + m(T/M)$ becomes

$$\hat{x}\left(nT + m\frac{T}{M}\right) = \sum_{k=-\infty}^{\infty} x(kT)g_m(n-k)$$

$$= \sum_{k=-L}^{L} g_m(k)x(nT - kT) \quad \text{for} \quad m = 0, \ldots, M-1 \qquad (12.17)$$

where M is the number of phase increments between samples and subfilter m has the impulse response

$$g_m(n) = g\left(nT + m\frac{T}{M}\right) \quad \text{for} \quad m = 0, \ldots, M-1 \qquad (12.18)$$

Again, a delay of L samples must be introduced to make the filters physically realizable.

The range for m was chosen to be 0 to $M-1$ above while the range for d in Section 12.1 was selected to be -0.5 to 0.5. These choices were somewhat arbitrary. For example, if M is even, the range for m could also have been chosen to be $-0.5M$ to $0.5M - 1$.

For typical modem applications, a reasonable value for the number of phase increments, M, might be between 32 and 64. To get finer resolution, some manufacturers linearly interpolate between the outputs of adjacent subfilters based on the required value of d. Unlike the transmit shaping filter bank shown in Figure 11.3, only the output of the one subfilter for the selected value of m is computed between each input sample.

The phase shifting method of Section 12.1 is efficient as far as data storage memory is concerned. However, when d is changing frequently, it effectively requires recomputation of the filter coefficients for each new d and is not computationally efficient. The quantized step phase shifter presented in this section is computationally efficient because all the filter coefficients are pre-computed and stored in data memory. However, it is not as efficient in terms of data memory usage. The choice between the two methods is a choice the designer must make based on system constraints.

12.3 Closing the Tracking Loop

A variable phase interpolator and the symbol clock tone generator presented in Section 11.5 can be combined into a phase-locked loop for tracking the symbol clock of a PAM signal as shown in Figure 12.1. In this figure, T is the sampling period and T_b is the symbol period. We will require T_b to be an integer multiple of T. In telephone line modems, T_b is typically 3 or 4 times T. In most cases, the transmitter and receiver symbol clocks are very close in frequency. For example, the ITU-T modem recommendations specify that the symbol clock should have an accuracy of $\pm 0.01\%$. Therefore, the symbol clock generated in the receiver drifts very slowly with respect to the transmitter clock without a tracking loop. However, transmissions may last for a long period of time and the accumulated phase shift can become large. Therefore, the receiver must adjust the frequency of its local clock to eliminate the drift and track the symbol clock embedded in the received signal.

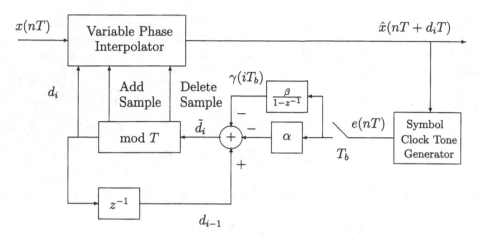

Figure 12.1: Symbol Clock Tracking Loop

The output of the Variable Phase Interpolator, $\hat{x}(nT + d_iT)$, is applied to the Symbol Clock Tone Generator. It is assumed that the normalized advance, d_i, changes very slowly. Suppose that $T_b = KT$. The output of the Tone Generator $e(nT)$ is sampled at the symbol rate to generate the phase error sequence $e(iKT) = e(iT_b)$. The goal of the loop is to adjust d_i to force $e(iT_b)$ to zero so the loop locks to the positive zero crossings of the generated clock tone. When the loop is nearly in lock, it can be seen from a sketch of $\sin x$ that when $e(nT) > 0$ the sampling instant is too late and the advance d_i should be reduced. Similarly, when $e(nT) < 0$, the sampling instant is too early and d_i should be increased. The increment for d_i is formed by scaling $e(iT_b)$ by a small positive constant α. The increments should be a small fraction of the sampling period for accurate clock tracking. The increments are then subtracted from the previous value of d along with a second-order correction to get the new initial value

$$\tilde{d}_i = d_{i-1} - \alpha e(iT_b) - \gamma(iT_b) \tag{12.19}$$

where

$$\gamma(iT_b) = \beta e(iT_b) + \gamma((i-1)T_b) \tag{12.20}$$

The accumulator generating $\gamma(iT_b)$ adjusts for a constant frequency offset just as in the loops presented in Chapters 6, 8, and 15. The positive constant β should be 50 to 100 times smaller than α for reasonable transient response.

It is quite important to have a very stable symbol clock reference in high speed modems. The clock phase should have little jitter. In Chapter 14 a nonlinear filter, called a random walk filter, is added to the symbol clock tracking loop for a QAM modem to further smooth the estimated clock phase.

If the transmitter and receiver clocks are exactly the same frequency, the value for $\tilde{d}_i$ should hover around a constant value. Each new sampling period, a new sample of $x(\cdot)$ is shifted into the interpolator's FIR filter and the output for the desired value of $\tilde{d}_i$ is computed. When there is a frequency offset between the transmitter and receiver symbol clocks, the value of $\tilde{d}_i$ will slowly drift in a positive or negative direction. In the continuously variable phase interpolator, its value was restricted to the range $[-0.5, 0.5)$. When $\tilde{d}_i$ falls outside this range some corrective action must be taken. We will assume it can only fall a small distance outside this range. If $\tilde{d}_i > 0.5$, the normal sample of $x(\cdot)$ should be shifted into the FIR filter delay line and, additionally, the next new sample in time should be shifted in. Then 1 should be subtracted from $\tilde{d}_i$. If $\tilde{d}_i < -0.5$, no new sample should be shifted into the delay line and 1 should be added to $\tilde{d}_i$. The mod T box in Figure 12.1 performs this corrective action. It generates the final phase advance value d_i and informs the interpolator if a sample should be added or deleted.

A similar strategy can be used for the quantized variable phase interpolator. It is convenient to use a number of subfilters that is a power of two, say, $M = 2^J$. The subfilter index can be stored in a 32-bit integer in a TMS320C67xx DSP with the top J bits, exclusive of the sign bit, actually used as the index. The phase increments can be added into the lower bits of the the word which performs the accumulation and averaging. When the upper J bits fall outside the allowed interval, corrective action similar to that described in the previous paragraph must be taken.

12.4 Changing the Sampling Rate by a Rational Factor

A situation arises in modem design where it is necessary to use interpolation to change from one sampling rate to another rationally related rate. Modems have been designed where the output samples are not converted to a continuous-time signal but are sent directly over a digital network like the Internet to bypass telephone charges. Sometimes a T1 connection is used and the samples have to be interpolated to an 8 kHz sampling rate and transformed into μ or a-law 8-bit codes. For example, the modem might normally have a symbol rate of 2400 baud with three samples generated per symbol resulting in a sampling rate of 7200 Hz. Then it is necessary to interpolate these samples to an 8000 Hz rate. The least common multiple of 7200 and 8000 is $72000 = 10 \times 7200 = 9 \times 8000$. A standard approach to performing the sampling rate conversion is to interpolate the 7200 Hz samples up to a 72000 rate with an FIR interpolation filter bank and then down-sample by a factor of 9 to get the 8000 Hz samples.

More generally, let the initial sampling rate be f_1 and the final rate be f_2. Suppose the ratio of f_1 and f_2 when reduced to lowest terms is

$$\frac{f_1}{f_2} = \frac{n_1}{n_2} \tag{12.21}$$

with n_1 and n_2 relatively prime. Then the intermediate sampling rate should be

$$f_3 = n_2 f_1 = n_1 f_2 \tag{12.22}$$

In the example above, $n_1 = 9$ and $n_2 = 10$. Down-sampling the f_3 rate sequence by a factor of n_1 gives the desired rate of f_2.

The first step in changing the sampling rate from f_1 to f_3 is to use the sampling theorem to express the continuous-time signal $x(t)$ in terms of its rate f_1 samples. The reconstruction formula is

$$x(t) = \sum_{k=-\infty}^{\infty} x(kT_1)h(t - kT_1) \tag{12.23}$$

where

$$h(t) = \frac{\sin \pi f_1 t}{\pi f_1 t} \tag{12.24}$$

is the impulse response of an ideal lowpass filter with cutoff frequency $f_1/2$ and amplitude response $T_1 = 1/f_1$. Letting $t = nT_1 + m\frac{T_1}{n_2}$ gives

$$\begin{aligned} x_m(nT_1) &= x\left(nT_1 + m\frac{T_1}{n_2}\right) = \sum_{k=-\infty}^{\infty} x(kT_1)h\left(nT_1 + m\frac{T_1}{n_2} - kT_1\right) \\ &= \sum_{k=-\infty}^{\infty} x(kT_1)h_m(n - k) \quad \text{for} \quad m = 0, \ldots, n_2 - 1 \end{aligned} \tag{12.25}$$

where subfilter m is

$$h_m(n) = h(nT_1 + m\frac{T_1}{n_2}) \quad \text{for} \quad m = 0, \ldots, n_2 - 1 \tag{12.26}$$

This formula shows how to interpolate n_2 points between each rate f_1 sample starting at time nT_1 to generate the rate $f_3 = n_2 f_1$ sequence. The approach is the same as the one used for implementing the baseband shaping filter discussed in Section 11.3. For each m, $x_m(nT_1)$ is generated by a discrete-time filter operating with sampling rate f_1. The interpolation formula is illustrated in Figure 12.2 as a bank of n_2 filters.

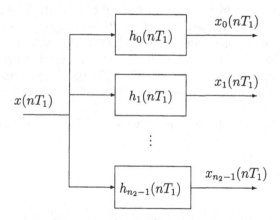

Figure 12.2: Interpolation Filter Bank

The next step is to down sample the rate f_3 sequence by a factor of n_1 to get the desired rate f_2 sequence. Suppose the output of subfilter m is chosen from the n_2 subfilter outputs generated at time nT_1. The next sample that should be selected is the output of subfilter $m + n_1$ if $m + n_1 < n_2$. If $m + n_1 \geq n_2$, the output of subfilter $\text{mod}(m + n_1, n_2)$ generated for the interval starting at time $(n + 1)T_1$ should be selected.

In practice, the impulse response $h(t)$ must be truncated to a finite time duration by, for example, a Hamming window. Then each subfilter is an FIR filter. The data samples used to calculate each subfilter output are the same, so only one "delay line" is required to store them. Each subfilter uses a different set of taps to convolve with the contents of this delay line. Also, for computational efficiency, the outputs of all n_2 subfilters should not be computed each T_1 period. Only the output of the subfilter for the required down-sampled output sample should be calculated.

The interpolation filter bank can be designed using the program `rascos.exe` in the directory `C:\digfil` discussed in Section 11.2.2. To approximate the ideal lowpass filter, set the excess bandwidth factor, α, to 0 and set the number of points per symbol to n_2.

12.5 Experiments for Variable Phase Interpolation

To continue along with the PAM experiments of Chapter 11, set the sampling rate to 16 kHz for the experiments of this section.

12.5.1 Experiment 12.1: Open Loop Phase Shifting Experiments

First, you will design and test a quantized variable phase interpolator. Perform the following tasks:

1. Using the sampling rate of $f_s = 16000$ Hz, generate the samples of a 2000 Hz cosine wave and send the resulting sequence to the left channel of the DAC. Connect the left channel line output to the left channel line input and the oscilloscope. The 2000 Hz cosine wave represents the dotting signal caused by alternating plus and minus input symbols to a two-level PAM transmitter when the symbol rate is 4000 baud.

2. Design a quantized step phase shifting filter bank as discussed in Section 12.2. Use a cutoff frequency of $f_c = f_s/2 = 8000$ Hz and let the number of phase steps between samples be $M = 8$.

3. First take the left channel ADC input samples and pass them thorough subfilter $m = 0$. Send the output sequence to the right DAC channel. Observe the right and left line outputs simultaneously on the oscilloscope. You will observe a phase shift caused by the delay introduced to make the subfilters physically realizable and by the system filters in the signal paths.

4. Once the $m = 0$ filter is working, test each of the subfilters for $m = 1$ through 7 to make sure they add the expected advance.

12.5.2 Experiment 12.2: Making a Symbol Clock Tracking Loop

Now you will make the symbol clock tracking loop shown in Figure 12.1. Do the following:

1. Write a program to implement the Symbol Clock Tone Generator. You should be able to use the program you created for Chapter 11. Continue to use the sampled 2000 Hz cosine wave as the input to your variable phase interpolator as in Experiment 12.1. Pipe the output of your Variable Phase Interpolator directly to the input of the Symbol Clock Tone Generator inside the DSP program. Do not close the loop yet. Send the output, $e(nT)$, of the tone generator to the right DAC and observe the result on the oscilloscope. You should see a clean 4000 Hz sine wave.

2. Once the clock tone generator is working, implement the rest of the loop. Check that it locks to the 4000 Hz symbol clock.

3. Change the frequency of the 2000 Hz input tone slightly and check that your loop tracks the frequency offset.

4. (Optional) Generate a PAM signal with a 4000 baud symbol rate either internally in the local DSP or, preferably, on another station. Connect the PAM signal to your clock tracking loop and check that it works.

12.6   Additional References

Discussions of quantized phase interpolation filter banks can be found in Crochiere and Rabiner [II.C.4, Section 3.3] and Proakis and Manolakis [II.C.15, Chapter 10]. The only reference for the continuously variable phase interpolator know to the author is the paper by Farrow [II.C.6]. Symbol clock tracking is crucial to good modem performance, but little is published in the open literature. Textbooks usually completely ignore the problem and assume the exact symbol instants are known at the receiver. Manufacturers seem to consider their methods to be trade secrets not to be disclosed.

Chapter 13

Fundamentals of Quadrature Amplitude Modulation

Quadrature amplitude modulation (QAM) is a widely used method for transmitting digital data over bandpass channels. It can be viewed as a generalization of PAM to bandpass channels. All current telephone line modems based on the ITU-T V series recommendations for transmission at rates of 2400 bps or more use QAM or include it as an option. These include recommendations V.22 through V.92. This series includes FAX modems. Recommendation V.90 modems normally use PAM in the downstream direction from the server to the client modem and always use QAM in the upstream direction from the client to the server. V.90 modems can choose to use QAM downstream if a digital link from the server to the codec in the local office on the client side does not exist. V.92 modems normally use PAM in the downstream and upstream directions but can choose to use QAM based on line conditions. QAM is also used in DSL telephone line, high speed cable, multi-tone wireless, microwave, and satellite systems. It is a popular choice because it uses bandwidth efficiently and linear channel distortions can be corrected by adaptive equalization at the receiver. In addition, QAM fits in nicely with a common combined coding and modulation scheme used for band limited channels called *trellis coded modulation* (TCM).

This chapter primarily deals with the QAM transmitter. However, a brief introduction to the QAM receiver is included. The subsystems required to construct a practical receiver are described in the following chapters.

13.1 A Basic QAM Transmitter

The block diagram of a basic QAM transmitter is shown in Figure 13.1. It has many similarities to the PAM transmitter shown in Figure 11.1. The transmitter input is a serial binary data stream d_n arriving at the rate of R_d bps. The Serial to Parallel Converter groups the input bits into J-bit binary words. Each J-bit word selects a channel symbol from a 2^J element alphabet resulting in a channel symbol rate of $f_s = R_d/J$ baud. As in PAM, $T = 1/f_s$ will be used to denote the symbol period. The alphabet consists of pairs of real numbers representing points in a 2-dimensional space and is called the *signal constellation*. More will be said about constellations later. It will be convenient to consider the 2-dimensional space to be the complex plane and represent the channel symbol sequence by the sequence of complex numbers $c_n = a_n + jb_n$. It is customary to call the real part, a_n, the *inphase* or I component and the imaginary part, b_n, the *quadrature* or Q component.

The inphase and quadrature symbol components are passed through separate PAM modulators identical to the one described in Chapter 11. Thus, they are first passed through impulse modulators resulting in the signals

$$a^*(t) = \sum_{k=-\infty}^{\infty} a_k \delta(t - kT) \tag{13.1}$$

and

$$b^*(t) = \sum_{k=-\infty}^{\infty} b_k \delta(t - kT) \tag{13.2}$$

These signals are passed through identical baseband transmit shaping filters, each with impulse response $g_T(t)$. The properties required for $g_T(t)$ are exactly the same as the ones required for the PAM shaping filter discussed in Chapter 11. The outputs $a(t)$ and $b(t)$ of the shaping filters are called the inphase and quadrature components of the continuous-time transmitted signal $s(t)$ and are given by the equations

$$a(t) = \sum_{k=-\infty}^{\infty} a_k g_T(t - kT) \tag{13.3}$$

and

$$b(t) = \sum_{k=-\infty}^{\infty} b_k g_T(t - kT) \tag{13.4}$$

The baseband shaping filter is typically a lowpass filter approximating the raised cosine or square-root of raised cosine response, so its cutoff frequency is somewhat greater than $f_s/2$. Consequently, $a(t)$ and $b(t)$ are lowpass signals with power spectra extending down to 0 Hz. In order to translate the spectra up to the passband of a bandpass channel, $a(t)$ and $b(t)$ are DSBSC-AM modulated by the quadrature carriers $\cos \omega_c t$ and $\sin \omega_c t$ and subtracted to form the transmitted QAM signal

$$s(t) = a(t) \cos \omega_c t - b(t) \sin \omega_c t \tag{13.5}$$

The carrier frequency ω_c must be greater than the shaping filter cutoff frequency to prevent spectral fold-over. For example, a typical voiceband telephone line channel has a passband extending from about 300 Hz to 3100 Hz, the symbol rate might be $f_s = 2400$ Hz, and the carrier frequency might be $f_c = 1800$ Hz.

By using (5.24), it can be shown that the pre-envelope of the QAM signal is

$$s_+(t) = s(t) + j\hat{s}(t) = [a(t) + jb(t)]e^{j\omega_c t} \tag{13.6}$$

Therefore, the transmitted QAM signal can be expressed as

$$s(t) = \Re\{s_+(t)\} = \Re\{[a(t) + jb(t)]e^{j\omega_c t}\} \tag{13.7}$$

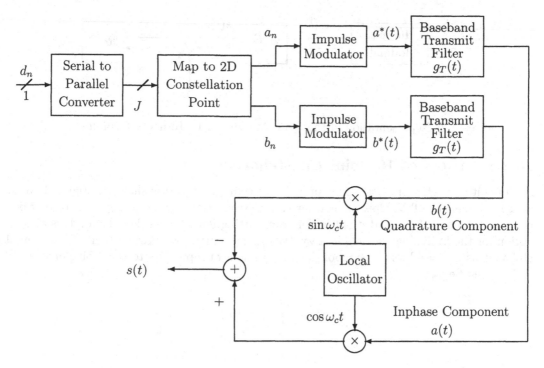

Figure 13.1: A Basic QAM Transmitter

The complex envelope of $s(t)$ is

$$\tilde{s}(t) = s_+(t)e^{-j\omega_c t} = a(t) + jb(t) \tag{13.8}$$

Combining (13.6), (13.3), and (13.4) yields

$$s_+(t) = \sum_{k=-\infty}^{\infty} (a_k + jb_k)g_T(t - kT) \; e^{j\omega_c t} = \sum_{k=-\infty}^{\infty} c_k g_T(t - kT) \; e^{j\omega_c t} \tag{13.9}$$

Therefore, the QAM modulator can be compactly represented in terms of these complex signals as shown in Figure 13.2. The complex envelope $\tilde{s}(t)$ is simply a complex PAM signal generated by the complex input symbols $c_n = a_n + jb_n$.

13.2   Two Constellation Examples

Examples of two common constellations are described in this section. The first is a rectangular 16-point constellation and the second is a 4-point subset of the first. A method for assigning data bits to the constellation points so that the system is transparent to 90° carrier ambiguities is described.

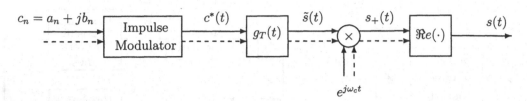

Figure 13.2: Representation of the QAM Modulator in Terms of Complex Signals

13.2.1   The 4×4 16-Point Constellation

A 16-point constellation with points on a 4×4 rectangular grid is shown in Figure 13.3. It is used in the ITU-T V.22bis modem for transmitting 2400 bps with a symbol rate of 600 baud and in the V.32 uncoded option for transmitting 9600 bps at 2400 baud. It is also an option for the individual carriers in a variety of multi-carrier wireless systems. This symbol alphabet uses $J = 4$ bits per symbol. The assignment of input bits to the 4-bit point labels is discussed below.

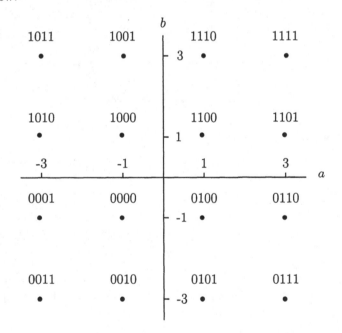

Figure 13.3: The 16-Point Rectangular QAM Constellation

with Labels $(Y1_n, Y2_n, Q3_n, Q4_n)$

The 4×4 constellation is invariant to 90° rotations. That is, a 90° rotation of the constellation results in the same set of points. It can be seen from (13.9) that a 90° carrier phase offset, that is, changing $\omega_c t$ to $\omega_c t + \pi/2$, has the effect of multiplying the constellation point $a_n + jb_n$ by j which rotates it by 90°. With this symmetry, carrier tracking loops in the receivers can only determine the correct phase to the nearest multiple of 90°. The system can be made transparent to 90° phase offsets by a combination of differentially encoding two of the input bits to specify the quadrant, and assigning the remaining two input bits to points within a quadrant so that a 90° rotation leaves them unchanged.

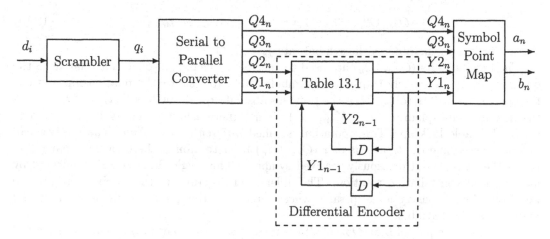

Figure 13.4: Mapping Input Bits to Constellation Points

A block diagram of the method for assigning input bits to constellation points is shown in Figure 13.4. The input data bits d_i arrive at R_d bps. This bit sequence is passed through a scrambler like the one described in Chapter 9 resulting in the scrambled sequence q_i. The reason for using the scrambler is to break up long strings of 1's or 0's in the input data sequence and cause the constellation points to be chosen pseudo-randomly. This randomization causes the transmitted spectrum to be distributed and have a shape like the transmit filter independent of the unscrambled input data sequence. Systems in the receiver like an adaptive equalizer, and carrier and symbol clock tracking loops require this symbol variation to operate properly. The V.22bis scrambler uses the difference equation

$$q_i = d_i \oplus q_{i-14} \oplus q_{i-17} \tag{13.10}$$

where $\oplus$ represents modulo 2 addition or the exclusive-or logical function. The corresponding connection polynomial is

$$h(D) = 1 + D^{14} + D^{17} \tag{13.11}$$

Actually, the V.22bis scrambler is slightly more complicated in that it contains a means for detecting a string of 64 1's at its output and complementing the next output bit. This prevents the all 1's scrambler lock-up condition when the input is all 1's. Since the lock-up probability is small, we will ignore this addition to the V.22bis scrambler. The V.32 calling modem uses the scrambler connection polynomial

$$h_c(D) = 1 + D^{18} + D^{23} \tag{13.12}$$

and the V.32 answer modem uses the connection polynomial

$$h_a(D) = 1 + D^5 + D^{23} \tag{13.13}$$

The scrambler output q_i is passed to a serial-to-parallel converter which groups the serial stream into the 4-bit words

$$(Q1_n, Q2_n, Q3_n, Q4_n) = (q_{4n}, q_{4n+1}, q_{4n+2}, q_{4n+3}) \tag{13.14}$$

These words are generated at the symbol rate $f_s = R_d/4$.

The first two bits $(Q1_n, Q2_n)$ are used to specify the change in quadrant relative to the quadrant of the previously transmitted symbol. Table 13.1 shows the relationship between the current input bits $(Q1_n, Q2_n)$, the previous absolute quadrant bits $(Y1_{n-1}, Y2_{n-1})$, and the new absolute quadrant bits $(Y1_n, Y2_n)$. This function is most easily implemented in a DSP by table look-up. This operation is called *differential encoding*. The receiver can uniquely determine the input bit pair $(Q1_n, Q2_n)$ by determining the quadrant change between the current and previously received symbols. This angle difference is unaffected by any constant constellation rotation. The differential decoding in the receiver can also be performed by a 16-entry look-up table whose inputs are the quadrants of the current and previous received symbol.

The remaining pair of bits $(Q3_n, Q4_n)$ are used to select a point in the quadrant specified by $(Y1_n, Y2_n)$. If you examine Figure 13.3, you will see that $(Q3_n, Q4_n)$ are assigned so that they do not change with 90° constellation rotations. The combination of the differential quadrant encoding and this bit assignment makes the overall system transparent to 90° rotations.

13.2.2   A 4-Point Four Phase Constellation

A 4-point constellation can be formed from the 4×4 constellation by selecting the subset of points with labels $\{(1101), (1001), (0001), (0101)\}$. Notice that these points lie on a circle and are separated by 90°. This constellation is sometimes called a 4PSK or QPSK constellation. PSK is an abbreviation for *phase shift keying*. This constellation is used in many low and high speed modems. In particular, it is used in the V.22bis modem for transmission at 1200 bps with a symbol rate of 600 baud and in the V.32 modem for transmission at 4800 bps with a symbol rate of 2400 baud. It is also used in single and multi-carrier wireless systems.

This constellation is also invariant to 90° rotations. It can be made transparent to these rotations by using the same differential quadrant encoding scheme as for the 4×4 constellation. The V.22bis and V.32 schemes for mapping input bits to constellation points using 2 bits/symbol is a simple modification of Figure 13.4. For these modems, the input data sequence d_i is again passed through the scrambler to generate the sequence q_i. The serial-to-parallel converter groups the scrambler output into the 2-bit blocks $(Q1_n, Q2_n) = (q_{2n}, q_{2n+1})$. These pairs are differentially encoded exactly as for the 4×4 constellation. To select the desired constellation points, the uncoded bits used in the 4×4 constellation are always set to $(Q3_n, Q4_n) = (01)$. Therefore, the inputs to the symbol point mapper are

$$(Y1_n, Y2_n, Q3_n, Q4_n) = (Y1_n, Y2_n, 0, 1) \tag{13.15}$$

Inputs		Previous Outputs		Quadrant	Outputs	
$Q1_n$	$Q2_n$	$Y1_{n-1}$	$Y2_{n-1}$	Phase Change	$Y1_n$	$Y2_n$
0	0	0	0	$+90°$	0	1
0	0	0	1	$+90°$	1	1
0	0	1	0	$+90°$	0	0
0	0	1	1	$+90°$	1	0
0	1	0	0	$0°$	0	0
0	1	0	1	$0°$	0	1
0	1	1	0	$0°$	1	0
0	1	1	1	$0°$	1	1
1	0	0	0	$+180°$	1	1
1	0	0	1	$+180°$	1	0
1	0	1	0	$+180°$	0	1
1	0	1	1	$+180°$	0	0
1	1	0	0	$+270°$	1	0
1	1	0	1	$+270°$	0	0
1	1	1	0	$+270°$	1	1
1	1	1	1	$+270°$	0	1

Table 13.1: Differential Quadrant Coding for V.22bis and V.32 Uncoded Options

The selection of these four points may seem odd at first, but the standards committee selected them to make the 4-point constellation an easily generated subset of the 16-point constellation and also to have about the same average power as the 16-point constellation.

13.3   A Modulator Structure Using Passband Shaping Filters

In this section, an alternative QAM modulator that uses passband shaping filters will be derived. This structure is slightly more efficient computationally than the one shown in Figure 13.1. As a starting point, (13.9) can be modified to

$$s_+(t) = \sum_{k=-\infty}^{\infty} \left(c_k e^{j\omega_c kT} \right) g_T(t - kT) e^{j\omega_c(t-kT)} \tag{13.16}$$

where $g_T(t)$ is the real baseband shaping filter impulse response. Let

$$h(t) = g_T(t)e^{j\omega_c t} = h_I(t) + jh_Q(t) \tag{13.17}$$

where

$$h_I(t) = g_T(t)\cos\omega_c t \ \text{ and } \ h_Q(t) = g_T(t)\sin\omega_c t \tag{13.18}$$

This filter is a bandpass filter with the frequency response $H(\omega) = G_T(\omega - \omega_c)$. Let

$$c'_k = c_k e^{j\omega_c kT} = a'_k + jb'_k \tag{13.19}$$

where

$$a'_k = \Re\{c'_k\} = a_k \cos \omega_c kT - b_k \sin \omega_c kT \qquad (13.20)$$

and

$$b'_k = \Im\{c'_k\} = a_k \sin \omega_c kT + b_k \cos \omega_c kT \qquad (13.21)$$

Substituting these definitions into (13.16) gives

$$s_+(t) = \sum_{k=-\infty}^{\infty} c'_k h(t - kT) \qquad (13.22)$$

and

$$s(t) = \Re\{s_+(t)\} = \sum_{k=-\infty}^{\infty} a'_k h_I(t - kT) - b'_k h_Q(t - kT) \qquad (13.23)$$

A block diagram for the modulator in terms of the complex signals is shown in Figure 13.5. The form based on (13.23), which is the equation that would be actually implemented, is shown in Figure 13.6.

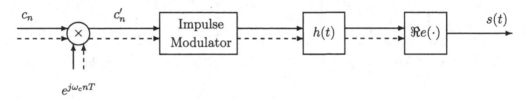

Figure 13.5: QAM Modulator Using a Passband Shaping Filter

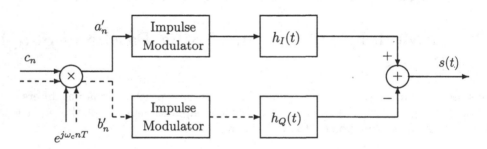

Figure 13.6: Expanded Block Diagram of the New QAM Modulator

The shaping filters, either baseband or passband, can be implemented by interpolation filter banks as described in Chapter 11. These banks generate L samples/baud. For example, V.32 modems use a symbol rate of 2400 baud, so L might be chosen to be 3 or 4 resulting in an output sampling rate of 7200 or 9600 samples/sec. In the original modulator structure using baseband shaping filters, each filter output sample must be multiplied by the appropriate inphase and quadrature carrier samples. In the modulator with passband shaping filters, the input symbols c_n must be rotated (modulated) before being applied to the passband shaping filters. This operation is required just once per symbol resulting in a slight computational savings.

13.4 Ideal QAM Demodulation

Two QAM demodulators will be described in this section using the assumption that the receiver has exact knowledge of the carrier and symbol clock phases and frequencies. Methods for tracking these signals will be presented in the next chapter. In addition it will be assumed that the channel is perfect and that all the shaping is performed at the transmitter by filters with no intersymbol interference.

A block diagram of the first demodulator is shown in Figure 13.7. It is based on (13.9). At the receiver input, the Hilbert transform of the received signal is formed to generate the pre-envelope $s_+(t)$. Then, according to (13.9)

$$\tilde{s}(t) = s_+(t)e^{-j\omega_c t} = \sum_{k=-\infty}^{\infty} (a_k + jb_k)g_T(t - kT) \tag{13.24}$$

If $g_T(t)$ has no intersymbol interference,

$$\tilde{s}(nT) = a_n + jb_n \tag{13.25}$$

which is exactly the transmitted symbol.

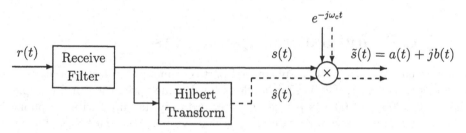

Figure 13.7: QAM Demodulator Using the Complex Envelope

A block diagram of the second demodulator is shown in Figure 13.8. It is based on (13.5) and uses a pair of DSBSC-AM coherent demodulators which are discussed in Chapter 6. The output of the product modulator in the upper branch is

$$s(t)2\cos\omega_c t = a(t) + a(t)\cos 2\omega_c t - b(t)\sin 2\omega_c t \tag{13.26}$$

Remember that $a(t)$ is a lowpass signal with a cutoff frequency around $f_s/2$. The second and third terms have spectra centered around $2\omega_c$ and do not overlap the baseband signal spectrum since the carrier frequency must be chosen to be greater than the cutoff frequency of the baseband signal. The unwanted highpass terms are eliminated by the lowpass post detection filter $F(\omega)$ which has a cutoff frequency selected to pass $a(t)$ and eliminate the unwanted terms. The output of the product modulator in the lower branch is

$$-s(t)2\sin\omega_c t = b(t) - b(t)\cos 2\omega_c t - a(t)\sin 2\omega_c t \tag{13.27}$$

Again, the undesired second and third terms can be eliminated by an identical post detection filter. The resulting inphase and quadrature components, $a(t)$ and $b(t)$ can then be sampled

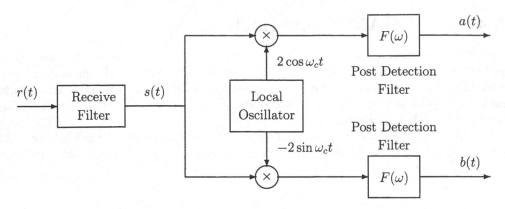

Figure 13.8: A Second Form of QAM Demodulator

at the symbol instants to recover the transmitted symbols. This second demodulator is not as popular in DSP implementations because it requires two post detection filters operating at a sampling rate dictated by the $2\omega_c$ terms. The first demodulator requires one filter to form the Hilbert transform and the $2\omega_c$ terms are automatically cancelled.

13.5   QAM Modulator Experiments

In this experiment, you will make a substantial part of a modem transmitter similar to the V.22bis calling transmitter. The V.22bis modem was designed primarily for full duplex data transmission at 1200 or 2400 bits per second over ordinary dial-up 2-wire telephone lines. The V.22bis modem uses a symbol rate of $f_s = 600$ symbols/sec. *Full duplex* means that one modem transmits to a second and the second transmits to the first simultaneously, and the transmissions in both directions are independent. Two modems communicate in what is called a *point-to-point* fashion. The modem that initiates the transmission by placing a call to the other modem is referred to as the *calling* or *originate* modem and the called modem is referred to as the *answer* modem. Because of the available sampling rates for the AIC23 codec on the TMS320C6713 DSK, the transmitter you make will use a symbol rate of 1000 baud and data rates of 2000 and 4000 bits per second. The carrier frequencies will also be changed.

Frequency division multiplexing is used to achieve the full duplex transmission. The V.22bis calling modem transmits a QAM signal using a carrier frequency of 1200 Hz and the answer modem transmits using a 2400 Hz carrier. We will use a calling modem carrier frequency of 2000 Hz and an answer modem carrier frequency of 4000 Hz. Notice that the carrier frequencies are equal to the data rates. This does not have to be the choice and is rarely true in general. This choice was convenient for full duplex transmission in the voice-band telephone line channel with a passband extending from about 300 to 3500. The V.22bis baseband shaping filters are specified to have a square-root of raised cosine frequency response with an excess bandwidth factor of $\alpha = 0.75$. Therefore, the spectrum of the V.22bis calling modem is nominally confined to the band $675 \leq f \leq 1725$ and the answer modem spectrum to $1875 \leq f \leq 2925$. For our transmitter, the spectrum of the calling modem will be confined to the band $1125 \leq f \leq 2875$ and the answer modem spectrum to $3125 \leq f \leq 4875$ Hz. Fifteen years ago, a typical voiceband telephone channel had a useful bandwidth extending from about 300 to 3100 Hz. The bandwidth is set by filters in the telephone plant. More recent voiceband channels may have a somewhat wider bandwidth. However, the PCM codecs in the local office use an 8 kHz sampling rate, so the upper cutoff frequency must be less than 4 kHz. Hybrids and bandpass filters are used to separate the transmit and receive signals at the receivers.

13.5.1   Steps to Follow in Making a Transmitter

Perform the following sequence of steps to make your transmitter. Be sure to use the highest compiler optimization level.

1. Write a C program to initialize the DSK as usual. Set the codec sampling rate to 16000 Hz. With a 16000 Hz sampling rate and a symbol rate of 1000 baud, you will have to generate $16000/1000 = 16$ output samples per symbol. You will be asked below to write the output samples to the McBSP1 data transmit register (DXR) with an interrupt service routine activated by the transmitter XRDY flag. You can generate the 1000 Hz symbol rate timing by counting interrupts in the interrupt service routine.

2. Implement the scrambler defined by (13.10). Set the initial shift register state to all 0's and use the input sequence $d_i = 1$ for all i. Check that your scrambler is working by computing an initial segment of the output sequence by hand and comparing it with your scrambler output. Your program should contain the options of generating two scrambled output bits per symbol for 2000 bps transmission or four scrambled output bits per symbol for 4000 bps transmission.

3. Implement the differential encoder shown in Figure 13.4 and Table 13.1. Your program should contain options for both the 1200 and 2400 bps V.22bis modes, which become 2000 and 4000 bps modes for our modem. At 2000 bps the output of your function should be $(Y1_n, Y2_n, 0, 1)$ and at 4000 bps it should be $(Y1_n, Y2_n, Q3_n, Q4_n)$.

4. Map the 4-bit differential encoder output to a constellation point by looking up the values for a_n and b_n in a table corresponding to Figure 13.3.

5. Now implement the modulator using passband shaping filters as shown in Figure 13.6. You should generate 16 output samples per symbol resulting in a 16000 Hz output sampling rate. Use the program C:\digfil\sqrtraco.exe along with (13.18) to generate the impulse response samples of your inphase and quadrature passband shaping filters. The filter impulse responses should be limited to the time interval $[-3T, 3T]$ where T is the symbol period. Since you are making a calling modem transmitter, use a carrier frequency of 2000 Hz. Notice that the symbol rotation shown at the input to the modulator of Figure 13.6 is not required in this case since

$$e^{j\omega_c nT} = e^{j2\pi n f_c/f_s} = e^{j2\pi n 2000/1000} = e^{j4\pi n} = 1 \tag{13.28}$$

For testing purposes, also generate the baseband shaping filter coefficients for a raised cosine response with the α and duration of the passband filters. You can use the program `C:\digfil\rascos.exe` to generate the impulse response.

Use the interpolation filter bank method presented in Section 11.3 to generate the $L = 16$ output samples from the inphase and quadrature passband shaping filters each symbol period. Combine the filter outputs to form 16 samples of the modulated signal $s(t)$ for the next symbol period and write the samples to a "mailbox". The samples put in the mailbox should be integers suitable for sending to the left channel of the DAC and they should be scaled so that the line output is limited to ± 0.5 v. The mailbox should be a 32-word array. One half of the array should contain the 16 output samples for the symbol currently being transmitted and the other half should contain the 16 new samples for the next symbol period. The halves of the array should be swapped after each symbol period. The mailbox can be implemented as a circular buffer.

6. Write an interrupt service routine to load output samples into the McBSP1 DXR. The interrupt should be triggered by the XRDY flag of McBSP1. The routine should contain a pointer to the next output sample in the mailbox. It should write the sample to the DXR and then increment the pointer. If the pointer is initialized to the start of the mailbox array, during the first symbol period it will be incremented through the first half of the array. During the next symbol it will be incremented through the second half of the array. Before the start of the next symbol, the pointer must be reset to the beginning of the array. More generally, the input pointer should be mod(output pointer value $+\,16, 32$) at the end of each symbol.

The interrupt service routine should also maintain a count of the number of interrupts that have occurred modulo 16 to provide the 1000 Hz symbol timing.

13.5.2 Testing Your Transmitter

Test your transmitter to verify that it is operating properly by performing the following steps:

1. First select the 2000 bps option. Clear the scrambler shift register and make its input $d_i = 1$ for all i. Set the coefficients of the inphase passband shaping filter equal to those of the baseband raised cosine shaping filter designed for this experiment. Make the quadrature component zero by setting the coefficients of the quadrature passband shaping filter to zero. Observe the eye diagram on the oscilloscope for the resulting signal. Explain the number of levels you observe in the eye diagram.

   Generate a sync signal for the eye diagram by sending a 1000 Hz clock to the right channel of the codec. You can do this by sending a constant like 16000 to the right channel for the first eight samples in a baud and -16000 for the last eight samples.

2. Next make the inphase passband shaping filter coefficients zero and make the quadrature passband shaping filter equal to the baseband raised cosine filter. Observe the resulting eye pattern on the oscilloscope.

3. Select the 4000 bps option and repeat the previous two steps.

4. Display the baseband signal constellation by sending the inphase constellation point a_n, scaled appropriately, to the left codec channel for each of the 16 samples in a symbol and send the quadrature component b_n to the right channel for each of the 16 samples in a symbol. Attach the left and right outputs to two oscilloscope channels and set it to the x-y display option. First set your transmitter to the 2000 bps mode and you should observe the 4-point constellation. Then set your transmitter to the 4000 bps option and you should observe the 16-point constellation.

5. Once you are convinced your basic program flow is correct, put in the correct inphase and quadrature passband shaping filter coefficients for square-root of raised cosine shaping. Observe the nature of the transmitted signal on the oscilloscope for the 2000 and 4000 bps options. If a spectrum analyzer is available, measure the spectrum of the transmitted signal and sketch the results. You could also use the spectrum analyzer you made for Chapter 4 if you did that chapter. Check that it has square-root of raised cosine shaping about the carrier frequency.

6. Select the 2000 bps option and make the differential encoder input $(Q1_n, Q2_n) = (1, 1)$ for all n. The ITU-T standard calls this the *unscrambled binary 1's* sequence and it is used by the answer modem in one segment of the handshaking sequence between the modems. The resulting sequence of transmitted constellation points continuously rotates by $-90°$. You should observe a periodic signal on the oscilloscope. Measure its fundamental frequency. Also measure the spectrum of the transmitted signal with the spectrum analyzer if one is available. Explain your results mathematically.

7. Repeat the previous step with the input to the differential encoder set to all 0's. This causes continuous $+90°$ phase shifts.

8. Another pattern called the S1 sequence is also used during handshaking. This pattern uses the 2000 bps constellation and alternates between two points separated by 90°. The two points are generated by making the differential encoder dibit inputs $(Q1_n, Q2_n)$ alternate between $(0, 0)$ which causes a $+90°$ phase change and $(1, 1)$ which causes a $-90°$ phase change. The exact pair of points used depends on the initial value $(Y1_0, Y2_0)$ of the absolute quadrant and is not specified in the V.22bis standard. For example, the S1 sequence could alternate between the points (0001) and (0101) shown in Figure 13.3. Make your transmitter send the S1 sequence continuously and observe the signal on the oscilloscope. Measure the spectrum with the spectrum analyzer if available. Determine the spectrum theoretically and compare your measured and theoretical results.

13.5.3   Generating a Startup Sequence

Before two modems begin transmitting customer data to each other, they must go through a startup sequence to agree on the transmission speed, to adjust their automatic gain controls (AGC), to train their symbol clock and carrier tracking loops, and to train their adaptive equalizers and echo cancellers. The startup sequence is also called the *handshake sequence*.

Now you will add a startup sequence to your transmitter. This sequence is similar to the one used for V.22bis modems. This startup sequence will be used when you make a receiver in the following chapters. Program your transmitter to generate the following three segment startup sequence:

1. First, send the S1 sequence described in item 8 of Section 13.5.2 for 100 ms.

2. Second, send scrambled binary 1 using the 2000 bps constellation for 700 ms. That is, clear the scrambler shift register and make its input identically 1.

3. Third, send scrambled binary 1 using the 4000 bps constellation for 200 ms.

After the startup sequence, continue to send scrambled binary 1 using the 4000 bps constellation.

13.6   Additional References

The book by Bingham [II.D.3] presents a good survey of the theory and practice of telephone line digital data modems, most of which use QAM modulation. More comprehensive presentations of the theory can be found in Gitlin, Hayes, and Weinstein [II.D.11, Chapter 5] and Lee and Messerschmitt [II.D.26, Sections 6.4 and 6.5]. These presentations include transmitters, receivers, constellations, and error probabilities for various constellations corrupted by additive Gaussian noise. For complete details of the ITU-T V series modem recommendations, see the CCITT Blue Book [II.D.4] for older printed recommendations. The name CCITT was changed to ITU-T and recent recommendations can be found at the ITU-T web site www.itu.ch. In particular, you may be interested in Recommendation V.32 [II.D.20], V.34 [II.D.21], V.90 [II.D.22], and V.92 [II.D.23]. Advanced modulation and coding techniques invented for V.34 modems including 4D constellations, shell mapping, nonlinear precoding, and trellis coding are discussed in Tretter [II.D.39].

Chapter 14

QAM Receiver I – General Description of Complete Receiver Block Diagram and Details of the Symbol Clock Recovery and Other Front-End Subsystems

In this chapter and the next you will make a QAM receiver. You should not do these experiments until you have completed Chapter 13 and have made a working QAM transmitter. First, the basic subsystems required in the receiver are briefly described. Then the receiver front-end components, in particular a symbol clock recovery method, are described in detail. These front-end subsystems are what you will implement in the experiments for this chapter.

14.1 Overview of a QAM Receiver

The block diagram of a QAM receiver is shown in Figure 14.1. We will call the top half of the figure the receiver front-end. The input signal $r_0(t)$ represents the signal at the receiver input which is the transmitted QAM signal distorted by the non-ideal frequency response of the channel and additive noise. This signal is passed through the Receive Filter which is a bandpass filter that passes the QAM signal and eliminates out-of-band noise. The Receive Filter can also be used in combination with the transmitter filters to perform the spectral shaping required for no intersymbol interference with a perfect channel. In transmission through a communications channel like a voiceband telephone circuit, the signal is often significantly attenuated. Therefore, the output of the Receive Filter is scaled by the automatic gain control (AGC) to increase its amplitude to a level that fully loads the ADC. This scaled signal $r(t)$ is sampled at a rate $f_0 = 1/T_0 = n_0/T$ that is n_0 times the symbol rate $f_s = 1/T$ and is at least twice the highest frequency component in the QAM signal to satisfy the the sampling theorem. The ADC output samples $r(nT_0)$ are used to adjust the AGC gain. These samples are also used by the Carrier Detect block to determine when a QAM signal is actually present at the receiver input and not just channel noise. Many of the receiver functions are not started until an input signal is detected. The proper sampling times for the ADC are determined from the ADC output samples by the Symbol Clock Recovery subsystem. The frequency and phase of the symbol clock must be tracked by this subsystem. The ADC block could be a converter with hardware capability for shifting the sampling phase, or an ADC with a fixed sampling phase combined with a variable phase interpolator implemented by the DSP as discussed in Chapter 12. Finally, the receiver front-end forms the pre-envelope $r_+(nT_0)$ of the received signal. The subsystem that forms the pre-envelope is often called a *phase splitter*.

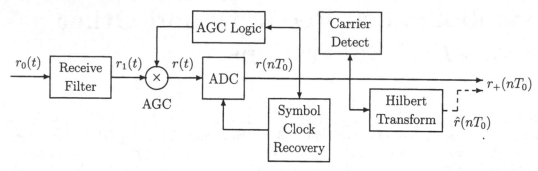

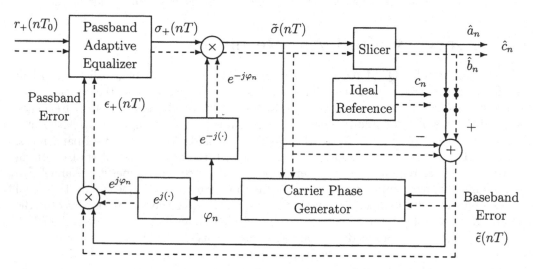

Figure 14.1: Block Diagram of a QAM Receiver

A real channel does not have a flat amplitude response and constant envelope delay and this causes intersymbol interference in the received signal. The Passband Adaptive Equalizer compensates for the channel response to minimize intersymbol interference. An adaptive filter is used because the exact frequency response of the channel is rarely known. For example, in the switched telephone network, a different channel can be selected each time a new call is made. The equalizer is an adaptive FIR filter that operates on samples spaced by T/n_1 and generates output samples spaced by the symbol period T. The constant

n_0 which determines the input sampling period $T_0 = T/n_0$ must be divisible by n_1. This is called a *fractionally spaced* equalizer. In our experiments we will use $n_1 = 2$. The equalizer input sequence $r_+(nT/n_1)$ is obtained by skip sampling the sequence $r_+(nT_0) = r_+(nT/n_0)$. Actually, this sampling rate reduction can be performed by having the FIR Hilbert transform filter operate on the T_0 spaced samples $r(nT_0)$ and computing its output only at the desired times nT/n_1. The equalizer is discussed in detail in the next chapter.

The equalizer output $\sigma_+(nT)$ is multiplied by the locally generated carrier reference $e^{-j\varphi_n}$ to demodulate it to the baseband signal $\tilde{\sigma}(nT)$. If all the system components were perfect, the baseband signal samples would be ideal constellation points. In practice, they deviate from the ideal points due to noise and intersymbol interference. The Slicer quantizes the baseband samples to the nearest ideal constellation points which are used as the receiver's estimates of the transmitted symbols. When the adaptive equalizer is working well and the carrier reference is good, the quantized output symbols will be the same as the transmitted symbols with high probability. It will be shown in the next chapter how the local carrier reference can be synchronized with the received signal's carrier by using the transmitted constellation sequence c_n and the baseband error $c_n - \tilde{\sigma}(nT)$ between the transmitted sequence and baseband equalizer output. The Carrier Phase Generator block performs this function. During an initial training period, the Ideal Reference generator is used to create a local replica of the known transmitted training sequence. After that, the outputs of the slicer are used as good estimates of the transmitted symbols. This is called *decision directed* operation. The equalizer coefficients are adjusted by a least mean-square error algorithm which uses the passband error. A significant portion of the initial training sequence is used to adjust the adaptive equalizer. The carrier recovery loop typically converges much faster than the equalizer.

The receiver described here is one of several approaches. Some modem designers prefer to remove the Hilbert Transform block and force an adaptive equalizer that operates on the real samples $r(nT_0)$ with two sets of real coefficients to perform both the equalization and phase splitting simultaneously [II.D.30][II.D.27].

14.2 Details About the Receiver Front-End Subsystems

In this section, more details about how to implement most of the receiver front-end subsystems are presented. With this information and some ingenuity you should be able to implement these blocks with the TMS320C6x.

14.2.1 Automatic Gain Control

The purpose of the automatic gain control (AGC) is to scale the analog input voltage to a level that almost fully loads the ADC but avoids clipping. Various combinations of strategies can be used. For example, the peak magnitude of the digitized samples can be monitored for a fixed time period and the analog gain can be adjusted to load the ADC converter to a desired level with some margin against clipping. This peak detection method can be combined with a scaling function that adjusts the average power of the sequence of samples to a desired level.

14.2.2   The Carrier Detect Subsystem

The purpose of the Carrier Detect subsystem is to determine when a QAM modem signal is present at the receiver input and not just channel noise. In the modem jargon, people say that a carrier has been detected when it is decided that a modem signal is being received. When no QAM signal is present, the receiver is kept in a default state waiting for a known training sequence to begin. When no carrier is detected, some of the things the receiver does are: (1) set the AGC to a higher gain, (2) set the EIA RS232 output levels for clear-to-send (CTS) and carrier detect (CD) to the off state (-12 v), (3) clamp the output data to steady marks (logical 1 or -12 v), (4) keep the equalizer taps cleared to zero and set the equalizer adaptation speed control to a fast value, (5) keep the frequency offset variable cleared in the carrier tracking loop, and (6) put the symbol clock tracking loop in a fast mode.

One approach to carrier detection is to form a running estimate of the received signal power. This power estimate can be formed by passing the squared ADC output samples through a first-order recursive lowpass filter with the transfer function

$$H(z) = \frac{1-c}{1-c\,z^{-1}} \tag{14.1}$$

The constant c is a number slightly less than 1. The closer c is to 1, the more narrow band the lowpass filter is, but the slower it is to reach steady state. The numerator $1-c$ was chosen to make the gain 1 at zero frequency. The resulting equation for the power estimate is

$$p(n) = (1-c)\,r^2(nT_0) + c\,p(n-1) \tag{14.2}$$

This is sometimes called *exponential averaging*. When the power estimate exceeds a predetermined threshold for a period of time, a received modem signal is declared to be present. Once a carrier is detected, the threshold should be reduced by 5 dB according to Recommendation V.22.

Another function of the Carrier Detect box is to detect when the received modem signal stops. This is called loss of carrier. Loss of carrier is declared when $p(n)$ falls below the reduced threshold for a period of time. The threshold hysteresis is used to avoid false detection of carrier loss caused by the random fluctuations of $p(n)$. According to Recommendation V.22bis, the carrier detect (CD) RS-232C connector signal should be turned off 40 to 65 ms after the power level of the received input signal falls below the lower threshold. It should be turned off in 10 to 24 ms for the V.22 modem.

14.2.3   Symbol Clock Recovery

At the receiver, the transmitter's symbol clock frequency is known quite accurately, but not perfectly. The clock phase is completely unknown and can be modeled as a random variable uniformly distributed over one symbol period. The fractionally spaced adaptive equalizer which will be discussed in the next chapter can automatically correct for the unknown clock phase. However, any error in the clock frequency will cause the equalizer timing reference to drift towards one end of its delay line and fall off that end at which point the receiver crashes. Therefore, the clock frequency must be tracked very closely.

A symbol clock recovery scheme described by Godard [II.D.13] is used in some commercial wireline modems. An idealized block diagram for this scheme is shown in Figure 14.2. The ADC samples the analog input signal $r(t)$ at the frequency $f_0 = 1/T_0 = n_0 f_s$ where $f_s = 1/T$ is the symbol rate and n_0 is chosen so the sampling frequency satisfies the Nyquist criterion. Thus, the sampling instants are $nT_0 + \tau$ where τ represents the clock phase. This phase varies with time because of clock frequency offsets between the transmitter and receiver and adjustments made by the receiver's tracking algorithm. The goal of the tracking loop is to adjust the sampling frequency so that it is n_0 times the true symbol frequency and then drive τ to zero.

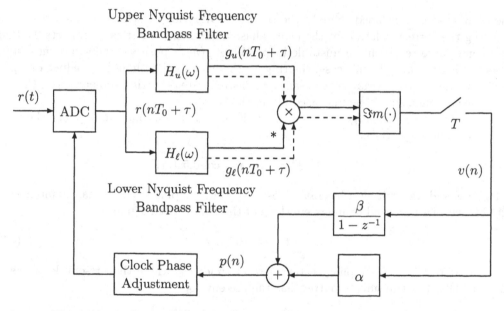

Figure 14.2: General Block Diagram of Symbol Clock Tracking Loop

The samples are applied to two bandpass filters operating at the sampling rate of $f_0 = n_0 f_s$. One filter is tuned to the upper Nyquist frequency $\omega_u = \omega_c + 0.5 f_s$ and the other to the lower Nyquist frequency $\omega_\ell = \omega_c - 0.5 f_s$. Let the responses of these filters over the Nyquist band $|\omega| < 0.5 f_0$ be

$$H_u(\omega) = \begin{cases} 2 & \text{for } |\omega - \omega_u| < B/2 \\ 0 & \text{elsewhere} \end{cases} \tag{14.3}$$

and

$$H_\ell(\omega) = \begin{cases} 2 & \text{for } |\omega - \omega_\ell| < B/2 \\ 0 & \text{elsewhere} \end{cases} \tag{14.4}$$

The bandwidth $B/2$ should be reasonably small, for example, 100 Hz for a $f_s = 2400$ baud modem. Notice that these filters only pass positive frequency components and have complex impulse responses.

The complex output of the upper Nyquist frequency bandpass filter is multiplied by the complex conjugate of the output of the lower Nyquist frequency bandpass filter. The imaginary part of this product is sampled at the symbol rate $f_s = 1/T$. It will be shown in an example below that the resulting sequence $v(n)$ gives an estimate of the timing phase error. Each symbol period, the clock phase is advanced or retarded by an amount α times the phase error plus an amount β times the accumulated phase error. That is, the phase advancement increment is

$$p(n) = \alpha v(n) + \beta \gamma(n) \tag{14.5}$$

where

$$\gamma(n) = v(n) + \gamma(n-1) \tag{14.6}$$

The accumulator is included to make the timing recovery loop track frequency offsets.

Using the correct polarity for the clock phase adjustment is critically important. Using the wrong phase results in an unstable loop. When $p(n)$ is positive, the sampling instants are occurring too late. In this case, the time to the next sample should be reduced by $p(n)$. Similarly, when $p(n)$ is negative, the sampling instants are occurring too early and the time to the next sample should be increased by $|p(n)| = -p(n)$.

To see how this system generates symbol clock control information, suppose the transmitted symbol sequence is

$$c_n = (-1)^n = \cos n\pi = \cos 0.5\omega_s nT \tag{14.7}$$

If the baseband transmit filter has raised cosine spectral shaping so that it has no intersymbol interference, the baseband complex envelope of the transmitted signal is

$$\tilde{s}(t) = \cos 0.5\,\omega_s t \tag{14.8}$$

Notice that the correct sampling instants are at times nT where the pre-envelope has the values $(-1)^n$. The transmitter output has the pre-envelope

$$s_+(t) = \cos(0.5\,\omega_s t)e^{j\omega_c t} = 0.5e^{j(\omega_c + 0.5\omega_s)t} + 0.5e^{j(\omega_c - 0.5\omega_s)t} \tag{14.9}$$

Thus, the transmitted signal is the sum of sinusoids at the upper and lower Nyquist frequencies. The output of the upper Nyquist frequency bandpass filter is

$$g_u(nT_0 + \tau) = e^{j(\omega_c + 0.5\omega_s)(nT_0 + \tau)} \tag{14.10}$$

and the output of the lower Nyquist frequency bandpass filter is

$$g_\ell(nT_0 + \tau) = e^{j(\omega_c - 0.5\omega_s)(nT_0 + \tau)} \tag{14.11}$$

The multiplier output is

$$q(nT_0 + \tau) = g_u(nT_0 + \tau)\bar{g}_\ell(nT_0 + \tau) = e^{j\omega_s(nT_0 + \tau)} = e^{j\omega_s(nT/n_0 + \tau)} \tag{14.12}$$

Replacing n by $n\,n_0$ to evaluate this signal once per symbol and taking the imaginary part gives

$$v(n) = \Im m\ q(nT + \tau) = \sin \omega_s \tau \qquad (14.13)$$

When $|\omega_s \tau| < \pi$, $v(n)$ has the same polarity as the sampling phase error τ. Also, when $|\omega_s \tau| \ll 1$, $\sin \omega_s \tau$ is closely approximated by the linear function $\omega_s \tau$.

The block diagram for a practical realization of the symbol clock tracking loop is shown in Figure 14.3. The upper Nyquist frequency bandpass filter is approximated by a filter with a single complex pole at $z = \nu e^{j\omega_u T_0}$. Its transfer function is

$$
\begin{aligned}
H_u(z) &= \frac{1}{1 - \nu e^{j\omega_u T_0} z^{-1}} = \frac{1 - \nu e^{-j\omega_u T_0} z^{-1}}{(1 - \nu e^{j\omega_u T_0} z^{-1})(1 - \nu e^{-j\omega_u T_0} z^{-1})} \\
&= \frac{1 - z^{-1} \nu \cos \omega_u T_0}{1 - z^{-1} 2\nu \cos \omega_u T_0 + \nu^2 z^{-2}} + j \frac{z^{-1} \nu \sin \omega_u T_0}{1 - z^{-1} 2\nu \cos \omega_u T_0 + \nu^2 z^{-2}} \qquad (14.14)
\end{aligned}
$$

The amplitude response of this filter has a peak value of $1/(1 - \nu)$ at the upper Nyquist frequency ω_u. The bandwidth of the filter is determined by the parameter ν which should be in the range $[0, 1)$. The closer ν is to 1, the narrower the bandwidth. The method used to compute the complex output from the real input $\rho(n) = r(nT_0 + \tau)$ is suggested by (14.14). The first step is to compute the intermediate real variable

$$\eta(n) = \rho(n) + 2\nu \cos(\omega_u T_0)\eta(n - 1) - \nu^2 \eta(n - 2) \qquad (14.15)$$

This recursion must be computed at the fast input sampling rate $f_0 = n_0 f_s$. The real and imaginary parts of the output are computed as

$$\Re e\{g_u(nT_0 + \tau)\} = \eta(n) - \nu \cos(\omega_u T_0)\eta(n - 1) \qquad (14.16)$$

and

$$\Im m\{g_u(nT_0 + \tau)\} = \nu \sin(\omega_u T_0)\eta(n - 1) \qquad (14.17)$$

The real and imaginary parts only have to be computed at the symbol rate f_s.

The lower Nyquist bandpass filter is implemented in a similar manner by simply replacing u by ℓ in the previous equations.

The imaginary part of the product $g_u(nT_0 + \tau)\bar{g}_\ell(nT_0 + \tau)$ is computed once per symbol as shown in Figure 14.3 to form the timing error signal $v(n)$. This signal has significant variability when random data is transmitted. The variability increases as the number of points in the constellation increases. The philosophy for adjusting the symbol clock sampling phase is similar to the approach used in the phase-locked loops in previous experiments. The variability in $v(n)$ is lowpass filtered by incrementing the clock phase by a small fraction α of $v(n)$ each symbol. In addition, $v(n)$ is accumulated to detect any DC component caused by a clock frequency offset, and a small fraction β of the accumulation is added to the clock phase increment. For good transient response, β should be a factor of 50 to 100 times less than α. The tracking loop becomes more narrow band as α is decreased.

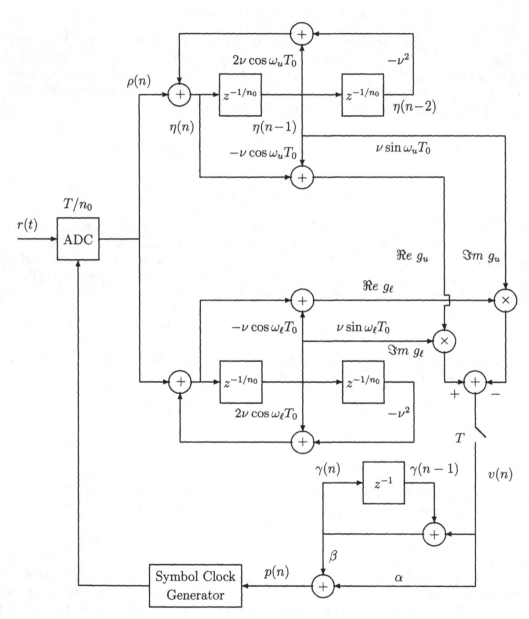

Figure 14.3: Practical Realization of Symbol Clock Tracking Loop

An Additional Suggestion for Rapid Symbol Clock Acquisition

The bandwidth of the clock tracking loop should be very small to make the clock jitter negligible. It also must be small to make the loop stable in light of significant delays in the loop from the time the phase updates are made to the time they propagate through the system components to the phase error measurements. This means the loop transient response will be very slow. If the local symbol clock is far off in phase from the clock in the received signal, the loop will take a long time to slew around to the correct position. The worst case is when the clocks are 180 degrees out of phase. During the initial training sequence for many modems a dotting sequence like the S1 sequence discussed in Chapter 13 is transmitted for a short period of time. The dotting sequence has strong components at the carrier frequency plus or minus half the symbol rate and the output of the Godard clock tone generation system is a strong tone at the symbol rate. This tone can be used to rapidly lock on to the symbol clock. The clock tracking loop updating can be turned off and the symbol rate samples of the complex multiplier output, $q(nT + \tau) = g_u(nT + \tau)\bar{g}_\ell(nT + \tau)$, can be observed for a period of time during the dotting sequence and their phase measured. Suppose the measured phase is θ_q. Then $q(\cdot)$ can be multiplied by $\exp(-j\theta_q)$ to form the complex signal

$$\tilde{q}(nT + \tau) = q(nT + \tau)e^{-j\theta_q} \tag{14.18}$$

which has an angle of zero. Geometrically, this is equivalent to rotating the complex clock tone sample to point in the positive real direction. The imaginary part of $\tilde{q}(nT + \tau)$ will then be zero. Then the loop updating can be turned on with $\Im m\{\tilde{q}(nT + \tau)\}$ used as the signal $v(n)$ that specifies the clock phase update increments. In effect, this fools the loop into thinking it is initially at the correct phase and it does not have to move from this position. The fractionally spaced equalizer will compensate for any fixed phase offset.

Including a Random Walk Filter in the Symbol Clock Tracking Loop

In practice, it has been found that the signal $p(n)$ shown in Figure 14.3 must be additionally filtered to reduce symbol clock jitter, particularly when the signal constellation contains many points. A technique called a *random walk filter* has been found to work well and is shown in Figure 14.4. First, the output of the Godard band edge filter cross-correlator is hard limited to form the signal

$$\check{v}(n) = \text{sign}\, v(n) = \begin{cases} 1 & \text{for}\ v(n) > 0 \\ 0 & \text{for}\ v(n) = 0 \\ -1 & \text{for}\ v(n) < 0 \end{cases} \tag{14.19}$$

The hard limiting provides a simple AGC action that keeps the loop gain constant independent of the input signal level. This signal is then passed through the same kind of second-order loop filter as shown in Figure 14.3 resulting in the signal $\check{p}(n)$.

The output of the second-order loop filter is then applied to the random walk filter. The random walk filter is basically an accumulator that gets reset when its output exceeds a positive or negative threshold. The random walk filter accumulator output is

$$y(n) = x(n) + \check{p}(n) \tag{14.20}$$

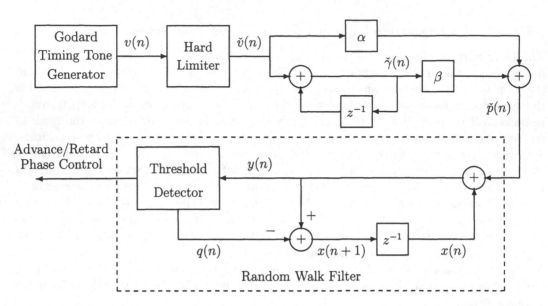

Figure 14.4: Including a Random Walk Filter in the Clock Tracking Loop

where

$$x(n+1) = y(n) - q(n) = x(n) + \check{p}(n) - q(n) \tag{14.21}$$

The signal q(n) is generated by the Threshold Detector and is zero most of the time. When the accumulator output $y(n)$ exceeds the thresholds of L or $-L$, the accumulator value is reset and the sampling phase of the ADC or variable phase interpolator is advanced or retarded. As long as the accumulator output remains between the thresholds, the sampling phase is not changed. This significantly reduces clock jitter. The exact rules describing the Threshold Detector and accumulator are:

$$q(n) = \begin{cases} 0 & \text{for} \quad |y(n)| < L \\ L & \text{for} \quad y(n) \geq L \\ -L & \text{for} \quad y(n) \leq -L \end{cases} \tag{14.22}$$

and

$$x(n+1) = \begin{cases} y(n) & \text{for} \quad |y(n)| < L \\ y(n) - L & \text{for} \quad y(n) \geq L \\ y(n) + L & \text{for} \quad y(n) \leq -L \end{cases} \tag{14.23}$$

When $y(n)$ exceeds the positive threshold L, this indicates that the sampling phase is too late, so the phase must be advanced. When $y(n)$ is algebraically less that the negative threshold, $-L$, the phase must be retarded. The phase is not changed while $y(n)$ remains between the thresholds.

14.3   Experiments for the QAM Receiver Front-End

In this experiment you will build the front-end for a modified V.22bis receiver operating in the answer mode. Use the modified V.22bis calling mode transmitter you made in Chapter 13 as the signal source. Set the transmitter to the 2000 bps data rate and 4-point constellation option. Run the transmitter on an adjacent station and connect its line output to the line input of the receiver station. Remember that the signal transmitted by the calling modem uses a 2000 Hz carrier and a symbol rate of 1000 baud. It uses 75% excess bandwidth square-root of raised cosine spectral shaping so the transmitted signal is theoretically band limited to the interval $(1125, 2875)$ Hz. Perform the following items to build and test your receiver front-end. Be sure to use the highest level of compiler optimization

1. Initialize McBSP1 as usual.

2. The sampling rate for the codec must be at least twice the highest input frequency component or $2 \times 2875 = 5750$ Hz. It also must be a multiple of the symbol rate and a frequency the AIC23 can generate. Use a sampling rate of 8000 Hz which meets all these requirements. Then, the input will be sampled eight times per symbol interval.

3. Forget about implementing Receive Filter and the AGC function. Just make sure the input signal level is adjusted so that no clipping occurs.

4. Write an interrupt service routine for the DSP triggered by the McBSP1 Receiver Ready Flag (RRDY) to read samples from the Data Receive Register (DRR). Count interrupts modulo 8 to determine the symbol timing. You can use this count to control when various receiver functions are performed during a symbol period. You will also need a cumulative interrupt or symbol count to time segments of the hand shaking sequence when you implement the rest of the receiver in the next chapter.

5. Implement the Carrier Detect function using the approach presented in Section 14.2.2. Keep the accumulator register in the symbol clock tracking loop cleared until a carrier is detected. You can write the output of the carrier detector to the DAC converter to check that is functioning properly.

6. Implement the symbol clock tracking system shown in Figure 14.3 and discussed in Section 14.2.3. Choose the bandpass filter parameter ν to achieve a 3 dB bandwidth of about 50 Hz. Use the $f_0 = 8000$ Hz sampling rate.

   Initially, do not close the loop and modify the sampling phase. For testing purposes, generate the imaginary part of the complex multiplier output at the 8 kHz sampling rate. Send the resulting samples to the DAC and observe the result on the oscilloscope when the transmitter is set to send the S1 sequence described in item 8 of Section 13.5.2 continuously. You should observe a 1000 Hz sine wave. Once this function is working, set the transmitter to send scrambled 1's and observe the result on the oscilloscope. You might also set the transmitter to the 4000 bps mode and observe the result. When you are sure the bandpass filters and complex multiplier are working properly, change the program to generate the product once per symbol. Generate the phase error signal $p(n)$

and close the loop by incrementing the time between samples appropriately. You can send $p(n)$ to the DAC to observe its behavior.

7. To complete the receiver front-end, design and implement the phase splitter. Design the Hilbert transform filter so its amplitude response is quite flat over the signal passband. The Hilbert transform FIR filter should operate on samples spaced by $T_0 = 1/f_0 = T/n_0$ where $n_0 = 8$. However, compute its output at the slower rate $2f_s = f_0/4 = 2000$ Hz, that is, compute two equally spaced output samples per symbol. Pass these pre-envelope samples on to the adaptive equalizer which is discussed Chapter 15.

14.4 Additional References

See Godard [II.D.13] for a detailed analysis of the symbol clock tracking method described in this experiment. He calls it the band edge component maximization (BECM) approach. Additional discussions of timing recovery can be found in Bingham [II.D.3, Chapter 7], Gitlin, Hayes, and Weinstein [II.D.11, Chapter 6], and Lee and Messerschmitt [II.D.26, Chapter 15].

 See Section 5.2.2 of Chapter 5 Amplitude Modulation for an introduction to Hilbert transforms and Section 5.3.4 for tools for designing FIR Hilbert transform filters.

Chapter 15

QAM Receiver II – The Passband Adaptive Equalizer and Carrier Recovery System

An important milestone in high speed data transmission over narrow band channels like the voice band telephone channel was the invention and commercialization of the FIR adaptive equalizer by R.W. Lucky at AT&T Bell Laboratories in the early 1960's [II.D.28]. The purpose of the adaptive equalizer is to remove the intersymbol interference caused by the amplitude and phase distortions of the channel. Adaptive filters are used because the frequency response of the channel is not known accurately in many situations. Lucky's original equalizers used the zero forcing algorithm. Other people soon replaced this algorithm by Widrow's [II.D.41] more powerful least-mean-square (LMS) algorithm. Another major influence has been the remarkable advances in VLSI technology. This has led to ever more powerful DSP's which allow complex algorithms to be implemented very inexpensively. For example, modems that include data rates of 300 bps, 1200 bps, and 2400 up to 56000 bps, as well as error correction, data compression capabilities, and FAX modes can be bought for less the $100.

In this chapter, you will complete the QAM receiver by making the adaptive equalizer and carrier recovery system. You will make two kinds of equalizers – the complex cross-coupled passband equalizer and the phase-splitting fractionally spaced equalizer. You will implement the LMS equalizer adjustment algorithm and a technique known as blind equalization.

15.1 The Complex Cross-Coupled Passband Adaptive Equalizer

A type of equalizer that operates on samples of the pre-envelope of the received signal is shown in Figure 15.1. The input to the equalizer is the sequence $r_+(nT/n_1)$ obtained by evaluating the output of the Hilbert transform filter in the receiver front end at the desired times. Thus, the equalizer operates on samples taken at the rate $f_1 = n_1 f_s$ where $f_s = 1/T$ is the symbol rate. The blocks in the figure containing z^{-1/n_1} represent complex signal delays of $T_1 = T/n_1$. Remember that the spectrum of the QAM pre-envelope is confined to the positive frequency interval $[f_c - 0.5(1 + \alpha)f_s \leq f \leq f_c + 0.5(1 + \alpha)f_s]$ where f_c is the carrier frequency and α is the excess bandwidth factor. The width of this interval is $(1 + \alpha)f_s$. The integer n_1 should be chosen to prevent aliasing of the pre-envelope. Since $0 \leq \alpha \leq 1$, n_1 must be greater than or equal to 2 so that no aliasing occurs. In this chapter, we will use $n_1 = 2$ and the resulting structure is commonly called a $T/2$ *spaced equalizer*. An equalizer which operates on samples spaced by less than the symbol period T is called a *fractionally spaced* equalizer. Early equalizers used $n_1 = 1$ which corresponds to T spaced samples. It was soon recognized that fractionally spaced equalizers performed better and, in particular, could act as interpolators and compensate for any fixed symbol clock timing phase offset.

The equalizer output at time nT/n_1 is

$$\sigma_+(nT/n_1) = \sum_{k=0}^{N-1} h_k r_+((n - k)T/n_1) \tag{15.1}$$

The equalizer coefficients $h_0, \ldots, h_{N-1}$ are complex numbers and are sometimes called the *equalizer tap values*. It will soon be shown how to adaptively adjust the tap values to minimize ISI. The Down Sampler selects every n_1th equalizer output sample to generate the symbol spaced sequence $\sigma_+(nT)$. Replacing n by $n n_1$ in (15.1) gives

$$\sigma_+(nT) = \sum_{k=0}^{N-1} h_k r_+ \left(nT - k\frac{T}{n_1} \right) \tag{15.2}$$

In practice, the fractionally spaced equalizer and down sampler are implemented by just evaluating (15.1) once per symbol period. Notice that this sum involves pre-envelope samples spaced by T/n_1. The remainder of the receiver operates on symbol spaced samples.

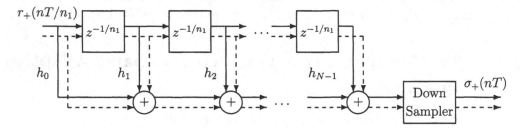

Figure 15.1: The Complex Cross-Coupled Passband Equalizer

15.1.1   The LMS Method for Adjusting the Equalizer Tap Values

The equalizer output samples are demodulated to baseband using the carrier angle φ_n generated by the carrier tracking system. For the time being, it will be assumed that the carrier phase is known exactly. The demodulated samples are

$$\tilde{\sigma}(nT) = \sigma_+(nT) \, e^{-j\varphi_n} \tag{15.3}$$

The goal of the equalizer is to make the baseband output samples as close as possible to a delayed version c_{n-n_d} of the transmitted input symbol sequence. With a perfect channel, this can be accomplished by setting h_{n_d} to 1 and all other taps to 0. The choice of n_d effectively sets the time reference for the receiver. The symbol at tap n_d is considered to be the current received symbol. The time reference n_d is usually selected to be near the center of the equalizer delay line. Then, the equalizer can be thought of as a non-causal system where the taps before n_d operate on future samples and the taps after n_d operate on past samples. With real telephone lines, it has been found experimentally that n_d should be chosen to be closer to $N - 1$ than 0. The optimum placement depends on the channel frequency response.

A mathematically tractable criterion for selecting the equalizer tap values is to choose them to minimize the mean-squared baseband or passband error. The instantaneous baseband error is

$$\tilde{\epsilon}(nT) = c_{n-n_d} - \tilde{\sigma}(nT) \tag{15.4}$$

and the instantaneous passband error is

$$\epsilon_+(nT) = \tilde{\epsilon}(nT)e^{j\varphi_n} = [c_{n-n_d} - \tilde{\sigma}(nT)]\,e^{j\varphi_n} = c_{n-n_d}\,e^{j\varphi_n} - \sigma_+(nT) \tag{15.5}$$

The mean-squared error to be minimized is

$$\Lambda = \mathrm{E}\{|\tilde{\epsilon}(nT)|^2\} = \mathrm{E}\{|\epsilon_+(nT)|^2\} = \mathrm{E}\{|c_{n-n_d} - \tilde{\sigma}(nT)|^2\} \tag{15.6}$$

where E denotes statistical expectation. Similar results are obtained when E is thought of as a sum over n. Let the complex tap values have the representation

$$h_k = h_{R,k} + j\,h_{I,k} \quad \text{where} \quad h_{R,k} = \Re e\{h_k\} \quad \text{and} \quad h_{I,k} = \Im m\{h_k\} \tag{15.7}$$

The optimum coefficients can be found by setting the derivatives of Λ with respect to the tap value components equal to zero. Since the mean-squared error is a quadratic function of the tap components, the error function is convex and a unique solution exists. The derivative with respect of $h_{R,m}$ is

$$\frac{\partial \Lambda}{\partial h_{R,m}} = \mathrm{E}\left\{ \frac{\partial}{\partial h_{R,m}}[\tilde{\epsilon}(nT)\overline{\tilde{\epsilon}(nT)}] \right\} = \mathrm{E}\left\{ \tilde{\epsilon}(nT)\frac{\partial\overline{\tilde{\epsilon}(nT)}}{h_{R,m}} + \overline{\tilde{\epsilon}(nT)}\,\frac{\partial\tilde{\epsilon}(nT)}{\partial h_{R,m}} \right\}$$

$$= 2\,\mathrm{E}\left\{ \Re e\left[\tilde{\epsilon}(nT)\frac{\partial\overline{\tilde{\epsilon}(nT)}}{\partial h_{R,m}} \right] \right\} = -2\,\mathrm{E}\left\{ \Re e\left[\tilde{\epsilon}(nT)e^{j\varphi_n}\overline{r_+(nT - mT/n_1)} \right] \right\} \tag{15.8}$$

In terms of the passband instantaneous error, this result can be written as

$$\frac{\partial \Lambda}{\partial h_{R,m}} = -2\,\mathrm{E}\left\{ \Re e\left[\epsilon_+(nT)\,\overline{r_+(nT - mT/n_1)} \right] \right\} \tag{15.9}$$

Similarly, it can be shown that

$$\begin{aligned}
\frac{\partial \Lambda}{\partial h_{I,m}} &= \mathrm{E}\left\{ \frac{\partial}{\partial h_{I,m}}[\tilde{\epsilon}(nT)\overline{\tilde{\epsilon}(nT)}] \right\} = 2\,\mathrm{E}\left\{ \Re e\left[\tilde{\epsilon}(nT)\frac{\partial\overline{\tilde{\epsilon}(nT)}}{\partial h_{I,m}} \right] \right\} \\
&= -2\,\mathrm{E}\left\{ \Re e\left[\tilde{\epsilon}(nT)(-j)e^{j\varphi_n}\overline{r_+(nT - mT/n_1)} \right] \right\} \\
&= -2\,\mathrm{E}\left\{ \Im m\left[\tilde{\epsilon}(nT)e^{j\varphi_n}\,\overline{r_+(nT - mT/n_1)} \right] \right\}
\end{aligned} \tag{15.10}$$

So, in terms of the passband error signal

$$\frac{\partial \Lambda}{\partial h_{I,m}} = -2\,\mathrm{E}\left\{\Im m\left[\epsilon_+(nT)\,\overline{r_+(nT - mT/n_1)}\right]\right\} \tag{15.11}$$

Let the "derivative" with respect to the complex tap value h_m be defined as

$$\frac{\partial \Lambda}{\partial h_m} \triangleq \frac{\partial \Lambda}{\partial h_{R,m}} + j\,\frac{\partial \Lambda}{\partial h_{I,m}} = -2\,\mathrm{E}\{\epsilon_+(nT)\,\overline{r_+(nT - mT/n_1)}\} \tag{15.12}$$

Thus, the derivative with respect to tap h_m is proportional to the average of the product of the instantaneous passband error $\epsilon_+(nT)$ and the complex conjugate of the passband data sample $r_+(nT - mT/n_1)$ sitting at tap m at time nT.

The optimum equalizer tap values must satisfy the equations

$$\frac{\partial \Lambda}{\partial h_m} = -2\,\mathrm{E}\{\epsilon_+(nT)\,\overline{r_+(nT - mT/n_1)}\} = 0 \quad \text{for} \quad m = 0, \ldots, N-1 \tag{15.13}$$

Substituting (15.5) for the passband error and rearranging yields the set of equations

$$\sum_{k=0}^{N-1} h_k \mathrm{E}\{r_+(nT - kT/n_1)\,\overline{r_+(nT - mT/n_1)}\} = \mathrm{E}\{c_{n-n_d} e^{j\varphi_n}\,\overline{r_+(nT - mT/n_1)}\}$$
$$\text{for} \quad m = 0, \ldots, N-1 \tag{15.14}$$

These are called the *normal equations* in estimation theory and the fact that the error sequence must be uncorrelated with the data samples is called the *orthogonality principle*. Assuming that the channel and baseband symbol sequence information required to compute the expectations is available, this is a set of N linear equations in the N unknown equalizer coefficients. Let the transpose of the N-dimensional coefficient column vector be

$$\mathbf{h}^t = [h_0, h_1, \ldots, h_{N-1}] \tag{15.15}$$

Let the $N \times N$ correlation matrix $\mathbf{R}$ for the received samples in the equalizer delay line have the elements

$$\mathbf{R}_{m,k} = \mathrm{E}\{r_+(nT - kT/n_1)\,\overline{r_+(nT - mT/n_1)}\} \quad \text{for} \quad k, m = 0, \ldots, N-1 \tag{15.16}$$

Also, let the $N \times 1$ column vector $\mathbf{p}$ of cross-correlations between the desired equalizer output and delay line samples have elements

$$\mathbf{p}_{m,1} = \mathrm{E}\{c_{n-n_d} e^{j\varphi_n}\,\overline{r_+(nT - mT/n_1)}\} \quad \text{for} \quad m = 0, \ldots, N-1 \tag{15.17}$$

Then, the linear set of equations can be written as the matrix equation

$$\mathbf{R}\mathbf{h} = \mathbf{p} \tag{15.18}$$

When $\mathbf{R}$ is nonsingular, the solution for the optimum tap values is

$$\mathbf{h} = \mathbf{R}^{-1}\mathbf{p} \tag{15.19}$$

and it can be shown that the resulting minimum mean-squared error is

$$\Lambda_{\min} = E\{|c_{n-n_d}|^2\} - (\bar{\mathbf{p}})^t \mathbf{R}^{-1} \mathbf{p} \tag{15.20}$$

In many real world applications like data transmission over voice band telephone channels in the switched telephone network, the channel frequency response and noise statistics are known only roughly at the transmitter and receiver. Therefore, the correlation matrices cannot be computed and the optimum tap values cannot be calculated by (15.19). A solution to this problem is to use an adaptive tap adjustment algorithm. The most popular algorithm is the *least-mean square* (LMS) or *stochastic gradient* algorithm. The basic philosophy is to iteratively minimize Λ by incrementing the tap values by small amounts in the directions opposite the derivatives given by (15.12). This is a form of gradient search algorithm known as *the method of steepest descent*. The expected value required to compute a derivative cannot be evaluated when the channel is unknown. However, a known training sequence is usually sent at the beginning of transmission, so $\epsilon_+(nT)$ and $r_+(nT - mT/n_1)$ are known to the receiver and a time average of the products of these quantities can be used as an unbiased estimate of the true expected value. These ideas suggest using the following tap adjustment formula:

$$h_m(n+1) = h_m(n) + \mu\,\epsilon_+(nT)\,\overline{r_+(nT - mT/n_1)} \ \ \text{for} \ \ m = 0, \ldots, N-1 \tag{15.21}$$

where $h_m(n)$ is the value of the m-th tap at time n and μ is a small positive constant. This is the LMS tap adjustment algorithm.

The parameter μ controls the speed and smoothness of the convergence of the taps to their optimum values. A large value of μ gives rapid initial convergence but large variations about the theoretically optimum final value because of the small averaging effect. A small value results in slow convergence but small tap variations around the optimum values. Very large values of μ cause the algorithm to become unstable, while very small values can result in arithmetic underflow which causes the adjustments to stop. In practice, the adaptation is often started with a moderately large value of μ to get rapid initial convergence for a period of time and then "gear shifted" to a small value for precise final adjustment.

The block diagram of a section of an adaptive passband equalizer illustrating the LMS algorithm for adjusting one tap is shown in Figure 15.2. The outputs of all the tap multipliers are summed in the box labeled "+" to form the passband output signal $\sigma_+(nT)$. The passband output is demodulated to the baseband signal $\tilde{\sigma}(nT)$ using the angle φ_n generated by the carrier tracking system. The Slicer quantizes its input to the closest ideal constellation point. During initial training, a known sequence c_n is transmitted and a delayed version c_{n-n_d} is generated in the receiver by the Ideal Reference block. The exact baseband error signal $\tilde{\epsilon}(nT)$ can be formed during the initial training period. This error signal is modulated to passband and correlated against the data sample at the tap being adjusted and scaled by μ to form the tap update increment. After the equalizer converges to the point where the baseband output symbols $\tilde{\sigma}(nT)$ are close to the ideal constellation points, the switch can be moved to the slicer output and $\hat{c}_{n-n_d}$ can be used as an accurate estimate of the delayed transmitted symbol sequence. This mode is called *decision directed* equalization. Decision

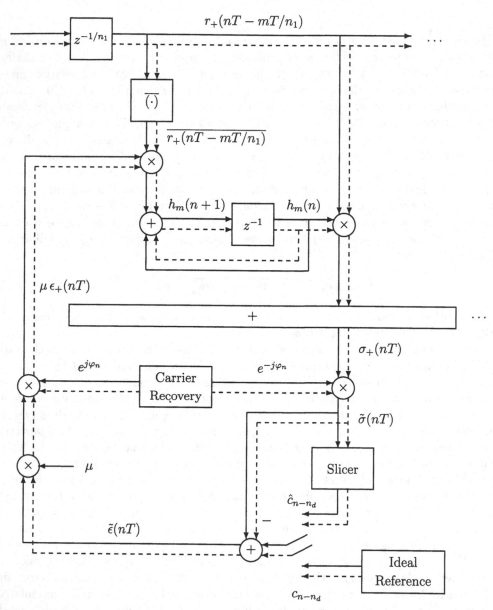

Figure 15.2: Block Diagram Illustrating the LMS Algorithm in a Passband Equalizer

directed equalization is required in practice because the receiver does not know the random symbol sequence transmitted during normal data transmission. If the majority of decisions are correct, the equalizer will converge because of the averaging effect of a small μ. Infrequent errors can not move the equalizer taps very far from their optimum values.

15.1.2 Theoretical Behavior of the LMS Algorithm

The behavior of the LMS algorithm has been extensively analyzed [II.D.11, 18, 26, 32, 41]. It has been shown by invoking an independence approximation that the behavior depends on the eigenvalues of the correlation matrix $\mathbf{R}$. Let these eigenvalues, arranged in order of increasing size, be $\{\lambda_1, \ldots, \lambda_N\}$ so $\lambda_1 = \lambda_{\min}$ is the smallest and $\lambda_N = \lambda_{\max}$ is the largest eigenvalue. Then it has been shown that the mean of a tap value error, $E\{h_m(n) - h_m\}$, is the sum of exponential modes of the form $(1 - \mu\lambda_i)^n$. Therefore, the mean tap values converge to the optimum tap values if

$$|1 - \mu\lambda_i| < 1 \quad \text{for} \quad i = 1, \ldots, N \tag{15.22}$$

or

$$0 < \mu < \frac{2}{\lambda_{\max}} \tag{15.23}$$

It is shown in Lee and Messerschmitt [II.D.26] that the value for μ that maximizes the speed of convergence is

$$\mu_{\text{opt}} = \frac{2}{\lambda_{\min} + \lambda_{\max}} \tag{15.24}$$

Then

$$|1 - \mu_{\text{opt}}\lambda_{\min}| = |1 - \mu_{\text{opt}}\lambda_{\max}| = \frac{\dfrac{\lambda_{\max}}{\lambda_{\min}} - 1}{\dfrac{\lambda_{\max}}{\lambda_{\min}} + 1} \tag{15.25}$$

Therefore, the maximum speed of convergence is determined by the ratio of the maximum and minimum eigenvalues. This ratio is called the *eigenvalue spread*. Maximum speed of convergence can be achieved when the eigenvalue spread is 1 which implies that all the eigenvalues are the same.

Convergence of the mean tap values does not imply convergence of the mean-squared equalizer output error. It is shown in Haykin [II.D.18, p. 329] that the mean-squared error converges if and only if

$$0 < \mu < \frac{2}{\displaystyle\sum_{i=1}^{N} \lambda_i} = \frac{2}{\text{trace } \mathbf{R}} \tag{15.26}$$

The denominator of this upper bound is

$$\text{trace } \mathbf{R} = \sum_{k=0}^{N-1} \mathbf{R}_{k,k} = \sum_{k=0}^{N-1} E\{|r_+(nT - kT/n_1)|^2\} \tag{15.27}$$

All of the eigenvalues of $\mathbf{R}$ are real and positive since it is a Hermitian matrix. Therefore, (15.26) is a tighter bound on μ than (15.23).

Physically, the denominator is the sum of the average power of the received samples in the equalizer delay line. When a T spaced equalizer ($n_1 = 1$) is used, these samples form a stationary random sequence and the sum is N times the average power of the received signal at the symbol instant nT. For fractionally spaced equalizers, the sequence is cyclostationary and the average powers vary periodically as a function of k and have period n_1. In the limit as the excess bandwidth factor α approaches zero, the fractionally spaced samples become a stationary sequence. In any case, the denominator increases monotonically and nearly linearly with the equalizer length N. Therefore, the stability constraint on the tap update factor μ becomes tighter as the equalizer becomes longer.

The mean-squared error does not converge to the theoretical minimum value with the LMS algorithm as a result of the noisy gradient estimates. These noisy estimates cause the taps to jitter about their optimum values in steady-state and this increases the mean-squared error. It is shown in Haykin [II.D.18, p. 327] that the mean-squared error converges to

$$\Lambda_{\mathrm{LMS}} = \Lambda_{\min} + \Lambda_{\mathrm{ex}} \tag{15.28}$$

where

$$\Lambda_{\mathrm{ex}} = \Lambda_{\min} \frac{\mu \sum_{i=1}^{N} \lambda_i / (2 - \mu\lambda_i)}{1 - \mu \sum_{i=1}^{N} \lambda_i / (2 - \mu\lambda_i)} \simeq \Lambda_{\min} \frac{\mu \sum_{i=1}^{N} \lambda_i}{2 - \mu \sum_{i=1}^{N} \lambda_i} \quad \text{if } |\mu\lambda_i| \ll 1 \text{ for all } i \tag{15.29}$$

is the *excess mean-squared error*. In practice, μ is usually switched to a very small value after an initial training period to minimize the tap jitter and excess mean-squared error.

15.1.3   Adding Tap Leakage to the LMS Algorithm

Convergence problems can occur with fractionally spaced equalizers because $\mathbf{R}$ is nearly singular. It can be shown that the eigenvalues of $\mathbf{R}$ are proportional to the amplitude spectrum of $r_+(nT/n_1)$ for large N. For $n_1 > 1$ and no additive channel noise, this spectrum is zero over a region around half the sampling rate. When $\mathbf{R}$ is singular, the normal equation (15.18) does not have a unique solution for the tap vector since the sum of any solution and a vector in the null space of $\mathbf{R}$ is another solution and all solutions have the same mean-squared error. In practice, the taps can slowly drift as a result of computational biases while the mean-squared error remains small. When a tap gets too large for its finite-word-length hardware representation and overflows, the system crashes. A solution to this problem is to modify the LMS algorithm to include *tap leakage*. The modified tap adjustment algorithm is

$$h_m(n+1) = (1 - \gamma)h_m(n) + \mu\, e_+(nT)\, \overline{r_+(nT - mT/n_1)} \quad \text{for } m = 0, \ldots, N-1 \tag{15.30}$$

where γ is a small positive constant. Thus, with each iteration, the current tap value is shrunk slightly before adding the estimated gradient increment.

15.2 The Phase-Splitting Fractionally Spaced Equalizer

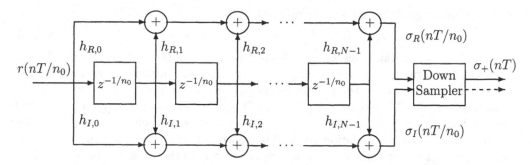

Figure 15.3: The Phase-Splitting Fractionally Spaced Equalizer

The phase-splitting fractionally spaced equalizer shown in Figure 15.3 is commonly used in current telephone line modems. Mueller and Werner [II.D.30] observed that the real and imaginary outputs of the cascade of the Hilbert transform FIR filter and complex cross-coupled equalizer are linear combinations of the input samples to the Hilbert transform filter. Therefore, they suggested combining the two functions into the structure shown in Figure 15.3. The phase-splitting equalizer is computationally more efficient than the complex cross-coupled equalizer because the complex products, which require four real multiplications, are replaced by two real products. Ling and Qureshi [II.D.27] show that the price paid is slower convergence.

When the phase-splitting equalizer is used, the Hilbert transform filter shown in the receiver front end block diagram on page 230 is removed and the signal $r(nT_0)$ is connected directly to the equalizer input. The input sampling rate $f_0 = 1/T_0$ must be at least twice the upper cutoff frequency of the received signal to satisfy the Sampling Theorem. As usual, we will use T to denote the symbol period and $f_s = 1/T$ to denote the symbol rate. It is convenient to let $T_0 = T/n_0$, and then $f_0 = n_0 f_s$. For example, a V.32 modem uses a carrier frequency of $f_c = 1800$ Hz, a symbol rate of $f_s = 2400$ Hz, and square-root of raised cosine shaping with $\alpha = 12\%$ excess bandwidth resulting in an upper cutoff frequency of about 3144 Hz. Therefore, a convenient choice is to let $n_0 = 3$ resulting in a sampling rate of $f_0 = 7200$ Hz. The z^{-1/n_0} blocks in Figure 15.3 represent delays of T_0.

The output of the upper filter in the phase-splitting equalizer at time nT is

$$\sigma_R(nT) = \Re e\{\sigma_+(nT)\} = \sum_{k=0}^{N-1} h_{R,k} r(nT - kT/n_0) \tag{15.31}$$

and the output of the lower filter at time nT is

$$\sigma_I(nT) = \Im m\{\sigma_+(nT)\} = \sum_{k=0}^{N-1} h_{I,k} r(nT - kT/n_0) \tag{15.32}$$

Therefore, the output of the Down Sampler block is

$$\sigma_+(nT) = \sigma_R(nT) + j\,\sigma_I(nT) = \sum_{k=0}^{N-1} h_k r(nT - kT/n_0) \tag{15.33}$$

where the complex equalizer taps are defined to be

$$h_k = h_{R,k} + j\,h_{I,k} \quad \text{for} \quad k = 0, \ldots, N-1 \tag{15.34}$$

Notice that the complex tap value h_k is multiplied by the real data sample $r(nT - kT/n_0)$ which requires only two real products. On the other hand, a complex tap is multiplied by a complex data sample in the complex cross-coupled equalizer and this requires four real products. The equalized passband samples $\sigma_+(nT)$ are processed in exactly the same way as shown in the bottom half of Figure 15.2. Each sample is demodulated to baseband and quantized to the nearest constellation point. The baseband error $\tilde{\epsilon}(nT)$ is formed using the ideal reference sequence or slicer decision and remodulated to the passband error $\epsilon_+(nT)$.

As in the case of the complex cross-coupled equalizer, the tap values can be chosen to minimize the mean-squared baseband or passband error defined by (15.6). Using the same approach as before, it can be shown that the derivatives of Λ with respect to the tap values are

$$\frac{\partial \Lambda}{\partial h_{R,m}} = -2\,\mathrm{E}\left\{\Re e[\epsilon_+(nT)]r(nT - mT/n_0)\right\} \quad \text{for} \quad m = 0, \ldots, N-1 \tag{15.35}$$

and

$$\frac{\partial \Lambda}{\partial h_{I,m}} = -2\,\mathrm{E}\left\{\Im m[\epsilon_+(nT)]r(nT - mT/n_0)\right\} \quad \text{for} \quad m = 0, \ldots, N-1 \tag{15.36}$$

Thus,

$$\frac{\partial \Lambda}{\partial h_m} \triangleq \frac{\partial \Lambda}{\partial h_{R,m}} + j\,\frac{\partial \Lambda}{\partial h_{I,m}} = -2\,\mathrm{E}\{\epsilon_+(nT)\,r(nT - mT/n_1)\} \quad \text{for} \quad m = 0, \ldots, N-1 \tag{15.37}$$

These derivatives suggest using the following LMS algorithm with tap leakage for adapting the taps:

$$h_m(n+1) = (1 - \gamma)h_m(n) + \mu\,\epsilon_+(nT)r(nT - mT/n_0) \tag{15.38}$$

where γ is a small positive leakage constant and μ is a small positive update scale factor. In terms of the individual tap components, this formula is equivalent to

$$h_{R,m}(n+1) = (1 - \gamma)h_{R,m}(n) + \mu\,\Re e\{\epsilon_+(nT)\}\,r(nT - mT/n_0) \tag{15.39}$$

and

$$h_{I,m}(n+1) = (1 - \gamma)h_{I,m}(n) + \mu\,\Im m\{\epsilon_+(nT)\}\,r(nT - mT/n_0) \tag{15.40}$$

Notice that the complex passband error is multiplied by a real data sample to update the complex tap and this requires two real products. In the complex cross-coupled equalizer, the complex error is multiplied by a complex data sample and this requires four real products.

A block diagram for LMS updating of the phase-splitting equalizer taps would be very similar to Figure 15.2. The main difference would be that the complex signal $r_+(nT - mT/n_1)$ would be replaced by the real signal $r(nT - mT/n_0)$ so the complex T/n_1 spaced delay line would become a real T/n_0 spaced delay line. The (complex × complex) products involving $r_+(\cdot)$ for updating the taps and computing the equalizer output would be replaced by (real × complex) products involving $r(\cdot)$. Also, the tap leakage is not shown in Figure 15.2.

15.3   Decision Directed Carrier Tracking

Up to this point, it has been assumed that the carrier phase is known exactly. An approach to estimating and tracking the carrier phase can be motivated by minimizing the mean-squared error Λ with respect to the parameters of the phase sequence generated by the receiver's Carrier Recovery block. Suppose this sequence has the form

$$\varphi_n = \omega_c nT + \theta \tag{15.41}$$

where ω_c is the carrier frequency and θ is a fixed unknown phase offset. Replacing $h_{R,m}$ by θ in (15.8), we find that the derivative of the mean-squared error with respect to the phase offset is

$$\frac{\partial \Lambda}{\partial \theta} = 2\,\mathrm{E}\left\{\Re e\left[\overline{\tilde{\epsilon}(nT)}\,\frac{\partial \tilde{\epsilon}(nT)}{\partial \theta}\right]\right\} \tag{15.42}$$

Remember that the baseband error is

$$\tilde{\epsilon}(nT) = c_{n-n_d} - \tilde{\sigma}(nT) = c_{n-n_d} - \sigma_+(nT)e^{-j(\omega_c nT + \theta)} \tag{15.43}$$

Therefore,

$$\frac{\partial \Lambda}{\partial \theta} = 2\,\mathrm{E}\left\{\Re e\left[\overline{\tilde{\epsilon}(nT)}\,j\,\sigma_+(nT)e^{-j(\omega_c nT + \theta)}\right]\right\} = -2\,\mathrm{E}\left\{\Im m\left[\overline{\tilde{\epsilon}(nT)}\,\tilde{\sigma}(nT)\right]\right\} \tag{15.44}$$

On replacing $\tilde{\epsilon}(nT)$ by $c_{n-n_d} - \tilde{\sigma}(nT)$, the following alternative formula for the derivative is obtained:

$$\frac{\partial \Lambda}{\partial \theta} = -2\,\mathrm{E}\left\{\Im m\left[\overline{c}_{n-n_d}\,\tilde{\sigma}(nT)\right]\right\} \tag{15.45}$$

This derivative has an interesting physical interpretation. Let

$$c_{n-n_d} = R_c e^{j\beta_c} \quad \text{and} \quad \tilde{\sigma}(nT) = R_\sigma e^{j\beta_\sigma} \tag{15.46}$$

be the polar form representations for these two complex numbers. Remember that the equalized baseband output sample $\tilde{\sigma}(nT)$ is supposed to be a close approximation to the ideal symbol c_{n-n_d}. Then

$$\Im m\left[\overline{c}_{n-n_d}\,\tilde{\sigma}(nT)\right] = R_c R_\sigma \sin(\beta_\sigma - \beta_c) \tag{15.47}$$

so

$$\sin(\beta_\sigma - \beta_c) = \frac{\Im m\left[\overline{c}_{n-n_d}\,\tilde{\sigma}(nT)\right]}{R_c R_\sigma} \tag{15.48}$$

This has the same sign as the phase error between the ideal constellation point c_{n-n_d} and the equalized baseband received point $\tilde{\sigma}(nT)$ if the phase error is not too large, and is nearly a linear function of the error for small phase errors.

   This phase error measure can be used in a phase-locked loop to iteratively adjust θ so the baseband equalized points are aligned in angle with the ideal constellation points. Changing θ by some angle has the effect of rotating the baseband equalized points by the negative of the angle. In practice, c_{n-n_d} and $\tilde{\sigma}(nT)$ become close when the equalizer converges so R_σ

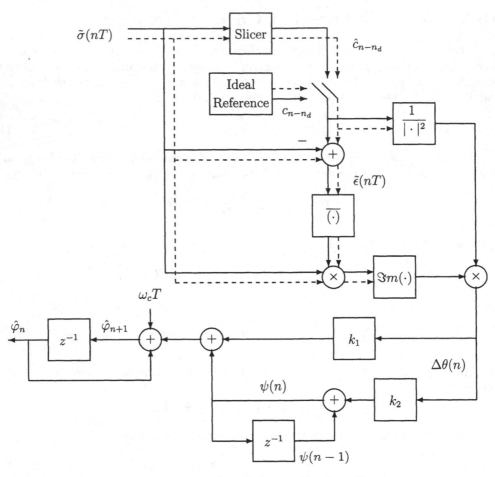

Figure 15.4: Decision Directed Carrier Tracking System

can be replaced by R_c in (15.48). These observations suggest adjusting θ according to the formula

$$\theta(n+1) = \theta(n) + k_1 \frac{\Im m\{\overline{\tilde{\epsilon}(nT)}\,\tilde{\sigma}(nT)\}}{|c_{n-n_d}|^2} \tag{15.49}$$

where k_1 is a small positive constant. A practical realization for a second-order carrier tracking loop based on this equation and including carrier frequency offset tracking is shown in Figure 15.4. First, an approximation to the phase error is computed from the baseband equalizer output sample $\tilde{\sigma}(nT)$ by the formula

$$\Delta\theta(n) = \frac{\Im m\{\overline{\tilde{\epsilon}(nT)}\,\tilde{\sigma}(nT)\}}{|c_{n-n_d}|^2} \tag{15.50}$$

During initial startup, a known symbol sequence is often transmitted and the Ideal Reference generator in the receiver replicates these symbols. After the equalizer and carrier tracking loop have converged, the outputs of the Slicer are substituted for the known sequence and the system operates in the decision directed mode. The phase estimate generated by the lower part of the block diagram is

$$\hat{\varphi}_{n+1} = \hat{\varphi}_n + \omega_c T + k_1 \Delta\theta(n) + \psi(n) \tag{15.51}$$

where

$$\psi(n) = \psi(n-1) + k_2 \Delta\theta(n) \tag{15.52}$$

Notice that $\omega_c T$ is the nominal change in the carrier phase angle between symbols. When $\Delta\theta(n)$ is zero for all n and the z^{-1} delay elements are initially cleared, the phase generated is

$$\hat{\varphi}_n = \omega_c n T \tag{15.53}$$

The philosophy behind the carrier tracking loop is to increment the phase angle predicted for the next symbol instant, $\hat{\varphi}_n + \omega_c T$, by a small fraction, k_1, of the current phase error estimate $\Delta\theta(n)$. In addition, a fraction, k_2, of the phase error is accumulated to measure any bias caused by a frequency offset, and added to the phase increment. The system is a second-order phase-locked loop similar in behavior to the ones discussed in previous chapters. It will track a constant phase and frequency offset with zero final error. The ratio k_1/k_2 should be in the order of 100 for good transient response.

15.4   Blind Equalization

In most cases, the adaptive equalizer in a QAM receiver is initially trained during handshaking with a known ideal reference sequence. There are times when ideal reference training is not possible or desirable. For example, if a tributary modem in a multi-drop network goes off line because of a power failure or for repairs, it would be desirable to bring the modem back online without having to retrain all the tributary modems on the network. With a simple constellation like the four-phase V.22 constellation, it is sometimes possible to achieve equalization using the decision directed mode. However, with 16 or more constellation points, starting LMS adaptation in the decision directed mode almost always fails with typical channels. A variety of algorithms called *blind equalization* have been discovered that use only very general knowledge of the transmitted constellation and not the exact transmitted sequence. These algorithms converge much slower than the LMS algorithm with ideal reference training and the final output constellation remains fuzzy, but the underlying ideal constellation becomes clearly apparent. After the constellation emerges with blind equalization, the equalizer can be switched to LMS decision directed training for fine equalization.

Blind equalization was first disclosed by Sato [II.D.33] in 1975 for the special case of one-dimensional multilevel PAM. Godard [II.D.14] in 1980 was the first to disclose and thoroughly analyze a class of blind equalization algorithms for QAM. Treichler and Agee independently developed a similar approach in the classified world and their work was published in the open literature in 1983 [II.D.37]. This method has become known as the *constant modulus algorithm* (CMA). Another approach is presented by Benveniste in [II.D.2] and is called a *reduced constellation algorithm* (RCA). It is a generalization of Sato's method to two dimensions. See the last section in this chapter for additional references.

Godard's 2nd order blind equalization algorithm is the one most commonly used in commercial modems. Jablon [II.D.24] experimentally compared this 2nd order CMA algorithm with the RCA algorithm and concluded that the CMA algorithm converges faster when carrier phase and symbol timing are also being estimated. Only Godard's algorithm will be presented in this chapter.

Godard's class of CMA algorithms operates directly on the passband equalizer output $\sigma_+(nT)$ shown in Fig.'s 14.1, 15.1, and 15.3. The equalizer taps are adjusted to minimize the *dispersion of order p* which is defined as

$$\mathcal{D}^p = \mathrm{E}\left\{(R_p - |\sigma_+(nT)|^p)^2\right\} \tag{15.54}$$

where p is a positive integer and R_p is a positive constant for which a formula will be given soon. Since the demodulated equalizer output is $\tilde{\sigma}(nT) = \sigma_+(nT)\,e^{-j\phi_n}$, the dispersion can also be expressed as

$$\mathcal{D}^p = \mathrm{E}\left\{(R_p - |\tilde{\sigma}(nT)|^p)^2\right\} \tag{15.55}$$

Notice that the dispersion is independent of the locally generated carrier phase. Therefore, the constellation can emerge rotated by any angle as the algorithm converges. The job of the carrier recovery loop is to rotate it to the correct angle. The carrier recovery loop is effectively a one-tap complex equalizer with the tap constrained to have magnitude 1. An equalizer with a small number of taps can adapt faster than one with many taps, so the carrier recovery loop can converge much faster than the equalizer. The fact that the CMA algorithm does not have to rotate the constellation to the correct angle may account for the better performance Jablon observed for the CMA algorithm over the RCA algorithm.

15.4.1 Blind Equalization with the Complex Cross-Coupled Equalizer

The equalizer taps can be adapted iteratively by taking small steps in the direction opposite to the gradient of $\mathcal{D}^p$ with respect to the taps. With some work, it can be shown that in the case of the complex cross-coupled passband equalizer

$$\frac{\partial \mathcal{D}^p}{\partial h_m} \triangleq \frac{\partial \mathcal{D}^p}{\partial h_{R,m}} + j\,\frac{\partial \mathcal{D}^p}{\partial h_{I,m}}$$

$$= -2p\,\mathrm{E}\left\{(R_p - |\sigma_+(nT)|^p)\,|\sigma_+(nT)|^{p-2}\sigma_+(nT)\,\overline{r(nT - mT/n_1)}\right\} \tag{15.56}$$

Godard shows that the value of R_p required to give the correct constellation size is

$$R_p = \frac{\mathrm{E}\left\{|c_n|^{2p}\right\}}{\mathrm{E}\left\{|c_n|^p\right\}} \tag{15.57}$$

Remember that c_n is the baseband symbol sequence randomly selected from the ideal constellation.

The statistical expectation in (15.56) is not known in practice but can be approximated by time averaging the product inside the expectation operator. This leads to the following stochastic gradient algorithm for adapting the equalizer taps:

$$h_m(n+1) = h_m(n) + \mu \left(R_p - |\sigma_+(nT)|^p\right) |\sigma_+(nT)|^{p-2} \sigma_+(nT) \overline{r(nT - mT/n_1)}$$
$$\text{for } m = 0, \ldots, N-1 \tag{15.58}$$

where μ is a small positive constant as in the LMS algorithm.

Even after convergence, the term $R_p - |\sigma_+(nT)|^p$ in the estimated gradient will fluctuate significantly and the blind equalized constellation will be somewhat fuzzy. The constant μ must be chosen significantly smaller for the CMA algorithm than for the LMS algorithm. After the equalizer converges with the blind equalization algorithm, a switch to decision directed LMS training can be made to achieve precise equalization.

The most commonly implemented algorithm is the $p = 2$ case. Then, the tap update formula reduces to

$$h_m(n+1) = h_m(n) + \mu \left(R_p - |\sigma_+(nT)|^2\right) \sigma_+(nT) \overline{r(nT - mT/n_1)}$$
$$\text{for } m = 0, \ldots, N-1 \tag{15.59}$$

and the required value of R_2 is

$$R_2 = \frac{\mathrm{E}\left\{|c_n|^4\right\}}{\mathrm{E}\left\{|c_n|^2\right\}} \tag{15.60}$$

The computational complexity for this adaptation algorithm is essentially the same as for the LMS algorithm. In both cases, an error signal is multiplied by the complex conjugate of the received signal sample at the tap being adjusted. In this case the error signal is

$$\epsilon(n) = \left(R_p - |\sigma_+(nT)|^2\right) \sigma_+(nT) \tag{15.61}$$

Godard derives some convergence properties for the CMA algorithms. Let $g(0)$ be the channel impulse response sample with the largest magnitude. Then for the $p = 2$ case, a sufficient but not necessary condition for convergence is that all the equalizer taps be initially set to zero except for the reference tap at some position L near the center of the delay line. This tap must satisfy the inequality

$$|h_L|^2 > \frac{\mathrm{E}\left\{|c_n|^4\right\}}{2|g(0)|^2 \left(\mathrm{E}\left\{|c_n|^2\right\}\right)^2} \tag{15.62}$$

15.4.2 Blind Equalization with the Phase-Splitting Equalizer

The CMA algorithm for the phase-splitting equalizer is almost identical to the one for the complex equalizer. It can be shown that the gradient of the dispersion with respect to tap m for the phase-splitting equalizer is

$$\frac{\partial \mathcal{D}^p}{\partial h_m} \triangleq \frac{\partial \mathcal{D}^p}{\partial h_{R,m}} + j \frac{\partial \mathcal{D}^p}{\partial h_{I,m}}$$
$$= -2p\, \mathrm{E}\left\{\left(R_p - |\sigma_+(nT)|^p\right) |\sigma_+(nT)|^{p-2} \sigma_+(nT)\, r(nT - mT/n_0)\right\} \tag{15.63}$$

This suggests the following stochastic gradient tap adjustment formula:

$$h_m(n+1) = h_m(n) + \mu \left(R_p - |\sigma_+(nT)|^p\right) |\sigma_+(nT)|^{p-2}\sigma_+(nT)\, r(nT - mT/n_0)$$
$$\text{for } m = 0, \ldots, N - 1 \tag{15.64}$$

The value required for R_p is again specified by (15.57).

For $p = 2$, the tap update formula reduces to

$$h_m(n+1) = h_m(n) + \mu \left(R_p - |\sigma_+(nT)|^2\right) \sigma_+(nT)\, r(nT - mT/n_0)$$
$$\text{for } m = 0, \ldots, N - 1 \tag{15.65}$$

At the start of blind equalization, the real taps should all be set to zero except for the reference tap at position L. The imaginary taps should be set to approximate a Hilbert transform with the reference tap at position L. The scaling should be adjusted to satisfy the convergence condition for the complex equalizer.

15.5   Complex Cross-Coupled Equalizer and Carrier Tracking Loop Experiments

Now it is time to complete the QAM receiver you began in Chapter 14 by building an adaptive equalizer and carrier tracking loop. Be sure to use the maximum compiler optimization level. You will be directed through a step-by-step approach to completing the receiver. First, you will make a 4-point and, optionally, a 16-point slicer. Then you will make the carrier tracking loop with the equalizer bypassed. Once the carrier loop is operating, you will build a complex cross-coupled equalizer. Finally, you will make the descrambler and check that all 1's are received when they are transmitted. Optionally, you can implement the phase-splitting passband equalizer after successfully completing the complex equalizer. You can also experiment with blind equalization.

This is a long experiment involving the implementation of a number of subsystems. It would be reasonable to team up with one or more other groups and work jointly on this experiment. You will have to decide how to partition the tasks among the team members, make sure the programs fit together, and manage the progress of subgroups. These are things you will have to do as an engineer on a product design and development team in industry.

Continue to use the modified V.22bis transmitter you made for Chapter 13 as the signal source. Program the transmitter to initially send the S1 alternating sequence described in item 8 of Section 13.5.2 for 1000 symbols (1 second) to allow the symbol clock recovery loop to lock up. Then continuously send scrambled 1's using the 4-point V.22 constellation. Optionally, after sending 4000 (4 seconds) 4-point symbols, switch to sending scrambled 1's using the 16-point V.22bis constellation shown in Figure 13.3.

15.5.1   Implementing the Slicer

The Slicer outputs are required to generate the baseband error sequence used by both the adaptive equalizer and carrier tracking loop in the decision directed mode. Therefore, the first logical step in completing the QAM receiver is to implement the slicers.

Making the 4-Point Slicer

The V.22 constellation consists of the 4 points shown in Figure 13.3 with the coordinates $(3, 1)$, $(-1, 3)$, $(-3, -1)$, and $(1, -3)$. These points lie on a circle and are separated by 90° but are rotated by an odd angle relative to the axes. Make a 4-point slicer to quantize each baseband equalizer output sample to the nearest ideal constellation point. To simplify the quantization operation, rotate the received baseband samples $\tilde{\sigma}(nT) = x_n + j\, y_n$ so that the ideal points lie on 45° lines. You can do this by using the transformation

$$
\begin{aligned}
x'_n + j\, y'_n &= (x_n + j\, y_n)(2 + j) \\
&= (2x - y) + j(x + 2y)
\end{aligned}
\tag{15.66}
$$

This transformation rotates the ideal points to the nearest 45° line and scales them to the four points $\pm 5 \pm j\, 5$. Then the slicing operation amounts to determining which quadrant the transformed point lies in, and this can be determined by simply examining the sign bits of x'_n and y'_n.

In your slicer function, also do the following tasks:

1. Form the unrotated baseband error $\tilde{\epsilon}(nT) = c_n - \tilde{\sigma}(nT)$ where c_n is the unrotated ideal constellation point corresponding to the slicer decision.

2. Generate and store the differentially encoded data bits $(Y1_n, Y2_n)$.

3. Differentially decode the data bits by using the inverse of Table 13.1.

Test your slicer by using Code Composer Studio to single-step through your code with a made up sequence of baseband input samples.

Making the 16-Point Slicer

As an optional task, make a 16-point slicer for the V.22bis constellation. The slicer should quantize the baseband equalizer output sample to the nearest ideal constellation point. The slicer program should contain a table that has a record for each of the 16 constellation points. Each record should include the coordinates of the ideal point, the data bits $(Y1_n, Y2_n, Q3_n, Q4_n)$ associated with the point, and the quantity $1/|c_n|^2$ for use in the carrier tracking algorithm. In addition, the slicer should perform the following tasks:

1. Form the baseband error.

2. Generate and store the differentially encoded bits $(Y1_n, Y2_n)$.

3. Differentially decode the data bits by using the inverse of Table 13.1.

Test your slicer by using Code Composer Studio to single-step through your code with a made up sequence of baseband input samples.

15.5.2   Making a Demodulator and Carrier Tracking Loop

As the next step in building the QAM receiver, implement the demodulator and carrier tracking system shown in Figures 14.1 and 15.4. Design your software to include a 16-tap $T/2$ complex cross-coupled equalizer. To reduce the interaction between subsystems for debugging purposes, effectively bypass the equalizer. To do this, set up the necessary circular buffers to hold the complex passband samples for the equalizer delay line coming from the front end phase splitter and create the arrays for the complex equalizer coefficients $h_0, \ldots, h_{15}$. Set all the equalizer coefficients to $0 + j0$ except make $h_8 = 1 + j0$. Implement the equalizer convolution sum with the taps frozen at these values so the equalizer input samples will simply be passed to its output with a delay. Always load samples from the phase splitter into the equalizer delay line regardless of whether or not the Carrier Detect system indicates the presence of a received modem signal. The Godard clock recovery system implemented in Chapter 14 should make the times at which its cross-correlator outputs are computed nearly optimum times at which to sample the received symbols. The equalizer output should be computed at these symbol spaced times.

Monitor the Carrier Detect signal generated by the receiver front end and when a carrier is detected begin incrementing a counter once per symbol period. Keep the 2nd order accumulator state variable $\psi(n-1)$ stored in the z^{-1} block shown in Figure 15.4 cleared until the count reaches 1008. At this time, the S1 sequence should be finished and the equalizer delay line should be filled with received samples of the 4-point scrambled 1's signal. In this part of the experiment, the equalizer taps should be frozen at the initial values specified above independent of the Carrier Detect signal.

Write a program to implement the carrier tracking loop. Before a received modem signal is detected, let the carrier loop phase free run. That is, let $k_1 = k_2 = \psi(n-1) = 0$ so that

$$\hat{\varphi}_{n+1} = \hat{\varphi}_n + \omega_c T \qquad (15.67)$$

To test your program, scale and convert the real and imaginary parts of the baseband equalized sample $\tilde{\sigma}(nT)$ into 16-bit integers. Send this pair of integers to left and right codec DAC channels to create a constellation display. Always make a constellation display independent of the state of the Carrier Detect signal. If everything is working properly, you should observe four distinct clouds of points separated by 90° on the constellation display.

You were instructed above to make all the equalizer taps zero except for h_8. It is possible because of a sample delay unaccounted for in your program that this is the wrong tap to pick as a time reference and your constellation display samples are being taken half way between the correct sampling instants. In this case, you probably will not observe four distinct clouds of points. In any case, test for the best sampling time by setting h_7 to $1 + j0$ and all the rest of the taps to zero. This advances the sampling instants by $T/2$. Observe the resulting constellation display. In the rest of this experiment, use the initial equalizer tap setting that gives the best results.

Once the carrier tracking loop is basically working, experiment using different values of k_1 and k_2. When k_1 is large, the constellation will appear to jitter due to the large random perturbations. When k_1 is very small, convergence will be slow.

15.5.3 Making a Complex Cross-Coupled Adaptive Equalizer

Now implement the LMS with leakage equalizer tap adaptation algorithm defined by (15.30). Keep the taps frozen at their initial values until the Carrier Detect count reaches 1008 symbols by making γ and μ zero until this time. Then turn on the update algorithm by setting μ to a small number like 0.001. The leakage constant γ should be very small so the approximate gradient updates significantly outweigh the tap reduction by the factor $1 - \gamma$. Observe the equalizer convergence on your constellation display. If the equalizer and carrier tracking loop are working properly, the clouds caused by intersymbol interference should converge to four tight points. Experiment with several values of the update constant μ and observe the convergence speed and tightness of the steady-state constellation.

15.5.4 Bit-Error Rate Test

Once the equalizer and carrier tracking loop are working, add the descrambler to your receiver. The descrambler output should be steady 1's. To check this, you can write the descrambled bits to an array and use Code Composer Studio to examine the array when you halt the DSP program. A probe point can not be used because it halts the DSP while reading so the program does not run in real-time. You can also use the converter box designed for the RS232 experiments and send the descrambled bit stream out through McBSP0 in real-time.

Add Gaussian noise samples to the ADC input samples using the method presented in Appendix A. Observe the dispersion of the constellation caused by the noise for different SNR.

Experimentally, generate a plot of the bit-error rate *vs.* SNR. Do this by waiting until the equalizer has converged and then counting the number of 0's in the descrambled bit stream for a fixed duration at each SNR. Make sure the count duration is long enough to give a statistically reliable estimate of the bit-error rate.

Derive a theoretical formula for the symbol error probability *vs.* SNR and plot the results. Compare this plot with your experimental bit-error rate plot. The general shape should be the same but the curves will be different because one is for symbol errors and the other is for bit errors. Also, the self-synchronizing descrambler generates three output bit errors for each isolated error in its input bit stream.

15.5.5 Optional Experiment – Receiving the 16-Point V.22bis Constellation

As an optional experiment, program the transmitter to send the S1 sequence for 1000 symbols, followed by the 4-point constellation for scrambled 1's for 4000 symbols, followed by the 16-point constellation for scrambled 1's indefinitely. Use a counter as above to start the equalizer and carrier tracking loop 1008 symbols after carrier detection using the 4-point slicer. After an additional 4400 symbols, switch to the 16-point slicer. If your timing is off by a few symbols, it should have a negligible effect when the carrier tracking loop and equalizer update constants are small.

Perform a bit-error rate test for the 16-point constellation and plot the results. Compare the plots for the 4 and 16-point constellations. Find in the literature or derive a theoretical formula for the symbol error rate and plot the results. Compare the results for the 4 and 16-point constellations.

15.5.6   Optional Experiment – Ideal Reference Training

You were directed to set the scrambler to the all 0 state at the start of scrambled 1's transmission. Therefore the transmitted training sequence is known and ideal reference training can be performed. Make a replica of the scrambler and constellation point selector in the receiver. About 1000 symbols after carrier detect, start generating the ideal reference symbol sequence and use it to train the equalizer and carrier loop. Initialize all the equalizer taps to zero for ideal reference training. Once the equalizer is trained, halt the receiver and examine the equalizer tap magnitudes using Code Composer Studio. Adjust the time at which you start the ideal reference training to center the largest tap in the equalizer delay line. If your initial timing phase is too far off, the main tap will want to be off one end or the other of the delay line and the equalizer will not converge. After the equalizer has converged, switch back to decision directed adaptation.

One of the virtues of a fractionally spaced equalizer is that it can adjust for any symbol timing phase offset. Start the ideal reference training $T/2$ seconds (half a symbol period) later than before and observe the system convergence.

15.6   Optional Phase-Splitting Fractionally Spaced Equalizer Experiment

If you are interested in pursuing adaptive equalizers further, build a phase-splitting fractionally spaced equalizer for the modified V.22bis receiver. First strip the phase splitter out of the receiver front end. Continue to use the sampling rate $f_0 = 8f_s = 8000$ Hz. Then, the phase-splitting equalizer must be a $T/8$ equalizer. The 16-tap $T/2$ complex cross-coupled equalizer spans eight symbols. A $T/8$ phase-splitting equalizer must have $8 \times 8 = 64$ complex taps to span eight symbols. However, remember that the complex equalizer must be preceded by a Hilbert transform filter which gives the cascade a longer memory span. In any case, make a 64 complex tap phase-splitting equalizer. The carrier tracking loop and slicer do not have to be changed from the previous design.

The equalizer taps must be initially set to make it a phase splitter when the decision directed startup mode is used. Do this by initially setting all the real taps $\{h_{R,n}\}$ to zero except for one near the center of the delay line which should be set to 1. Design a 63-tap Hilbert transform filter using remez87.exe or window.exe and set the imaginary taps $\{h_{I,n}\}$ to these values. Align the center tap of the Hilbert transform filter with the nonzero real tap.

The position you pick for the initial main tap (the one with real part equal to 1) may not give the optimum sampling phase. There are 8 choices for the symbol spaced sampling phase. Determine the optimum phase in a way similar to what you did for the $T/2$ complex equalizer. That is, freeze the equalizer taps at their initial values. Set the transmitter to send 1000 symbols of S1 followed by the 4-point constellation for scrambled 1's indefinitely. Observe the constellation display for each of the 8 phases when the 4-point constellation is transmitted. Choose the phase that results in the tightest constellation.

Also experiment with ideal reference training. Set all the equalizer taps to zero initially in this mode. The taps should automatically adjust for any timing phase offset and converge to a set of values that perform the equalization and phase-splitting functions. After a period of time when the equalizer has converged, switch to the decision directed mode.

You could also test your receiver with the 16-point constellation. As before, begin the training with the 4-point constellation and then switch to the 16-point constellation after the equalizer has converged.

Perform a bit-error rate test for your receiver. Compare the tightness of the constellation and the bit-error rate performance of your complex cross-coupled and phase-splitting equalizers.

15.7   Optional Blind Equalization Experiment

Use the modified V.22bis transmitter you made for Chapter 13 as the signal source for the blind equalization experiments of this subsection. You might want to pass the signal through the commercial telephone channel simulator to introduce additional intersymbol interference by selecting a bad line. Perform the following tasks:

1. Program the transmitter to first send the S1 alternations for 1000 symbols to allow your symbol clock tracking loop to lock up. Then switch to sending scrambled 1's using the 16-point constellation.

2. Compute the required value for R_2 using (15.60) to make the equalized constellation have the correct scaling for your 16-point slicer.

3. Modify your complex cross-coupled equalizer to use Godard's $p = 2$ tap update algorithm given by (15.59). Initialize all the taps to zero except for the reference tap near the center of the delay line that you experimentally selected in the previous exercises. Set the reference tap to a value you think will satisfy (15.62). Note that if all the taps are set to zero, the equalizer output will be zero, the tap update increments will all be zero, and adaptation will not occur.

4. Wait for 1008 symbols after carrier detect and then begin CMA updating of the equalizer taps. Turn on the carrier tracking loop at this time also. The 16-point constellation with the correct rotation should emerge. If the constellation display does not converge to the expected picture but to some different stable pattern, try adjusting the initial value of the reference tap. You might want to wait to turn on the 2nd order integrator in the carrier loop for some time after adaptive equalization is started so random slicing errors do not drive its output far from the correct value. If this happens the constellation will spin rapidly.

5. Repeat the previous task with the carrier loop free-running at the nominal carrier frequency. The constellation should emerge but spin at the carrier offset frequency.

6. Devise and implement a strategy for switching from CMA to LMS training and watch the received constellation converge to 16 tight points.

7. If you are interested in experimenting with blind equalization further, implement it for the phase-splitting fractionally spaced equalizer.

15.8   Additional References

The use of adaptive transversal equalizers in data modems was first investigated and commercialized by R.W. Lucky [II.D.28] at AT&T Bell Labs in the mid 1960's. This approach is also discussed in Lucky, Salz, and Weldon [II.D.29, Chapter 6] and is based on a *zero forcing* criterion. These two references are primarily of historical interest since the zero forcing approach was soon replaced by the more powerful LMS algorithm. For detailed presentations of the complex cross-coupled LMS algorithm for adaptive equalization, see Gitlin, Hayes, and Weinstein [II.D.11, Chapter 8], Haykin [II.D.18, Chapter 9], Lee and Messerschmitt [II.D.26, Chapter 9], Proakis [II.D.32, Chapter 6], Treichler, Johnson, and Larimore [II.D.38, Chapter 4], and Widrow and Stearns [II.D.41, Chapter 10].

The tap leakage algorithm is proposed and investigated in Gitlin, Meadors, and Weinstein [II.D.12]. Treichler, Johnson, and Larimore [II.D.38, Section 4.2.6.2] also analyze the behavior of the LMS algorithm with tap leakage.

Mueller and Werner [II.D.30] and Ling and Qureshi [II.D.27] are good references for the fractionally spaced phase-splitting equalizer. The equivalence of the cross-coupled complex equalizer and phase-splitting equalizer is presented in Tretter [II.D.39, Chapter 11] along with a derivation of the transfer function of the optimum linear fractionally spaced equalizer.

Discussions of carrier recovery techniques for QAM systems can be found in Gitlin, Hayes, and Weinstein [II.D.11, Section 6.4] and Lee and Messerschmitt [II.D.26, Chapter 14]. An addition to the carrier tracking loop presented in this chapter that allows tracking of sinusoidal phase jitter is discussed in Sugar and Tretter [II.D.35].

Blind equalization was first disclosed by Sato [II.D.33] for one dimensional multilevel PAM. The next major publication was by Godard [II.D.14] which presented a thorough analysis of a class of blind equalization algorithms and carrier tracking for QAM receivers. Treichler and Agee [II.D.37] independently discovered the 2nd order case of Godard's class of algorithms. Benveniste and Goursat [II.D.2] presented additional blind equalization algorithms. More recent papers on blind equalization include Foschini [II.D.6] and Picchi and Prati [II.D.31]. Books with discussions of blind equalization include Gitlin, Hayes,and Weinstein [II.D.11], Haykin [II.D.18, Chapter 20], Proakis [II.D.32, Chapter 20], and Treichler, Johnson, and Larimore [II.D.38, Chapter 6].

Chapter 16

Echo Cancellation for Full-Duplex Modems

An important advance in the design of high speed voice-band telephone line modems for the dial network was the introduction of echo cancelers to achieve full-duplex data transmission over 2-wire circuits. This technique was studied in the early 1980's and then widely introduced in commercial products in the mid 1980's when the CCITT V.32 recommendation for a 9600 bps modem was approved. A few years later, the V.32bis recommendation for 14.4 kbps modems was approved, and the V.34 recommendation for rates up to 33.6 kbps was approved in June 1994. These also use echo cancelers. The recent V.90 and V.92 modems that use PCM downstream use echo cancelers to. Echo cancelers are also used in some high speed digital subscriber lines at data rates of 64 kbps or more. Line echo cancelers were used with analog voice transmission to eliminate annoying talker echo prior to the inclusion of echo cancellation in digital data modems. The voice echo cancelers are placed at different points in the telephone circuit than the ones for data transmission and are disabled during data transmission by a special signal in the modem handshake sequence. The technique is also used in speaker phones to eliminate annoying acoustic reflections from the speaker to the microphone and then back to the far end talker.

In this chapter, you will learn the fundamentals of echo cancellation for voice-band modems. You will build an echo canceler for near-end echo and observe its behavior without a far end transmitted signal. You will also build a far-end canceler with frequency offset correction. Suggestions for additional references on echo cancellation are included at the end of this chapter.

You should do Chapter 13 before this experiment because the modified V.22bis transmitter you make there will be used as the signal source.

16.1   The Echo Sources in a Dialed Telephone Line Circuit

A typical full-duplex dialed telephone line circuit is shown in Figure 16.1. The Transmitter, Receiver, and Hybrid boxes shown on the left are all contained in the left modem and the ones on the right are in another modem. The left transmitter sends to the right receiver while the right transmitter simultaneously sends to the left receiver. The transmissions in both directions are independent of each other. Both modems are connected to ordinary 2-wire voice-band telephone line circuits which go from the modem site to the local office and are called *2-wire local loops*.

A modem transmits and receives over the same pair of wires. The function of the Hybrid box is to isolate the transmit and received signals at the 4-wire to 2-wire interfaces. An ideal hybrid routes the signal on the transmit terminal pair on the 4-wire side only to the 2-wire loop and not to the receive pair on the 4-wire side. Similarly, the received signal on the 2-wire loop is routed only to the receive terminal pair on the 4-wire side. Customers can lease 4-wire circuits at additional cost and eliminate the need for echo cancelers.

Signals are transmitted between offices within the telephone plant using 4-wire circuits. One pair is used to transmit in one direction and the other pair in the opposite direction. Again, hybrids are used at the 2-wire to 4-wire interfaces. The 4-wire circuit can consist of combinations of ground lines, fiber optic cables, microwave links, and satellite circuits. A very large percentage of the transmissions between offices is in digital form in the US.

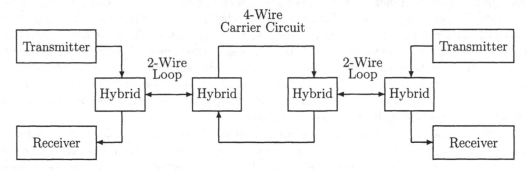

Figure 16.1: Typical Full-Duplex Dialed Telephone Line Connection Without Echo Cancellation

Real-world hybrids are not perfect. They are designed to work with a nominal impedance on the 2-wire loop and actual loops differ from the nominal. Therefore, some of the transmitted signal at a local modem leaks through the hybrid back to the local transmitter. This is known as *near-end echo*. When the impedance imbalance is large, the near-end echo power can be 30 dB above the power of the desired signal received from the transmitter of the far-end modem. The impulse response of the near-end echo path typically has a duration of from 5 to 18 ms. Another source of echo is leakage through the hybrid at the far end of the 4-wire circuit and is called *far-end echo*. The far-end echo is delayed by the propagation time through the echo path. This delay can range from a few milliseconds up to as much as 1.5 seconds when the transmission path contains a couple of satellite hops. The duration of the impulse response of the far-end echo path, ignoring the bulk delay, is typically no more than 20 ms. The level of the far-end echo is typically at least 15 dB below that of the near-end echo. In addition, there can be frequency offsets on the far-end echo when carrier circuits are in the path. Additional echos can circulate around the 4-wire circuit but are usually attenuated to the point of being negligible.

Techniques used for eliminating the echos depend on whether the modems are low or high speed. The voice-band circuit has a usable bandwidth extending from about 300 to 3200 Hz. Low speed modems like the V.22bis separate the transmit and receive paths by frequency division multiplexing and bandpass filtering. Higher speed modems like the V.34 use a symbol rate of 2400 Hz or more so the signal spectrum fills most of the available bandwidth and frequency division multiplexing cannot be used. The near and far-end modems use the same carrier frequency, so the transmitted and received spectra completely overlap. Echo cancellation was chosen as the solution to this problem.

16.2 The Data-Driven, Nyquist, In-Band Echo Canceler

16.2.1 General Description

The architecture for a practical echo canceler is shown in Figure 16.2. The transmitter uses the passband shaping filter approach presented in Section 13.3 and shown in Figure 13.6. As in Chapter 13, $c_n = a_n + j\, b_n$ is the sequence of constellation points selected by the input data and

$$c'_n = c_n e^{j\omega_c nT} = a'_n + j\, b'_n \tag{16.1}$$

is called the *rotated* symbol sequence. The echo canceler is basically an adaptive tapped delay-line filter with near and far-end sections. It is driven by the rotated symbols and synthesizes a replica of the real passband echo. For these reasons, it is called a data-driven, in-band canceler.

Since the synthesized echo is subtracted directly from the received signal, echo samples must be generated and the received signal sampled at a rate that is at least twice the highest frequency component in the received signal so the Nyquist criterion is satisfied. The received signal is typically sampled at L times the symbol rate $f_s = 1/T$ as indicated in the ADC block in the figure. For example, the symbol rate for a V.32 modem is 2400 Hz with a carrier frequency of 1800 Hz and uses 12% excess bandwidth square-root of raised cosine shaping, so the upper cutoff frequency is 3144 Hz. The smallest choice for L that satisfies the Nyqist criterion is $L = 3$ resulting in a sampling rate of 7200 Hz and is the most efficient choice in terms of minimizing the required number of computations per second.

The canceler also includes a far-end frequency offset correction phase-locked loop. Without frequency offset correction, the received constellation appears to pulsate at the offset frequency and the error rate performance is severely degraded. Modern links where the transmission between local offices is entirely digital do not have any frequency offset.

The synthesized echo samples are subtracted from the received signal samples to give what is called the *residual* signal sequence. The remaining echo component in the residual is typically required to be at least 30 dB below the received data signal component for adequate error rate performance. The echo canceler and received signal sampling instants are synchronized to the transmitter symbol timing. The signal received from the far-end

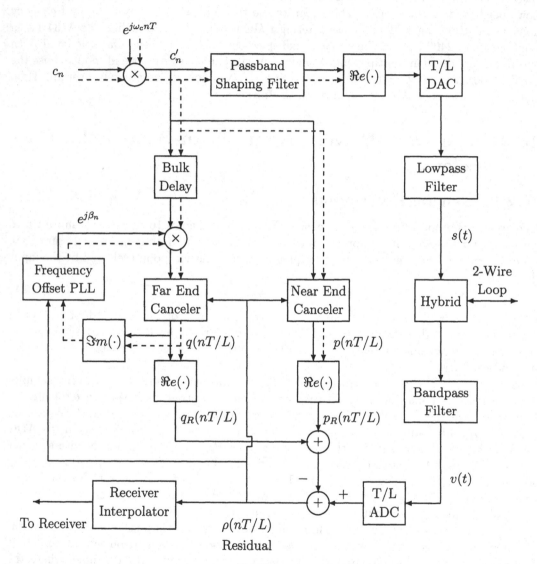

Figure 16.2: A Data-Driven, Nyquist, In-Band, Echo Canceler

modem has the same nominal symbol rate. However, it will have a slightly different frequency because of minor hardware differences. Therefore, the residual sampling instants are not synchronized with the desired receiver symbol timing. This problem is solved by the Receiver Interpolator block which generates output samples synchronized with the receiver timing. The interpolator can be implemented by converting the residual samples back to a continuous-time signal and re-sampling the analog lowpass reconstruction filter output at times synchronized with the receiver timing. Modem analog front end chips are commercially available that perform the interpolation using numerical DSP techniques. These chips also include the ADC, DAC, analog lowpass transmit and bandpass receive filters, and clock generation circuits. The interpolation can also be performed numerically using the variable phase interpolator methods presented in Chapter 12. The interpolated residual sequence is sent to the receiver which has the same structure as described in Chapters 14 and 15.

The echo canceler problem is simpler than the equalizer problem in one way. The modem's transmitter and receiver are located in the same box and may even be implemented as subroutines in the same DSP. Therefore, the local receiver has exact knowledge of the locally transmitted symbol sequence and the exact error signal, the residual, is formed. The adaptive equalizer must use decision directed adaptation algorithms after the initial handshaking sequence. The canceler problem is more difficult in another way. In the equalizer problem, the received signal to noise ratio is usually rather high. In the echo canceler problem it is usually low because the desired received signal from the far-end modem can be at a significant level relative to the echos that need to be cancelled. Therefore, a large amount of averaging must be performed in the adaptation algorithm to suppress the far end signal which looks like noise to the canceler. When both modems are transmitting simultaneously, they are said to be *double talking*. In practice, the local and remote echo cancelers are usually separately trained in the half-duplex mode during an initial handshake sequence. The data signals transmitted in both directions are statistically independent. Different scramblers are used in the near and far end modems to ensure this independence. This allows the echo impulse response to be tracked during double talking with enough averaging.

16.2.2 The Near-End Echo Canceler

The inputs to the near-end echo canceler are the rotated symbols c_n' which occur at the symbol rate $f_s = 1/T$. The canceler must generate outputs at a rate $L f_s$ that satisfies the Nyquist criterion for sampling the received signal $v(t)$. The canceler can be implemented by generalizing the interpolation filter bank structure discussed in Section 11.3. The canceler requires L FIR subfilters each of which is called a *subcanceler*. Suppose each subcanceler has N_1 complex taps and the taps for the m-th subcanceler are denoted by

$$A_m(n) = \alpha_{R,m}(n) + j\,\alpha_{I,m}(n) \quad \text{for} \quad n = 0, \dots, N_1 - 1 \quad \text{and} \quad m = 0, \dots, L - 1 \quad (16.2)$$

The subcancelers all use the same delay-line which stores the present and $N_1 - 1$ past rotated symbols. After each new rotated symbol c_n' arrives, the L subcanceler outputs are computed as

$$p_R(nT + mT/L) = \Re\left\{ \sum_{k=0}^{N_1-1} A_m(k) c_{n-k}' \right\}$$

$$= \sum_{k=0}^{N_1-1} \alpha_{R,m}(k)a'_{n-k} - \alpha_{I,m}(k)b'_{n-k} \quad \text{for} \quad m = 0,\ldots, L-1 \quad (16.3)$$

The residual for the m-th subcanceler is

$$\rho(nT + mT/L) = v(nT + mT/L) - p_R(nT + mT/L) - q_R(nT + mT/L) \qquad (16.4)$$

where $q_R(nT + mT/L)$ is the far-end echo canceler output. As with adaptive equalizers, choosing the subcanceler coefficients to minimize the L mean-squared residuals has been found to be a good strategy. The derivatives of the m-th residual with respect to the i-th real and imaginary tap components are

$$\frac{\partial}{\partial \alpha_{R,m}(i)} \rho^2(nT + mT/L) = -2\rho(nT + mT/L)a'_{n-i} \qquad (16.5)$$

and

$$\frac{\partial}{\partial \alpha_{I,m}(i)} \rho^2(nT + mT/L) = 2\rho(nT + mT/L)b'_{n-i} \qquad (16.6)$$

Therefore, using the LMS strategy presented in Chapter 15, the coefficients of the m-th subcanceler can be adjusted by the following formulas:

$$\alpha_{R,m}(i; n+1) = \alpha_{R,m}(i,n) + \mu\rho(nT + mT/L)\,a'_{n-i} \quad \text{for} \quad i = 0,\ldots, N_1-1 \qquad (16.7)$$

$$\alpha_{I,m}(i; n+1) = \alpha_{I,m}(i; n) - \mu\rho(nT + mT/L)\,b'_{n-i} \quad \text{for} \quad i = 0,\ldots, N_1-1 \qquad (16.8)$$

where μ is a small positive constant. These formulas can be expressed in terms of the complex coefficients and rotated data symbols by the single equation

$$A_m(i; n+1) = A_m(i; n) + \mu\rho(nT + mT/L)\,\overline{c'_{n-i}} \quad \text{for} \quad i = 0,\ldots, N_1-1 \qquad (16.9)$$

Some simple stability and rate of convergence formulas can be derived for this LMS algorithm. Suppose the constellation points are selected with equal likelihood, are an uncorrelated sequence, and are located so that

$$\sigma^2 = \mathrm{E}\{a_n^2\} = \mathrm{E}\{b_n^2\} \qquad (16.10)$$

Then, it can be shown [II.D.40] that this LMS adaptation algorithm is stable if the update scale factor μ satisfies the bounds

$$0 < \mu < \frac{1}{N_1\sigma^2} \qquad (16.11)$$

The scale factor that gives the most rapid convergence of the mean-squared error is

$$\mu_0 = \frac{1}{2N_1\sigma^2} \qquad (16.12)$$

Let $\xi(nT + mT/L)$ be the portion of the received signal $v(nT + mT/L)$ exclusive of the echos. It includes the received far-end data signal and channel noise. Then, the steady-state mean-squared residual value, assuming the far-end echo is not present, is

$$R_\infty = \frac{\dfrac{1}{L} \sum_{m=0}^{L-1} \mathrm{E}\{\xi^2(nT + mT/L)\}}{1 - \mu N_1\sigma^2} \qquad (16.13)$$

This equation shows that as μ is reduced to zero, the steady-state residual power reduces to the power of the interference signal. When μ is chosen to give the fastest speed of convergence, the steady-state residual power is double that of the interference signal.

During double talking, μ must be made very small to average out the effects of the interference from the far-end data signal. This can result in an underflow problem and cause the adaptation to stop, particularly with fixed point DSP's. A solution to this problem is to sum several past products to form the gradient estimate and is called the *block least mean-square* (BLMS) algorithm. The formula for adjusting the i-th complex tap of the m-th subfilter using a block depth of M is

$$A_m(i; n + M) = A_m(i; n) + \mu \sum_{k=0}^{M-1} \rho(nT + kT + mT/L)\,\overline{c'_{n+k-i}} \tag{16.14}$$

Notice that this reduces to the standard LMS algorithm for $M = 1$. With the BLMS algorithm, the taps are adjusted every M samples. The stability and convergence rate formulas are the same as for the LMS algorithm.

It is very likely with current DSP technology, that all the echo canceler taps can not be updated during one symbol period, especially when the DSP is performing other transmitter and receiver tasks in addition. A standard solution to this problem is to update just a portion of the taps each symbol period.

16.2.3 The Far-End Echo Canceler

The impulse response of the echo path usually consists of an initial period of activity caused by the near-end echo, followed by a period of silence, followed by a period of activity caused by the far-end echo. No echo canceler filter taps are required to model the silent period caused by the round-trip propagation time of the far-end echo. It is only necessary to include a *bulk delay* to model the round-trip delay. A portion of the handshake sequence is used to measure the round-trip delay in V.32, V.32bis, and V.34 modems. The far-end echo may have a frequency offset in addition to the delay. The frequency offset is corrected by multiplying the output of the Bulk Delay by $e^{j\beta_n}$. The details of the offset correction are presented in the following subsection. The structure of the far-end echo canceler FIR filter is exactly the same as for the near-end canceler except that it will have more taps because the far-end echo is usually dispersed more than the near-end echo.

Suppose the measured bulk delay is d_0 symbols. Then, the input to the far-end echo canceler is

$$c''_n = c'_{n-d_0}\,e^{j\beta_n} = a''_n + j\,b''_n \tag{16.15}$$

Let the taps for the m-th far-end echo subcanceler be

$$G_m(n) = \gamma_{R,m}(n) + j\,\gamma_{I,m}(n) \quad \text{for} \quad n = 0, \ldots, N_2 - 1 \quad \text{and} \quad m = 0, \ldots, L - 1 \tag{16.16}$$

Then, the m-th subcanceler output is

$$q_R(nT + mT/L) = \Re e\{q(nT + mT/L)\} = \Re e\left\{ \sum_{k=0}^{N_2-1} G_m(k)c''_{n-k} \right\}$$

$$= \sum_{k=0}^{N_2-1} \gamma_{R,m}(k)a''_{n-k} - \gamma_{I,m}(k)b''_{n-k} \qquad (16.17)$$

Using the same approach as for the near-end canceler, the LMS tap adjustment formula for the m-th subcanceler is found to be

$$G_m(i; n+1) = G_m(i; n) + \mu\rho(nT + mT/L)\,\overline{c''_{n-i}} \quad \text{for} \quad i = 0, \ldots, N_2 - 1 \qquad (16.18)$$

16.2.4   Far-End Frequency Offset Compensation

With older telephone circuits using analog frequency division multiplexing, it has been found that it is important to compensate for the far-end echo frequency offset to achieve satisfactory modem performance. As a first approximation, suppose this effect can be removed by letting

$$\beta_n = \theta + \omega_0 nT \qquad (16.19)$$

Then, the complex output of the m-th far end echo subcanceler is

$$
\begin{aligned}
q(nT + mT/L) &= \sum_{k=0}^{N_2-1} G_m(k)c'_{n-d_0-k}e^{j[\theta + \omega_0(n-k)T]} \\
&= e^{j(\theta + \omega_0 nT)} \sum_{k=0}^{N_2-1} G_m(k)e^{-j\omega_0 kT}c'_{n-d_0-k} \qquad (16.20)
\end{aligned}
$$

The derivative of the squared residual with respect to θ is

$$
\begin{aligned}
\frac{\partial}{\partial\theta}\rho^2(nT + mT/L) &= -2\rho(nT + mT/L)\frac{\partial}{\partial\theta}\Re e\{q(nT + mT/L)\} \\
&= -2\rho(nT+mT/L)\,\Re e\left\{\frac{\partial}{\partial\theta}q(nT+mT/L)\right\} = -2\rho(nT+mT/L)\,\Re e\{j\,q(nT+mT/L)\} \\
&= 2\rho(nT + mT/L)\,\Im m\{q(nT + mT/L)\} \qquad (16.21)
\end{aligned}
$$

where

$$\Im m\{q(nT + mT/L)\} = \sum_{k=0}^{N_2-1} \gamma_{I,m}(k)a''_{n-k} + \gamma_{R,m}(k)b''_{n-k} \qquad (16.22)$$

Let

$$\Delta_n = -\sum_{m=0}^{L-1} \rho(nT + mT/L)\Im m\{q(nT + mT/L)\} \qquad (16.23)$$

Then, a second-order phase-locked loop update formula for the far-end echo phase correction angle suggested by these results is

$$\beta_{n+1} = \beta_n + k_3\Delta_n + \Gamma_n \qquad (16.24)$$

where the second-order accumulator output is

$$\Gamma_n = \Gamma_{n-1} + k_4\Delta_n \qquad (16.25)$$

and k_3 and k_4 are small positive constants with $k_3/k_4 \approx 100$. The structure for this PLL is essentially the same as the lower part of the QAM carrier tracking loop shown in Figure 15.4 with $\omega_c T = 0$.

In order to save computation time, the phase update can be performed once per symbol period using the residual from just one subcanceler. For example, suppose the residual for subcanceler $m = 0$ is used. Then, the only change required in the formulas is to replace Δ_n by

$$\Delta_n = -\rho(nT)\Im m\{q(nT)\} \tag{16.26}$$

16.3   Echo Canceler Experiments

You will be asked to make a near-end echo canceler and a far-end echo canceler with frequency offset correction in these experiments. No interfering data signal from a far-end modem will be included. Make sure to use the highest compiler optimization level.

16.3.1   Making a Near-End Echo Canceler

Use the modified V.22bis transmitter you made in Chapter 13 as a starting point for your near-end echo canceler program. Send the output samples to the left channel of the DAC. Connect the DAC output to the ADC input to simulate the near-end echo. Your transmitter uses a passband shaping filter with an impulse response that spans 8 symbols. Therefore, your near-end echo canceler will have to span at least 8 symbols which is equivalent to 8 ms at 1000 baud. Use an 8000 Hz sampling rate for the echo canceler. This requires using $L = 8000/1000 = 8$ samples per symbol or 8 subcancelers.

You will have to figure out a strategy for interleaving the transmitter and echo canceler subfilter computations each symbol. Consider using the interrupts for transmitted samples somehow to generate the required timing.

Implement the echo canceler with 8 subcancelers using (16.7) and (16.8) to update the taps. Pass the residual signal to the right DAC channel and observe it on the oscilloscope. Experiment with different values of the update scale factor μ and observe the convergence rates and steady-state residual variations. Measure the power of the received sequence before the echo canceler subtraction node and the power of the residual. and calculate the ratio of the input to residual power in dB.

16.3.2   Making a Far-End Echo Canceler with Frequency Offset Correction

Simulate a frequency offset far-end echo by multiplying the complex output of the transmitter passband shaping filter by $e^{j2\pi f_d nT_0}$ before sending its real part to the DAC. The parameter T_0 is the filter bank sampling period and f_d is the frequency offset. Choose a frequency offset of $f_d = 0.25$ Hz. For simulation purposes, assume the far-end propagation time is zero and ignore the Bulk Delay block. Add the frequency correction functions to your near-end echo canceler and pretend it is a far-end canceler. Do not implement a near-end canceler also. Update the far-end frequency offset tracking loop only once per symbol when processing subcanceler $m = 0$ using (16.24), (16.25), and (16.26).

Send the residual to the right DAC channel and observe it on the oscilloscope. Experiment with the system parameters and observe the residual behavior. When you have found a good set of parameters, measure the input and residual powers and compute the ratio in dB.

16.4 Additional References

Werner [II.D.40] gives a very good presentation of echo cancelers for digital data modems. This work had a significant influence on the round trip delay estimation method chosen for the V.32 recommendation. Also see Gitlin, Hayes, and Weinstein [II.D.11, Chapter 9], and Lee and Messerschmitt [II.D.26, Chapter 18] for comprehensive theoretical treatments and alternative structures for echo cancelers.

Chapter 17

Multi-Carrier Modulation

This chapter explores the fundamental concepts of multi-carrier modulation. Multi-carrier modulation is important to know about today because it is used extensively in broadband digital wireline systems and in many broadband wireless systems. The systems are very complex and this chapter will only look at the fundamental concepts used for modulation and demodulation by implementing a simplified ADSL transmitter and receiver. The systems also include sophisticated error correcting codes, interleavers, and network protocols.

17.1 A Brief Overview of the History and Implementation of Multi-Carrier Modulation

Multi-carrier modulation is often selected for situations where the channel frequency response and noise interference vary significantly with frequency and/or time over the signaling bandwidth. Applications include fading wireless channels and broadband wireline digital subscriber lines. It is called *discrete multi-tone modulation* (DMT) in the wireline community and *orthogonal frequency division multiplexing* (OFDM) in the wireless community. The basic philosophy behind multi-carrier modulation is to partition the channel frequency band into multiple non-overlapping narrow-band sub-channels with nearly constant frequency response over each sub-channel. Data is transmitted by modulating carriers centered in each sub-channel using a narrow-band modulation method like QAM. The individual carriers are often called *tones* or *subcarriers*. At the receiver, the QAM constellation for each subcarrier is essentially scaled by a complex constant corresponding to the channel frequency response at the subcarrier frequency and can be easily equalized by multiplying by the reciprocal of the constant. A block of many data bits is taken from the data source and subsets of bits are assigned to the different subcarriers based on sub-channel signal-to-noise ratio measurements. The subcarriers are modulated and all transmitted for some time T and then another data block is taken and the process is repeated. The sum of modulated subcarriers transmitted for an input data block is called a multi-carrier symbol. Since many bits are assigned to a symbol, the symbol rate $1/T$ is small compared to the data rate or, in other words, the symbol period is long compared to the input bit period.

Multi-carrier modulation is used extensively is in broadband digital wireline systems using existing copper twisted pair telephone line cables. These telephone line systems are known as *digital subscriber lines*. The American National Standards Institute published a standard for a broadband multi-carrier wireline system in 1998 [E.1] called an *asymmetric digital subscriber line* (ADSL) . The International Telecommunication Union has approved a number of multi-carrier wireline recommendations including ITU-T G.992.3 [E.10], G.992.4 [E.11], and G.992.5 [E.12]. Multi-carrier modulation is called *discrete multi-tone modulation* (DMT) in the wireline world.

Voice-band telephone line modems starting with the Western Electric 110 baud Bell 103 binary FSK (frequency shift keyed) modems of the 1960's, followed by other proprietary commercial modems, and ones conforming to the international standard ITU-T V-series recommendations up to the ITU-T V.34 modem approved in 1998 have all used single-carrier modulation. The voice-band channel is well modeled by a linear time-invariant filter followed by additive noise. Almost all of the modems operating at data rates of 1200 bits per second or more have used quadrature amplitude modulation (QAM). A few early high speed modems used single sideband modulation. The ITU-T V.90 modem approved in September 1998 always uses single-carrier V.34 modulation upstream from the customer to the central office but pulse code modulation (PCM) downstream from the central office to the customer if channel conditions allow it. The V.92 modems approved in 2000 use PCM upstream and downstream but can fall back to V.34 in either direction. Chapters 13, 14, 15, and 16 present the theory and algorithms required to implement a single-carrier voice-band telephone line modem.

A few multi-carrier voice-band telephone line modems have been produced but none were commercially successful or included in international standards. The manufacturers claimed to make optimum use of the telephone channel by adjusting carrier amplitudes and number of bits assigned to each carrier based on capacity formulas from Information Theory. One reason for not being accepted is that they introduced large delays at voice-band data rates. Another reason is that their transmitted signal had a large peak-to-average ratio requiring amplifiers and circuits with a large dynamic range and small non-linearities. The performance of multi-carrier modems was also degraded by carrier phase jitter which was still present in telephone plant analog carrier circuits. These have now been almost entirely replaced by digital PCM systems. In addition, equivalent results could be obtained with single-carrier modems using adaptive precoding.

The situation changed dramatically when the telephone companies were forced by government regulation and competition from cable and wireless companies to provide broadband data transmission. This required direct access to the copper twisted pair local loops without going through the lowpass voice-band filters with a cut-off frequency of around 4 kHz. It was found that the local loops could support signals with spectral components up to 1.1 MHz or more, in some cases, allowing data rates up to several megabits per second. The useful bandwidth depends on the distance of the customer from the central office, bridge taps into the cable, and ambient noise. See Golden [E:3] for extensive discussion of the cable properties. The frequency response of the cables and noise interference varies greatly over these bandwidths. The noise includes AM radio signals and cross-talk from other cables in the same bundle. Experts decided that the adaptive equalization technique used in single-carrier systems was expensive to implement at high symbol rates, did not allow flexibility in using the channel, and that moderate delay was not a problem for applications using broadband systems like web surfing or video on demand. They chose to use the multi-carrier approach with the subcarrier frequencies uniformly spread over the usable bandwidth. In ADSL, upstream transmission uses subcarrier frequencies at multiples of 4.3125 kHz from about 25 kHz to 138 kHz and can provide data rates up to 896 kbit/s. Downstream transmission uses frequencies at multiples of 4.3125 kHz usually in the band from 138 kHz to 1.104 MHz but can use the full 25 kHz to 1.104 MHz band if echo cancellation is implemented providing data rates up to 8 Mbit/s. QAM modulation with a large range of constellation sizes is used for each of the subcarriers. The frequencies below 25 kHz are not used so the *plain old telephone system* (POTS) 0 to 4 kHz voice-band channels can be frequency division multiplexed with the ADSL signals onto the same cable. Channel measurements of the cable frequency response and interfering noise are used to determine which subcarriers to use, their amplitudes, and the number of data bits assigned to each to try to achieve a data rate approaching the capacity predicted by Information Theory.

Multi-carrier modulation is used extensively in wireless digital data communication systems where the channel has fading characteristics caused by the transmitted signal propagating over multiple paths to the receiver as a result of reflections and diffraction. Multi-carrier modulation is called *orthogonal frequency division multiplexing* (OFDM) in the wireless community. It is essentially the same as DMT used in the DSL systems. These fading channels are also called *multi-path* channels. They are time varying because of motion of the reflectors and, possibly, the transmitter and receiver. When an unmodulated carrier is transmitted and all the paths have about the same power and introduce independent random phase shifts and there are many paths, the amplitude of the received signal can be shown to have a probability density function that is well approximated by the Rayleigh density and the phase is uniformly distributed over $[0, 2\pi)$. When there is a strong line-of-sight component, the probability density function for the amplitude of the received signal has a Rician pdf. See books on wireless communications like Goldsmith [E.4], Haykin [E.5], Schwartz [E.15], and Stüber [E.17] for the detailed theory. The frequency response of the channel from transmitter to the receiver can have an amplitude that varies significantly with frequency and can have deep nulls at some frequencies depending on the delays and amplitudes of the paths. These channels are said to have *frequency selective* fading. The peaks and nulls are caused by reinforcement and cancellation of the signals arriving along the different paths.

Multi-carrier modulation was used as long as fifty years ago for HF digital radio communication over long distances like from ship to shore. The HF signal reflects off layers of the ionosphere and can propagate over long distances bouncing around the Earth. The received signal experiences significant time varying multi-path fading as the ionospheric layers shift. The initial radios used analog filters which were difficult to keep tuned. In the late 1960's, the military began using expensive special purpose mini-computers and digital signal processing algorithms including the FFT and IFFT to perform the speech compression, encryption, modulation, forward error correction, and demodulation. The systems used around a 3 kHz bandwidth and in the order of 32 subcarriers. Small inexpensive modern DSP's can now be used to implement the HF modems.

Arguably, the most common current use of multi-carrier modulation is in the ubiquitous Wi-Fi networks. There are a number of IEEE standards for Wi-Fi. The IEEE 802.11a standard [E.6] was released in 1999. It uses OFDM in the 5 GHz band, has a maximum data rate of 54 Mbit/s with a typical throughput of 21 Mbit/s, an indoor range of about 25 meters, and an outdoor range of about 75 meters. The IEEE 802.11g standard [E.7] was released in 2003. It uses OFDM in in the 2.4 GHz band, has a maximum data rate of 54 Mbit/s and typical throughput of 20 Mbit/s, an indoor range of about 40 meters, and an outdoor range of about 95 meters. Each individual subcarrier can be modulated with BPSK, QPSK, 16-QAM, or 64-QAM. A new standard, 802.11n, is in the advanced stages and is expected to be released around September 2007. It includes provisions for multiple transmit and receive antennas (MIMO). The indoor environment is quite hostile with extensive multipath propagation and interference from other electronic devices transmitting in the Wi-Fi bands. Multi-carrier modulation allows the Wi-Fi systems to adaptively select subcarriers at frequencies where the channel is good.

By far, the most common Wi-Fi network configuration is to have multiple users connect to the network through a nearby transceiver called an *access point*. A multiple access protocol is used. In the wireline DSL systems, a single transceiver on one end of the line communicates with a single transceiver on the other end. This arrangement is called a *point-to-point* connection. It is also possible for two nearby Wi-Fi users to make a point-to-point connection.

Currently, multi-carrier modulation is almost always being selected and implemented for long distance broadband digital wireless communications with fixed and mobile users rather than code division multiple access (CDMA). These systems will have architectures similar to cellular telephone networks where the mobiles connect to a nearby base station and are automatically switched between stations as they move to adjacent cells. The standardized system that is currently getting the most attention and is in the process of being deployed by major carriers like Sprint is popularly called *WiMax*. This system was initially standardized in 2004 by IEEE Std 802.16 [E.8] as a fixed broadband system. In 2005 it was expanded to include mobile units by IEEE Std 802.16e [E.9]. The 802.16e systems were designed to operate below 11 GHz and have bandwidths ranging from 1.25 to 20 MHz. These standards contain a multitude of options. An industry wide committee called the WiMax Forum has been meeting to select a subset of options that manufacturers should include to make WiMax transceivers compatible with each other. Companies like Texas Instruments have groups furiously working on making DSP's with special peripherals to implement signal processing tasks required in WiMax transceivers and groups creating optimized software to implement the required algorithms. It is projected that mobiles as much as 30 miles away will be able to communicate with a base station. The WiMax systems are very flexible in terms of channel bandwidths, data rates, and number of users. They are also very complicated and use the latest modulation and coding methods.

WiMax uses a variant of OFDM called *orthogonal frequency division multiple access* (OFDMA). In OFDMA, different mobiles are allocated different subsets of subcarrier frequencies. Periodically, the base station collects groups of bits from the digital network to be sent to each of the mobiles. Each group of bits is mapped to a set of QAM constellation points, one for each subcarrier allocated to that mobile. QPSK (4-QAM), 16-QAM, or 64-QAM constellations can be used based on desired data rates, transmitted signal bandwidths, and measured channel conditions. Each constellation point is a complex number that specifies the DFT value for the corresponding subcarrier frequency or DFT index. The set of constellation points for all the mobiles is the DFT frequency domain representation for one transmitted symbol. Modulation is performed by taking the inverse fast Fourier transform (IFFT) of the set of constellation points. WiMax systems can use IFFT sizes of 128, 512, 1024, or 2048 points. The total number of subcarriers used for all the mobiles is somewhat less than each of these IFFT sizes because some subcarriers at the band edges are set to zero to confine the spectrum and other subcarriers are used to send pilot signals for channel estimation. The mobiles demodulate each symbol by using an FFT. A similar process happens in the upstream direction from the mobiles to the base station. Each mobile periodically collects a group of bits to be sent to the base station, maps them to constellation points for its assigned set of subcarriers, and performs modulation using an IFFT. The transmissions from the mobiles must be timed so that when they arrive at the base station their sum appears to the base station as a single OFDM symbol. The 802.16e standard includes procedures for initially acquiring and tracking this timing. The base station demodulates the received symbols by taking FFT's.

The European cellular telephone community is also working on multi-carrier broadband wireless data communication standards to compete with WiMax. The Third Generation Partnership Project (3GPP) is working on a standard called Long Term Evolution (LTE) [E.18]. OFDMA will be used in the down link from the base station to the mobiles and another variation of OFDM called *single carrier frequency division multiple access* (SC-FDMA) will be used uplink. Again, QPSK, 16-QAM, and 64-QAM constellations are proposed. The bandwidths will be scalable from 1.25 to 20 MHz. Maximum data rates are projected to be 100 Mbit/s down link and 50 Mbit/s uplink. Systems are expected to be rolled out in 2009.

Multi-carrier modulation is also used in a variety of other systems. It is used for digital audio and video broadcasting in Europe. It is also included as one option in the IEEE 802.15 ultra-wideband standard. A number of proprietary systems have been built and field tested. One of these is called FLASH-OFDM which was created by a spin-off from Bell Labs called Flarion which was recently bought by Qualcomm.

17.2 Asymmetric Digital Subscriber Line (ADSL) System Architecture

Figure 17.1 shows the high level block diagram of an ADSL link. At a telephone company central office, data for a remote customer premise is collected from a broadband network like the Internet and routed to an ADSL modem. The Network Interface box often contains a device called a *Digital Subscriber Line Access Multiplexer* (DSLAM) that performs the multiplexing task. In the xDSL community, modems are often called transceivers for transmitter/receiver. The modem in the central office is designated by ATU-C which stands for "ADSL transceiver unit, central office." The ATU-C transmitter transforms its input data into a sequence of multi-carrier symbols that are sent through a splitter in the central office and then over the copper twisted pair local loop to the customer premise. The ATU-C transmit spectrum for ITU-T Recommendation G.992.3 [E.10] is most often confined to a frequency band extending from 138 kHz to 1.104 MHz. There is also an option to use echo cancellation and let the spectrum go from 25 kHz to 1.104 MHz but this is rarely used. The maximum downstream data rate for ADSL2 is 12 Mbit/s but the actual rate depends on the quality of the local loop. An ordinary voice-band telephone channel from the Public Switched Telephone Network (PSTN) which has a lowpass spectrum with a nominal cutoff frequency of 4 kHz is also applied to the central office splitter and sent to the customer over the local loop along with the ADSL signal. Ordinary telephone channels have been given the name *Plain Old Telephone Service* (POTS). At the customer premise, the ADSL and POTS signals are separated by the highpass and lowpass filters in a splitter at the customer premise. A somewhat simpler and lower cost option called *splitterless* ADSL in ITU-T Recommendation G.992.4 [E.11] eliminates the customer premise splitter and requires the customer to install lowpass filters at his ordinary telephones. The ADSL modem at the customer premise is called the *remote* transceiver and is designated by ATU-R where R stands for remote. The ATU-R demodulates the received multi-carrier symbols and sends the received data to the customer premise network. CPE in the figure stands for "customer premise equipment" which could be a PC or network printer. Data and voice transmission upstream from the customer premise to the central office follows the same process as downstream except that the upstream ADSL2 spectrum is limited to a band extending from 25 kHz to 138 kHz with a maximum data rate of 1 Mbit/s. The reason for the asymmetry in the upstream and downstream data rates is that the designers envisioned that the major use of the broadband connection would be for surfing the Internet or video on demand where small messages would be sent from the customer to a server and large files would be down-loaded from the server to the customer.

17.3 Components of a Simplified ADSL Transmitter

The block diagram of an ADSL transmitter that is a simplified version of the one in the ANSI T1.413 standard [E.1] is shown in Figure 17.2. The ITU-T G.993.3 transmitter [E.10] is very similar. These standards include extensive descriptions of how data is collected from a number of input streams and put into frames to form the "Data Source" shown in Figure 17.2 for a wide variety of data rates that are multiples of 32 kbit/s. The data rates are adaptively determined from measurements of the local loop frequency response and noise and are computed based on equations from Information Theory to maximize the rate of reliable transmission. The standards include a parallel path into the "Map block" box that does not include the interleaver to give a "fast" path with much less delay. They also include the option for a trellis encoder between the "Map block" and IFFT block.

17.3.1 The Cyclic Redundancy Check Generator

The cyclic redundancy check generator (CRC) computes a check byte for each successive block of k bits from the Data Source. Because of the flexibility of how the input sources are multiplexed and variety of data rates, these blocks can be anywhere from 67 up to around 14,875 bytes long. Let the block of data bits be the row vector

$$\mathbf{M} = [m_{k-1}, m_{k-2}, \ldots, m_1, m_0]$$

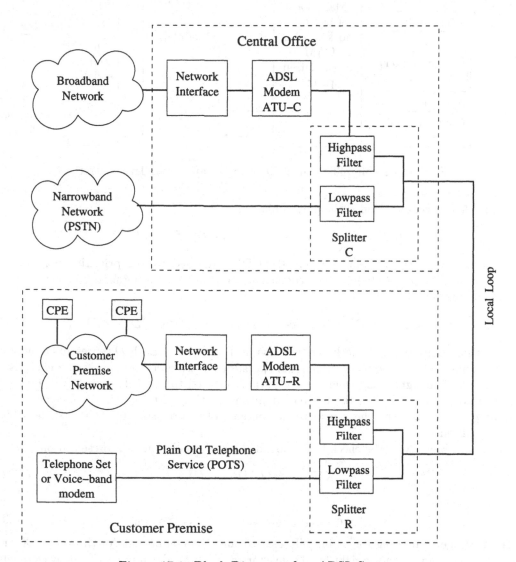

Figure 17.1: Block Diagram of an ADSL System

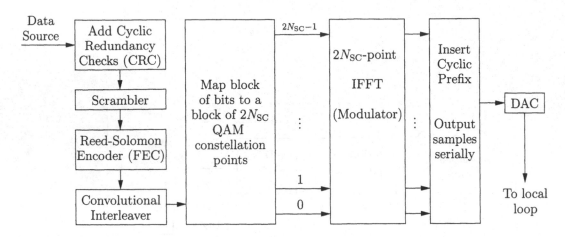

Figure 17.2: Simplified Block Diagram of an ADSL Transmitter

An equivalent representation is the polynomial

$$M(D) = m_{k-1} + m_{k-2}D + \ldots + m_1 D^{K-2} + m_0 D^{k-1}$$

The check byte is the remainder when $M(D)D^8$ is divided by the primitive polynomial $g(D) = D^8 + D^4 + D^3 + D^2 + 1$ where coefficients are computed using modulo 2 arithmetic. Let this check byte be

$$c(D) = \text{mod}[M(D)D^8, g(D)] = c_7 + c_6 D + \ldots + c_2 D^5 + c_1 D^6 + c_0 D^7 \qquad (17.1)$$

A circuit for computing the check byte is shown in Figure 17.3. Each D represents a one-bit delay element. The adders perform modulo two addition which is the same as an exclusive-or function. The eight delay elements are initially cleared to zero. Then the data bits are shifted into the right side of the feedback register starting with m_0, the coefficient of the highest power of D. The check byte is the contents of the register after the last data bit, m_{k-1} is shifted in.

The data block with the check byte appended has the polynomial representation

$$X(D) = M(D)D^8 + c(D) \qquad (17.2)$$

or vector form

$$\mathbf{X} = [c_7, c_6, \ldots, c_1, c_0, m_{k-1}, m_{k-2}, \ldots, m_1, m_0] \qquad (17.3)$$

If the CRC generator register is initially cleared and the components of the vector $\mathbf{X}$ are shifted into the adder on the right-hand side, starting with m_0 down to c_7, the delay elements will end up being all 0. If $Y(D) = X(D) + e(D)$ where $e(D)$ is an error pattern which is not divisible by $g(D)$, the register contents will not be zero when $Y(D)$ is shifted into the CRC generator. Any error pattern which is divisible by $g(D)$ is an undetectable error pattern.

Adding the CRC byte allows the receiver to make a final overall check for transmission errors. CRC failures are typically used by the system operators to check for equipment failures.

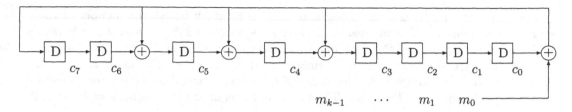

Figure 17.3: Cyclic Redundancy Check (CRC) Generator

17.3.2 The Scrambler

The bit stream from the CRC generator is then passed through a self-synchronizing scrambler like the one discussed in Chapter 9 and shown in Figure 9.1 to break up long strings of 1's and 0's. The connection polynomial is $h(D) = 1 + D^{18} + D^{23}$ which is the same as the one used in the V.32 and V.34 calling modems. Let the input to the scrambler be $x(n)$. Then, the scrambler output is

$$y(n) = x(n) + y(n-18) + y(n-23) \tag{17.4}$$

where "+" is modulo two addition. The input can be computed from the output, that is, descrambled by computing

$$x(n) = y(n) + y(n-18) + y(n-23) \tag{17.5}$$

17.3.3 The Reed-Solomon Encoder

Successive blocks of bits from the scrambler are operated on by the Reed-Solomon encoder which adds redundant bits to the blocks so that transmission errors can be corrected at the receiver. This process is often called *forward error correction* (FEC). The detailed theory of these codes is beyond the scope of this course. See books on error correcting codes like Peterson and Weldon [F.12] and Wicker [F.19] for complete details. A brief description of the encoder will be given here to give you a hint of what it does.

The ADSL Reed-Solomon encoder is based on the primitive polynomial $f(D) = D^8 + D^4 + D^3 + D^2 + 1$ which was used in the CRC generator. This polynomial has roots in the Galois field GF(2^8). Let α be one of the roots. Then all the 256 elements of GF(2^8) can be represented in the form

$$\beta = b_0 + b_1\alpha + \ldots + b_6\alpha^6 + b_7\alpha^7$$

where the b_i's are 0 or 1. An element can also be represented by the byte $[b_0, b_1, \ldots, b_7]$. The ADSL Reed-Solomon encoder considers code symbols to be elements of GF(2^8) or bytes. A codeword can be at most 255 bytes long including the information and check bits. The integer 255 is the natural length of codewords, but some information symbols can be set to zero and not used. A variety of generator polynomials can be used and have the form

$$g(D) = \prod_{i=0}^{R-1} (D - \alpha^i) \tag{17.6}$$

where R is a positive integer. The degree of $g(D)$ is R which is also the number of check symbols in a codeword. The number of information symbols is $255 - R$ and the code rate is $(255 - R)/255$. Up to $t = \lfloor R/2 \rfloor$ symbol errors can be corrected in a codeword. Anywhere from one to eight bit errors in a code symbol (byte) is a single symbol error as far as the Reed-Solomon code is concerned. The choice of R allows a tradeoff between error correction capability and code rate. The ANSI T1.413 standard requires implementations for $R = 2i$ for $i = 0, 1, \ldots, 8$.

A block of K message bytes $[\mathbf{m}_0, \mathbf{m}_1, \ldots, \mathbf{m}_{K-1}]$ can be represented by the polynomial

$$\mathbf{M}(D) = \sum_{i=0}^{K-1} \mathbf{m}_i D^{K-1-i} \tag{17.7}$$

The check symbols are the remainder when $\mathbf{M}(D)D^R$ is divided by $g(D)$ using GF(2^8) arithmetic, that is,

$$\mathbf{C}(D) = \mathrm{mod}[\mathbf{M}(D)D^R, g(D)] = \sum_{i=0}^{R-1} \mathbf{c}_i D^{R-1-i} \tag{17.8}$$

and the complete codeword is $x(D) = \mathbf{M}(D)D^R + \mathbf{C}(D)$. The highest order K symbols are the message bytes and the lower order R symbols are the check symbols. The codeword length is $N = K + R$.

17.3.4   The Convolutional Interleaver

Impulse noise is one of the major disturbances in ADSL systems. An impulse can corrupt several multi-carrier symbols and cause bursts of errors in the received data streams if no special precautions are taken. ADSL systems protect against bursts in two ways. First, the Reed-Solomon codes discussed in the previous section naturally can correct multiple bit errors in a single code symbol, which consist of a string of eight bits, since any pattern of eight or less bits is treated as a single GF(2^8) code symbol error. Second, a technique that has been used for many years in other systems called interleaving is required by the standards for the data streams needing low error rates. Detailed discussions of interleaving are presented in Ramsey [F.13] and Forney [F.5]. Basically, an interleaver shuffles multiple codewords together before transmission so symbols from an individual codeword are separated in time when the interleaved codewords are transmitted over the channel. Suppose the symbols from one codeword occur in the interleaved stream at multiples of an integer L called the *interleaving depth*. Then an error burst of length B on the channel will cause about B/L errors in the word obtained by deinterleaving at the receiver. The exact number of errors in the deinterleaved word depends on the burst phasing and length.

The ADSL standards specify the use of a type of interleaver called a *convolutional* interleaver. Suppose we start entering codewords of length N bytes into the interleaver at time 0. Then, entering of the jth codeword starts at time jN. Let this codeword be designated by

$$\mathbf{X}_j = [B_0^j, B_1^j, \ldots, B_{N-1}^j]$$

The symbols are entered into the interleaver starting from the left with B_0^j at time jN and ending on the right with B_{N-1}^j at time $jN + N - 1 = (j + 1)N - 1$. Symbol B_i^j is entered at time $jN + i$. The ADSL interleaver delays B_i^j by $i(L - 1)$ symbols for $i = 0, 1, \ldots, N - 1$. The integer L is called the interleaving depth. The ANSI standard requires support of the values 1, 2, 4, 8, 16, 32, and 64 for L which are all powers of 2. The symbol B_i^j emerges from the interleaver at time

$$T_i^j = jN + i + i(L - 1) = jN + iL \tag{17.9}$$

When N and L are relatively prime, each code symbol emerges from the interleaver at a distinct time and every time slot is filled with a code symbol. When N is odd and L is a power of 2, N and L are relatively prime.

One possible interleaver implementation is shown in Figure 17.4. Each D represents a delay of one time unit for a code symbol, that is, a byte, so D^L is a delay of L time units. We will call the cascade of delay elements a shift register. The output of each adder is the bit-wise exclusive-or of the two input bytes. At time jN the symbols for codeword $\mathbf{X}_j$ appear in parallel at the inputs to the "and" gates. The output of a gate is the input byte when the "Load" signal is 1, and a 0 byte when Load is 0. The Load signal becomes 1 at time $n = jN$ so B_0^j appears at the output labeled I_n. The register is clocked once and B_0^j is clocked out of the interleaver into the next system component. At the same time, the rest of the B_i^j's are loaded into the D^L boxes. Then Load is set to 0 to make the outputs of the and gates 0 bytes and the registers are clocked $N - 1$ more times. This process is repeated for each new codeword. The Load signal is 0 except at times jN when a new codeword arrives. To summarize, the N code symbols of a new codeword are loaded into the register stages separated by L time units and then N bytes are clocked out of the register. This process is repeated every N clock pulses for each new codeword. It can be shown that when N and L are relatively prime, new code symbols are entered into empty stages in the register and all positions are filled with bytes from some codeword. At the start of each new codeword, that is, at time jN, the horizontal inputs to the adders are all zero bytes that have been loaded into the shift register during the $N - 1$ times when Load is 0. This structure is equivalent to the Ramsey Type II [F.13] interleaver.

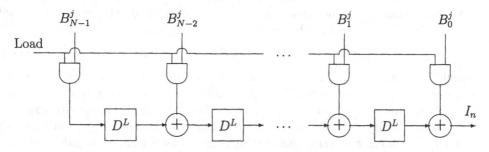

Figure 17.4: Shift Register Implementation of the ADSL Interleaver

The block diagram of a deinterleaver is shown in Figure 17.5. Assuming that the interleaver output is directly connected to the deinterleaver input, starting at time 0, symbols of the interleaved sequence are shifted into an $(N-1)L$ stage shift register with taps every L stages as shown in the figure. After a delay of $(N-1)L$ symbols, B_0^0 appears at the right side of the shift register. Then N symbols are combed out of the shift register at the taps and collected into the original codeword transmitted starting at time 0. The figure shows these N symbols being loaded in parallel into an N-symbol shift register by the Load signal. The re-assembled codeword can then be shifted out of this register serially. It could also just be written into an array in a RAM memory. This process is then repeated continually with N symbols shifted into the register and N combed out. Figure 17.5 shows the situation when the jth codeword is ready to be combed out of the register. Each input symbol to the interleaver experiences a delay of $(N-1)L$ symbols before it is combed out of the deinterleaver, ignoring the other system delays.

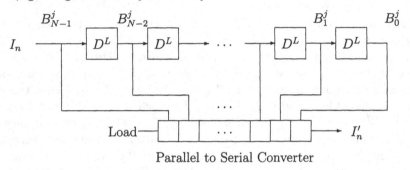

Parallel to Serial Converter

Figure 17.5: A Deinterleaver

The interleaver can also be implemented using a RAM organized as a two-dimensional array with N columns and L rows. An example for $N = 5$ and $L = 2$ is illustrated in Figure 17.6. A row of N interleaved symbols is read out of the array starting from column 0 on the left. Then the symbols of a new codeword, say, $\mathbf{X}_j$, are entered into the array from row j down in a manner that will now be described. The symbol B_i^j should appear at the interleaver output at time $T_i^j = jN + iL$ for $i = 0, \ldots, N-1$. Therefore symbol B_0^j should be written into row j and column 0 of the array. The next symbol, B_1^j should be written into column L of row j if $L < N$. Symbol B_i^j is written into column iL of row j as long as $iL < N$. In general, iL can be divided by N to get quotient, q_i, and remainder r_i with $0 \le r_i \le N - 1$ and $iL = q_iN + r_i$. Therefore,

$$T_i^j = jN + (q_iN + r_i) = (j + q_i)N + r_i \qquad (17.10)$$

and B_i^j should be written into column r_i of row $j + q_i$. In words, it should be written q_i rows down from row j and r_i columns over. After $\mathbf{X}_j$ is entered into the array, the N symbols in row j are read out from left to right. The process is then repeated starting with row $j + 1$, and so on. Deinterleaving can be performed by writing the received rows into a memory array similar to the interleaver array and combing the symbols from each codeword out of the array.

The maximum number of rows down occurs for $i = N - 1$ and is, assuming $N > L$,

$$q_{N-1} = \left\lfloor \frac{(N-1)L}{N} \right\rfloor = \left\lfloor L - \frac{L}{N} \right\rfloor = L - 1 \qquad (17.11)$$

Therefore, the symbols of codeword $\mathbf{X}_j$ are distributed over L rows starting with row j. After row j is read out, its contents are no longer needed. Consequently, the row can be cleared and the row containing the last few symbols for codeword $\mathbf{X}_{j+1}$ can be written into the row just read out. Thus the memory can be used as a circular array of L rows.

	0	1	2	3	4
$j-1$	$\vdots$	$\vdots$	$\vdots$	$\vdots$	$\vdots$
j	B_0^j	B_3^{j-1}	B_1^j	B_4^{j-1}	B_2^j
$j+1$	B_0^{j+1}	B_3^j	B_1^{j+1}	B_4^j	B_2^{j+1}
$j+2$	B_0^{j+2}	B_3^{j+1}	B_1^{j+2}	B_4^{j+1}	B_2^{j+2}
$j+3$	$\vdots$	B_3^{j+2}	$\vdots$	B_4^{j+2}	$\vdots$

Figure 17.6: RAM Interleaver Implementation Example for $N = 5$ and $L = 2$

17.3.5   The Map and IFFT Modulator Blocks

Successive blocks of bits are collected from the interleaver every T_0 seconds and assigned to subcarriers in the transmitted symbols. For ADSL, the nominal data symbol rate is $f_0 = 1/T_0 = 4000$ symbols per second. The number of bits assigned to a subcarrier is determined by measurements of the channel frequency response and noise at the subcarrier frequency made during the transceiver initialization sequence. The channel characteristics can also be monitored during data transmission and the bit assignments adjusted. See Golden, et al., [E.3, pp. 204–206] for a discussion of bit allocation algorithms. Each group of bits assigned to a subcarrier is mapped to a complex number $Z_k = X_k + jY_k$ corresponding to a point in a QAM constellation. The number of bits assigned to a subcarrier can range from 2 through 15, that is, the constellation size can vary from $2^2 = 4$ up to $2^{15} = 32,768$ points according to the ADSL standard. The constellation points Z_k for $0 \leq k \leq N/2 - 1$ are considered to be the first half of the elements of an N-point DFT. Actually, the standard requires that $Z_0 = 0$. The remaining half of the elements are set to

$$Z_{N-k} = \overline{Z}_k \text{ for } k = 1, \ldots, N/2 - 1 \text{ and } Z_{N/2} = \overline{Z}_{N/2} \qquad (17.12)$$

This implies that $Z_{N/2}$ must be real. The standard specifies that it should not be used for data transmission. With this complex conjugate symmetry, it can be shown that the N-point time sequence $z_n = \text{IDFT}\{Z_k\}$ is real. Conversely, when z_n is real, its DFT must have this complex conjugate symmetry. The central office ATU-C transmitter uses $N = 512$ and the remote ATU-R transmitter uses $N = 64$. In Figures 17.2 and 17.8, $N_{SC} = N/2$.

Using an IDFT for Modulation

Modulation is performed by taking an IDFT of the sequence of constellation points. Since N is a power of 2, the IDFT can be efficiently computed using an inverse fast Fourier transform (IFFT). The time sequence computed by the IDFT, and more efficiently by an IFFT, is

$$z_n = \frac{1}{N} \sum_{k=0}^{N-1} Z_k e^{j\frac{2\pi}{N}kn} = \frac{1}{N} \sum_{k=0}^{N-1} Z_k e^{j2\pi\left(k\frac{f_1}{N}\right)nT_1} \quad \text{for} \quad n = 0, \dots, N-1 \qquad (17.13)$$

where $f_1 = 1/T_1$ is the sampling frequency. Let $Z_k = A_k e^{j\theta_k}$ where $A_k = |Z_k|$ and $\theta_k = \arg Z_k$. Using the complex conjugate symmetry property of the Z_k sequence and assuming $Z_0 = Z_{N/2} = 0$, it follows that

$$
\begin{aligned}
z_n &= \frac{1}{N} \sum_{k=1}^{\frac{N}{2}-1} \left[Z_k e^{j2\pi k \frac{f_1}{N}nT_1} + Z_{N-k} e^{j2\pi(N-k)\frac{f_1}{N}nT_1} \right] \\
&= \frac{1}{N} \sum_{k=1}^{\frac{N}{2}-1} A_k \left[e^{j\left(2\pi k\frac{f_1}{N}nT_1 + \theta_k\right)} + e^{-j\left(2\pi k\frac{f_1}{N}nT_1 + \theta_k\right)} \right] \\
&= \frac{2}{N} \sum_{k=1}^{\frac{N}{2}-1} A_k \cos\left(2\pi k\frac{f_1}{N}nT_1 + \theta_k \right)
\end{aligned}
\qquad (17.14)
$$

Therefore, the time sequence z_n is the sum of cosines at the subcarrier frequencies $f^{(k)} = kf_1/N$ for $k = 1, \dots, N/2-1$ sampled with period T_1. The frequency $f_1/2$ corresponding to $k = N/2$ is called the *Nyquist frequency*.

Algorithm for Creating Constellations

If b bits are assigned to a subcarrier, its constellation must have 2^b points. Let the bits assigned to a subcarrier be the binary vector $(v_{b-1}, v_{b-2}, \dots, v_1, v_0)$ which has the decimal value $d = \sum_{i=0}^{b-1} v_i 2^i$. In the ANSI T1.413-1998 ADSL standard [E.1], the X and Y coordinates of a constellation point when b is even are, to within a scale factor, the odd integers with the 2's complement representations

$$X_k \leftrightarrow (v_{b-1}, v_{b-3}, \dots, v_1, 1) \quad \text{and} \quad Y_k \leftrightarrow (v_{b-2}, v_{b-4}, \dots, v_0, 1) \qquad (17.15)$$

Constellations for $b = 2$ and 4 are shown if Figure 17.7. The ADSL standard also gives an algorithm for constellations with b odd. In practice, the actual constellations are scaled to give desired power levels for each subcarrier.

The ANSI ADSL standard [E.1, Section 6.8] also includes an option for using a 4-dimensional, 16-state, Wei [F.18] trellis code for converting input data bits to sequences of constellation points. This code is a slight variation of the one used in V.34 modems. The details of this code will not be described here and are left for interested readers to pursue.

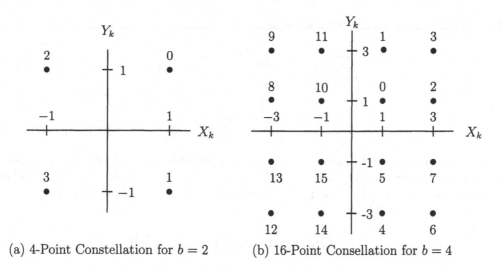

(a) 4-Point Constellation for $b = 2$ (b) 16-Point Consellation for $b = 4$

Figure 17.7: Constellations with 4 and 16 Points

The Cyclic Prefix

The impulse response of a local loop lasts for many samples. However, it changes very slowly over time, so the local loop can be considered to be a linear time-invariant system, for all practical purposes. The impulse response of a wireless channel can also last for many samples as a result of multi-path propagation and this is referred to as *delay spread*. The wireless channel can change relatively quickly with time as a result of motion of the transmitter, receiver, and reflecting objects and must be treated as a time-varying channel. However, multi-carrier modulation and linear time-invariant methods can be used if the channel changes insignificantly over the duration of a symbol and channel estimates are continually updated. In either case, the channel impulse response causes intersymbol interference (ISI) between adjacent multi-carrier symbols. Suppose the impulse response, $h(n)$, can be nonzero only for $n = 0, 1, \ldots, L$ and a symbol, z_n, starting at time $n = 0$ ends at time $n = M$ and the symbol duration is longer than the impulse response duration, that is, $M > L$. The ISI from this symbol into the next is

$$I(n) = \sum_{k=n-L}^{M} z_k h(n-k) \quad \text{for} \quad n = M+1, \ldots, M+L \tag{17.16}$$

as $h(n-k)$ slides off the right side of the non-zero portion of z_k. A similar initial transient occurs as $h(n-k)$ slides onto the next symbol during the same time period.

The ISI and initial transient problem has been solved for wireline and wireless systems by increasing the symbol duration to $N' = N + \nu$ samples by prepending a cleverly chosen sequence of ν samples which is longer than the impulse response duration. This provides a guard time between symbols for the ISI to disappear. The prepended samples are the last ν samples of the symbol, that is, $(z_{N-\nu}, \ldots, z_{N-2}, z_{N-1})$, so the augmented symbol sequence is $(z_{N-\nu}, \ldots, z_{N-2}, z_{N-1}, z_0, z_1, \ldots, z_{N-1})$. Suppose z_0 occurs at time $n = 0$. Then the channel output for $0 \leq n \leq N - 1$, ignoring noise, is

$$r(n) = \sum_{k=n-L}^{n} z_{\mathrm{mod}(k,N)} h(n-k) = \sum_{k=0}^{N} z_k h_{\mathrm{mod}(n-k,N)} \qquad (17.17)$$

where

$$h_n = \begin{cases} h(n) & \text{for } 0 \leq n \leq L \\ 0 & \text{for } L+1 \leq n \leq N-1 \end{cases}$$

This formula can be thought of as the convolution of $h(n)$ with a periodically repeated version of the N symbol samples computed by the IFFT, or the *circular* or *cyclic* convolution of the N-point sequences z_n and h_n. The DFT of the cyclic convolution of two N-point signals is the product of their DFT's, that is

$$R_k = \mathrm{DFT}\{r_n\} = \mathrm{DFT}\{z_n\}\mathrm{DFT}\{h_n\} = Z_k H_k \qquad (17.18)$$

The $N' = N + \nu$ points of the augmented sequence are sent to the DAC starting with the cyclic prefix.

The ATU-C and ATU-R both use cyclic prefixes of length $\nu = N/16$. Thus the length of the cyclic prefix for the ATU-C is $\nu_C = 512/16 = 32$ and $N'_C = 544$. For the ATU-R, $\nu_R = 64/16 = 4$ and $N'_R = 68$.

The Actual Sampling Rates and Subcarrier Frequencies

The ATU-C and ATU-R insert a known N'-point sync sequence in their symbol streams after every 68 data symbols. To maintain a data symbol rate of 4000 data symbols per second, the channel symbol rate is increased by the factor 69/68. Therefore, the sampling rates are

$$f_1 = \frac{69}{68} \times 4000 \times (N + \nu) = \begin{cases} \frac{69}{68} \times 4000 \times 544 & = 2.208 \text{ MHz} & \text{for ATU-C} \\ \frac{69}{68} \times 4000 \times 68 & = 276 \text{ kHz} & \text{for ATU-R} \end{cases} \qquad (17.19)$$

The subcarrier frequencies in both cases are multiples of $f_1/N = 4.3125$ kHz.

17.3.6 Some Signals Used for Initialization and Synchronization

The ADSL standards define a complicated procedure to initialize the central office and remote transceivers for full duplex data transmission. Interested readers should consult the standards for all the details. A few of the signals defined in the ANSI standard will be described in this section so they can be used for the experiments.

One obvious signal is called QUIET. This signal is just 0 volts for a period of time. It is used before transmission starts and the transceivers are idle as well as between several segments of the initialization sequence.

A signal transmitted by the ATU-C is called C-REVERB. This signal allows the ATU-R to adjust its automatic gain control (AGC), synchronize its timing, measure noise on the channel, measure the channel amplitude response, and adjust its adaptive equalizer. The data bit sequence for a DMT C-REVERB symbol is generated starting at time $n = 1$ by the following rule:

$$d_n = 1 \qquad \text{for} \quad n = 1, \ldots, 9$$
$$d_n = d_{n-4} + d_{n-9} \quad \text{for} \quad n = 10, \ldots 512 \qquad (17.20)$$

where "$+$" is modulo 2 addition. The bits d_1 through d_9 are initialized to all 1's for each C-REVERB symbol, so the C-REVERB symbols all use the same data sequence. This is the rule for a 9-stage PN sequence generator as discussed in Chapter 9 and shown in Figure 9.1 except the output is taken from the delay element on the right side of the generator and all the delay elements are initially set to 1. The connection polynomial is the primitive polynomial $h(D) = 1 + D^4 + D^9$. The period of this sequence is $2^9 - 1 = 511$, so $d_{512} = d_1$. Subcarriers 0 and 256 are set to 0 and bits d_1 and d_2 are not used. Bit pairs (d_{2k+1}, d_{2k+2}) are used to QPSK modulate subcarriers $k = 1, \ldots, 255$ as shown up to a scale factor in Table 17.1. Bits d_{129} and d_{130} which modulate subcarrier 64 are overwritten by bits $(0, 0)$ to generate the constellation point $(X_{64}, Y_{64}) = (+1, +1)$ to form a *pilot carrier*. The 512 C-REVERB symbol time samples are transmitted without the cyclic prefix. The C-REVERB symbol is repeated anywhere from 512 up to 1536 times depending on the several positions it occupies in the initialization sequence.

Another signal called C-SEGUE is the 180 degree phase reversal of C-REVERB for 10 symbols and follows right after the end of C-REVERB. This phase reversal can be detected by the receiver and used as a timing mark.

Table 17.1: Mapping of Two Data Bits to QPSK Constellation Points

d_{2k+1}	d_{2k+2}	Decimal Label	X_k	Y_k
0	0	0	$+1$	$+1$
0	1	1	$+1$	-1
1	0	2	-1	$+1$
1	1	3	-1	-1

The ATU-R transmits similar signals. The bits for R-REVERB are generated by a 6-stage PN sequence generator according to the rule:

$$d_n = 1 \qquad \text{for} \quad n = 1, \ldots, 6$$
$$d_n = d_{n-5} + d_{n-6} \quad \text{for} \quad n = 7, \ldots 64 \qquad (17.21)$$

Bits d_1 through d_6 are initialized to all 1's for each R-REVERB symbol, so they all use the same data sequence. This sequence has period 63 so $d_{64} = d_1$. Subcarriers 0 and 32 are set to 0 and bits d_1 and d_2 are not used. Bit pairs (d_{2k+1}, d_{2k+2}) are used to QPSK modulate subcarriers $k = 1, \ldots, 31$ as shown up to a scale factor in Table 17.1. Bits d_{33} and d_{34} which modulate subcarrier 16 are overwritten by bits $(0, 0)$ to generate the constellation point $(X_{16}, Y_{16}) = (+1, +1)$ to form a *pilot carrier*. R-REVERB is transmitted with no cyclic prefix and repeated many times. An R-SEGUE similar to C-SEGUE is used to generate a timing mark.

The ATU-C and ATU-R both send a synchronization symbol after every 68 data symbols. These symbols are identical to C-REVERB and R-REVERB except that the cyclic prefixes are included.

17.4   A Simplified ADSL Receiver

The block diagram of a simplified ADSL receiver is shown in Figure 17.8. Signals from the far end transmitter are first applied to an analog-to-digital converter (ADC). The sampling rate for the signal received at the ATU-R receiver from the ATU-C transmitter is typically $f_1 = 2.208$ MHz which is twice the Nyquist frequency for the downstream signal. The sampling rate for the signal received at the ATU-C receiver from the ATU-R transmitter is typically $f_1 = 276$ kHz. The remaining blocks in the figure, except for the Frequency Domain Equalizer, perform the inverse operations of those in the transmitter.

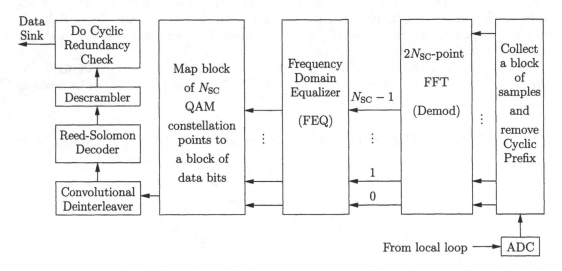

Figure 17.8: Simplified Block Diagram of an ADSL Receiver

One block that has not been shown is called a *time-domain equalizer* (TEQ). The impulse response of a local loop is often longer than the lengths of the cyclic prefixes specified in the standards. This problem has been solved by putting an adaptive filter between the ADC and "Collect a block of samples" block. The job of this filter is to shorten the channel impulse response to an acceptable length. It is trained during the initialization procedure. This type of filter is usually not used in wireless systems because the signal parameters are chosen to make the delay spread no longer than the cyclic prefix length. Also, the filter would have to be updated as fast as the wireless channel changes. We will not implement the TEQ in this course. See Chapter 11 of Golden, et al., [E.3] for details.

Other blocks that have not been shown explicitly are those for sampling clock and symbol clock acquisition and tracking. These functions are critical in making a transceiver actually work in the real world but are usually overlooked in textbooks. They are implicitly included in the "Collect a block of samples" block. Approaches to timing acquisition and tracking will be discussed in Sections 17.4.2 and 17.4.3.

17.4.1   Demodulation and Frequency Domain Equalization

The samples for a DMT symbol are collected and a sequence of $N = 2N_{SC}$ consecutive samples are extracted by removing the cyclic prefix. Let the resulting sequence for symbol i be $r_{n,i}$ for $n = 0, \ldots, N-1$. The FFT demodulator output is

$$R_{k,i} = \sum_{n=0}^{N-1} r_{n,i} e^{-j\frac{2\pi}{N}nk} = Z_{k,i}H_k + V_{k,i} \quad \text{for} \quad k = 0, \ldots, N-1 \qquad (17.22)$$

where H_k is the channel frequency response at subcarrier k, $S_k = \mathrm{E}\{|Z_{k,i}|^2\}$, and $V_{k,i}$ is a zero-mean noise sample with variance $\sigma_k^2 = \mathrm{E}\{|V_{k,i}|^2\}$. The effect of H_k is to scale and rotate the received constellation. Only the lower half of the points are needed because of the complex conjugate symmetry of the transform of a real sequence.

The channel frequency response can be estimated during initialization when a know sequence like REVERB is transmitted for many symbols. One method for estimating H_k is to average $R_{k,i}/Z_{k,i}$ over a number of symbols. This assumes that the sample clock has already been acquired and drifts negligibly between symbols.

The initial channel frequency response estimates can be adaptively updated during regular data transmission to refine the initial estimates, track slow changes in the channel caused by environmental factors like daily variations in temperature, and compensate for small sample clock drifts. The LMS algorithm, discussed for single carrier modems in Chapter 15, with a one-tap equalizer can be used for each subcarrier. The resulting set of equalizers is called a *frequency domain equalizer* (FEQ). Let the equalizer coefficient for subcarrier k be W_k, the equalized received point be $\tilde{Z}_{k,i} = W_k R_{k,i}$, and the estimation error be $\Delta_{k,i} = Z_{k,i} - W_k R_{k,i}$. Under noisy conditions, W_k can be chosen to minimize

$$\Lambda_k = \mathrm{E}\left\{|\Delta_{k,i}|^2\right\} = \mathrm{E}\left\{(Z_{k,i} - W_k R_{k,i})\overline{(Z_{k,i} - W_k R_{k,i})}\right\} \qquad (17.23)$$

If H_k is known, the closed form solution is

$$\hat{W}_k = \frac{S_k}{S_k|H_k|^2 + \sigma_k^2}\,\overline{H}_k \qquad (17.24)$$

and the corresponding mean-squared error is

$$\hat{\Lambda}_k = \sigma_k^2 \frac{S_k}{S_k|H_k|^2 + \sigma_k^2} \qquad (17.25)$$

Since H_k is not actually known, the iterative LMS algorithm can be used. Let A_k and B_k be the real and imaginary parts of W_k so that $W_k = A_k + jB_k$. Taking derivatives as in the LMS derivation in Chapter 15 gives

$$\frac{\partial \Lambda_k}{\partial A_k} = -2\,\mathrm{E}\{\Re e\{\Delta_{k,i}\overline{R}_{k,i}\}\} \qquad (17.26)$$

and

$$\frac{\partial \Lambda_k}{\partial B_k} = -2\operatorname{E}\{\Im m\{\Delta_{k,i}\overline{R}_{k,i}\}\} \tag{17.27}$$

so

$$\frac{\partial \Lambda_k}{\partial W_k} \triangleq \frac{\partial \Lambda_k}{\partial A_k} + j\frac{\partial \Lambda_k}{\partial B_k} = -2\operatorname{E}\{\Delta_{k,i}\overline{R}_{k,i}\} \tag{17.28}$$

This suggests the approximate update formula

$$W_k(i+1) = W_k(i) + \mu\Delta_{k,i}\overline{R}_{k,i} = W_k(i) + \mu[Z_{k,i} - W_k(i)R_{k,i}]\overline{R}_{k,i} \tag{17.29}$$

where μ is a small positive constant. It was stated in Chapter 15 that in order for the mean-squared error to converge, μ must satisfy the stability constraint

$$0 < \mu < 2/\operatorname{E}\{|R_{k,i}|^2\} = 2/(S_k|H_k|^2 + \sigma_k^2) \tag{17.30}$$

Since the product $\Delta_{k,i}\overline{R}_{k,i}$ is a rough estimate of its expected value, the coefficient estimate converges to the correct mean value with excess fluctuations. Making μ small reduces the excess variance but slows the convergence time. A formula for the excess mean-squared error is given by (15.29).

The true value of $Z_{k,i}$ is not known at the receiver. However, after initial channel estimation, the equalizer output $\tilde{Z}_{k,i} = W_k R_{k,i}$ will be close to the true value. Then an accurate estimate of $Z_{k,i}$ can be made by quantizing $\tilde{Z}_{k,i}$ to the closest ideal constellation point. The quantization procedure results in very accurate decisions during initialization symbol sequences when the 4-point QPSK constellation is used. This "decision directed" approach works very well in practice even for constellations of large size after the equalizer coefficient has converged and μ is made small.

17.4.2 Sample Clock Acquisition and Tracking

The sample clocks in a transmitter and remote receiver will have some frequency offset. This offset will be small because the standards place tight tolerances on the clock frequency. Nevertheless, a small frequency error will accumulate to a large phase error over time leading to significant degradation of performance. The receiver must lock its clock to the transmitter clock based on timing information derived from its received signal. The offset can be estimated in an open loop manner during initialization when known signals are transmitted, and updated in a closed loop tracking mode during normal data transmission. Most systems use a phase-locked loop (PLL) to generate and track the sampling times. Some systems emulate a phase-locked loop in software using numerical interpolation or buffers with sample stuffing or deletion to eliminate the cost of the hardware PLL.

Effect of Sample Clock Offset on the Output of the Receiver's FFT

Let $T_1 = 1/f_1$ be the sampling period of the transmitter. Consider the symbol with samples, z_n, given by (17.13) starting at time $n = 0$ with a ν sample cyclic prefix longer than the channel impulse response prepended for time $-\nu, -\nu+1, \ldots, -1$. When this $N + \nu$ point sequence is applied to the DAC, the resulting continuous-time signal is

$$z(t) = \frac{1}{N} \sum_{\ell=0}^{N-1} Z_\ell e^{j2\pi\ell\frac{f_1}{N}t} \tag{17.31}$$

Let the sampling period at the receiver be $T_2 = (1 + \epsilon)T_1$ where ϵ is a small constant. Then, the samples observed at the receiver are

$$r_{n,0} = \frac{1}{N} \sum_{\ell=0}^{N-1} H_\ell Z_\ell e^{j2\pi\ell\frac{f_1}{N}n(1+\epsilon)T_1} = \frac{1}{N} \sum_{\ell=0}^{N-1} H_\ell Z_\ell e^{j\frac{2\pi}{N}(1+\epsilon)n\ell} \tag{17.32}$$

where H_k is the channel frequency response at subcarrier k. The receiver computes the DFT of this sequence to get

$$
\begin{aligned}
R_{k,0} &= \sum_{n=0}^{N-1} r_{n,0}\, e^{-j\frac{2\pi}{N}nk} = \sum_{n=0}^{N-1} \frac{1}{N} \sum_{\ell=0}^{N-1} H_\ell Z_\ell e^{j\frac{2\pi}{N}(1+\epsilon)n\ell} e^{-j\frac{2\pi}{N}nk} \\
&= \sum_{\ell=0}^{N-1} H_\ell Z_\ell \frac{1}{N} \sum_{n=0}^{N-1} e^{j\frac{2\pi}{N}n[(1+\epsilon)\ell-k]} = \sum_{\ell=0}^{N-1} H_\ell Z_\ell\, Q(k,\ell,\epsilon)
\end{aligned}
\tag{17.33}
$$

where

$$Q(k,\ell,\epsilon) = \frac{1}{N} \sum_{n=0}^{N-1} e^{j\frac{2\pi}{N}n[(1+\epsilon)\ell-k]} = \frac{1}{N} \frac{e^{j2\pi[(1+\epsilon)\ell-k]} - 1}{e^{j\frac{2\pi}{N}[(1+\epsilon)\ell-k]} - 1} \tag{17.34}$$

By factoring out e to half the exponent in the numerator and denominator, this can be put into the following more informative form:

$$Q(k,\ell,\epsilon) = e^{j\frac{2\pi}{N}[(1+\epsilon)\ell-k]\frac{N-1}{2}} \frac{\sin \pi[(1+\epsilon)\ell - k]}{N \sin \frac{\pi}{N}[(1+\epsilon)\ell - k]} \tag{17.35}$$

Equation (17.33) shows that an effect of timing offset is to introduce crosstalk from one subchannel into another. The component proportional to Z_k in $R_{k,0}$ is

$$H_k Z_k Q(k,k,\epsilon) = H_k Z_k e^{j\frac{2\pi}{N}k\epsilon\frac{N-1}{2}} \frac{\sin \pi k\epsilon}{N \sin \frac{\pi}{N}k\epsilon} \tag{17.36}$$

The timing frequency offset causes a phase rotation of $H_k Z_k$ by the angle $\beta_k = k\frac{2\pi}{N}\epsilon\frac{N-1}{2}$ that increases linearly with frequency index and scales its amplitude. The amplitude scaling is negligible for small $|\epsilon|$ and computation shows that the crosstalk components are small when the timing offset is within the tolerances required by the standards. Therefore,

$$R_{k,0} \approx H_k Z_k e^{j\frac{2\pi}{N}k\epsilon\frac{N-1}{2}} \quad \text{for } |\epsilon| \ll 1 \tag{17.37}$$

Open Loop Estimation of the Sample Clock Period Error

Now suppose the symbol including the cyclic prefix is repeated starting at time N. The transmitted signal is $z_1(t) = z[t - (N + \nu)T_1]$ for $NT_1 \leq t \leq (2N + \nu)T_1$. The cyclic prefix for this repeated symbol extends over $NT_1 \leq t \leq (N + \nu)T_1$. Skipping over what it thinks is the received cyclic prefix, the receiver collects the N-point sequence

$$
\begin{aligned}
r_{n,1} &= z_1[(n + N + \nu)T_2] = z_1[(n + N + \nu)(1 + \epsilon)T_1] = z[n(1 + \epsilon)T_1 + (N + \nu)\epsilon T_1] \\
&= \frac{1}{N} \sum_{\ell=0}^{N-1} H_\ell Z_\ell e^{j\frac{2\pi}{N}\ell\epsilon(N+\nu)} e^{j\frac{2\pi}{N}(1+\epsilon)n\ell} \quad \text{for } n = 0, \ldots, N-1
\end{aligned}
\tag{17.38}
$$

and computes the DFT

$$
R_{k,1} = \sum_{\ell=0}^{N-1} H_\ell Z_\ell e^{j\frac{2\pi}{N}\ell\epsilon(N+\nu)} Q(k, \ell, \epsilon)
\tag{17.39}
$$

Again, the subcarrier crosstalk is small if $|\epsilon| \ll 1$ and

$$
R_{k,1} \approx H_k Z_k e^{j\frac{2\pi}{N}k\epsilon\frac{N-1}{2}} e^{j\frac{2\pi}{N}k\epsilon(N+\nu)}
\tag{17.40}
$$

At this point in the initialization procedure, the channel frequency response H_k may not be known yet. The phase effects of this response and the constellation point Z_k can be removed by forming the product

$$
\Gamma_{k,1} = R_{k,1}\overline{R}_{k,0} = |H_k Z_k|^2 e^{j\frac{2\pi}{N}k\epsilon(N+\nu)}
\tag{17.41}
$$

which has the angle

$$
\varphi_{k,1} = \arg \Gamma_{k,1} = \frac{2\pi}{N}k\epsilon(N + \nu)
\tag{17.42}
$$

The preceding derivations have neglected the effects of noise. An accurate estimate of ϵ can be obtained when noise is present by averaging $\Gamma_{k,i} = R_{k,i}\overline{R}_{k,i-1}$ over several symbols for a range of k, computing phase estimates $\hat{\varphi}_k$ from the averages, and then choosing ϵ to form a least-squares fit of a straight line to the phase estimates over the range of k. Suppose the range is $k_1 \leq k \leq k_2$. Then the least-squares fit estimate is

$$
\hat{\epsilon} = \frac{N}{2\pi(N + \nu)} \left(\sum_{k=k_1}^{k_2} k^2 \right)^{-1} \sum_{k=k_1}^{k_2} k\hat{\varphi}_k
\tag{17.43}
$$

The sum of squares can be computed by using the formula

$$
\sum_{n=1}^{N} k^2 = \frac{N(N + 1)(2N + 1)}{6}
\tag{17.44}
$$

Tracking the Sample Clock Period

Once initialization is finished, random DMT symbols generated by customer data are transmitted and the open loop method of estimating the timing period error just presented cannot be used. At the end of initialization, the frequency domain equalizer (FEQ) should be well adjusted and a good estimate of ϵ should have been made and inserted into the clock generator. The constellations observed for each subcarrier at the output of the receiver's FFT should be very close to the ideal transmitted constellations. If there is a small residual clock period error, the constellations will appear to slowly rotate if the FEQ is kept constant as shown by (17.42). These observations suggests that a decision-directed method can be used for tracking the sample clock. Let $\hat{Z}_{k,i} = |\hat{Z}_{k,i}|e^{j\alpha_{k,i}}$ be the ideal constellation point closest to the equalized point $\tilde{Z}_{k,i} = |\tilde{Z}_{k,i}|e^{j\psi_{k,i}}$ for the symbol at time i. The operation of finding this ideal point is often called quantizing or *slicing* $\tilde{Z}_{k,i}$ to the nearest ideal point. The angle between the equalized and estimated ideal constellation points can be determined as

$$\zeta_{k,i} = \arg(\tilde{Z}_{k,i}\overline{\hat{Z}}_{k,i}) = \arg\left(|\tilde{Z}_{k,i}|\,|\hat{Z}_{k,i}|e^{j(\psi_{k,i}-\alpha_{k,i})}\right) = \psi_{k,i} - \alpha_{k,i} \qquad (17.45)$$

In rectangular form

$$\tilde{Z}_{k,i}\overline{\hat{Z}}_{k,i} = |\tilde{Z}_{k,i}|\,|\hat{Z}_{k,i}|\cos\zeta_{k,i} + j|\tilde{Z}_{k,i}|\,|\hat{Z}_{k,i}|\sin\zeta_{k,i} \qquad (17.46)$$

The phase rotation from one symbol to the next will be small when the sample clock is being tracked well. Using the approximation $\sin x \approx x$ for $|x| \ll 1$ and the fact that $|\tilde{Z}_{k,i}| \approx |\hat{Z}_{k,i}|$ when the FEQ is well adjusted, gives

$$\Im m\{\tilde{Z}_{k,i}\overline{\hat{Z}}_{k,i}\} = |\tilde{Z}_{k,i}|\,|\hat{Z}_{k,i}|\sin\zeta_{k,i} \approx |\tilde{Z}_{k,i}|\,|\hat{Z}_{k,i}|\,\zeta_{k,i} \approx |\hat{Z}_{k,i}|^2\,\zeta_{k,i} \qquad (17.47)$$

which is proportional to the phase rotation. Computing $\Im m\{\tilde{Z}_{k,i}\overline{\hat{Z}}_{k,i}\}$ from the rectangular forms requires the difference of two real products and is an efficient method for implementation with DSP's. According to (17.42) $\zeta_{k,i}$ should be proportional to the frequency index k and ϵ if noise is ignored. A straight line can be fit to the phase rotations at different subcarriers to give the following formula similar to (17.43) for estimating the residual timing offset to within a constant:

$$\zeta_i = C \sum_{kused} k\zeta_{k,i} \qquad (17.48)$$

where C is an appropriate scale factor.

The phase rotation estimate can be used to update the sampling period of a hardware phase-locked loop. If $\zeta_i > 0$, the constellation has rotated counter clockwise (in the positive direction mathematically), and the sampling period is too large and should be reduced. This can be done in hardware by inserting extra high speed clock pulses into a divider chain in the PLL. If $\zeta_i < 0$ the sampling period is too small and should be increased. This can be done in hardware by inhibiting pulses to a divider chain.

The TMS320C6713 DSK does not have a codec that includes a hardware PLL for the sample clock, so we will have to use a different approach. Suppose the FEQ is perfectly adjusted at symbol time i and is held constant after that. At symbol time i there will be no constellation rotation. According to (17.42), the constellation at symbol time $i + 1$ will be rotated by $\varphi_{k,1} = \frac{2\pi}{N} k\epsilon(N + \nu)$. This can be corrected by rotating the equalized point in the opposite direction by $\varphi_{k,1}$. Mathematically, the counter-rotation can be accomplished by multiplying the equalized point $\tilde{Z}_{k,i+1}$ by $e^{-j\varphi_{k,1}}$. Each successive symbol the equalized point has to be counter-rotated by a larger angle. That is, at symbol $i + q$ the equalized point would have to be rotated by angle $-q\varphi_{k,1}$.

A second-order PLL can be implemented to generate the required rotation. Let the PLL output at symbol time i be τ_i. The corresponding constellation point should be rotated by the angle $-\frac{2\pi}{N} k\tau_i(N + \nu)$. First, the phase errors are accumulated using the equation

$$\delta_i = \delta_{i-1} + \beta\zeta_i \tag{17.49}$$

Then, the PLL output is computed as

$$\tau_i = \tau_{i-1} + \delta_i \tag{17.50}$$

where α and β are small constants. The idea is that when the phase rotation estimate ζ_i has a constant component, its accumulated value will converge in the closed loop to the constant required to increment τ_i each iteration to force the observed rotation error to zero. Adding a $\alpha\zeta_i$ term to the equation for τ_i would allow the loop to eliminate a constant constellation tilt. However, the FEQ already corrects for any constant tilt, so it is not necessary to add this term here. For this arrangement to work correctly, the FEQ must change its coefficients much more slowly than the timing PLL tracks.

The PLL output τ_i will grow linearly without bound unless some corrective action is taken. If no action is taken, the symbol timing can slip into the cyclic prefix or into an adjacent symbol. An approach is to keep $|\tau_i| \leq 0.5$ by stuffing or robbing a sample between symbols. When τ_i exceeds 0.5, this indicates that the equalized constellations are rotated in a counter-clockwise direction and the sampling period is too large. This can be corrected by advancing the timing by one sample. The advance can be implemented by "stuffing" an extra sample into the cyclic prefix to move the N points collected for the symbol forward in the sample buffer by one sample. Then τ_i should be replaced by $\tau_i - 1$. When τ_i becomes less than -0.5, this indicates the sampling is occurring too early and a one sample delay should be introduced. This can be accomplished by deleting or "robbing" a sample from the cyclic prefix to move the symbol collection time one sample later relative to the sample buffer and then τ_i should be replaced by $\tau_i + 1$. In this scheme, τ_i will vary between -0.5 and 0.5.

17.4.3 Symbol Alignment Acquisition and Tracking

The receiver must determine where a symbol starts so it can delete the cyclic prefix and select a sequence of N ISI free samples for the FFT demodulator. Before the initialization process begins, the transmitters are silent. A receiver can monitor the power level of its input signal, detect a jump above a preset threshold, and determine relatively quickly when an initialization signal has started. Special training signals like REVERB, which consists of repeated symbols without cyclic prefixes, are sent during initialization for specified time intervals. Symbol counting from the detection of the start of initialization can give a ball park estimate of when data symbols begin. To help the symbol synchronization process, the

SEGUE sequence is sent for 10 symbols at the end of REVERB. SEGUE symbols are the 180 degree phase shift of REVERB symbols. This phase reversal can be accurately detected and used as a timing mark.

Once regular data symbols are transmitted, a ν sample cyclic prefix is included in the symbols. The receiver can search symbols for the repetition of the prefix N samples later by some type of correlation process. For example, the receiver could compute the periodicity metric

$$p_{n,i} = \sum_{\ell=n-d}^{n} (r_{\ell,i} - r_{\ell-N,i})^2 = \sum_{\ell=n-d}^{n} r_{\ell,i}^2 + \sum_{\ell=n-d}^{n} r_{\ell-N,i}^2 - 2 \sum_{\ell=n-d}^{n} r_{\ell,i} r_{\ell-N,i} \qquad (17.51)$$

where d is somewhat less than the cyclic prefix length ν. This could also be averaged over several symbols. The metric will be close to zero when n is at the end of the symbol and the two samples subtracted are theoretically identical and gets large elsewhere. Of course, they will not be identical in the real world because of noise and ISI. The minimum of the metric indicates the location of the end of a symbol. A test for the location of the minimum is to check when for some appropriately chosen small positive constant γ

$$\sum_{\ell=n-d}^{n} r_{\ell,i}^2 + \sum_{\ell=n-d}^{n} r_{\ell-N,i}^2 - 2 \sum_{\ell=n-d}^{n} r_{\ell,i} r_{\ell-N,i} < \gamma \left(\sum_{\ell=n-d}^{n} r_{\ell,i}^2 + \sum_{\ell=n-d}^{n} r_{\ell-N,i}^2 \right) \qquad (17.52)$$

or

$$\sum_{\ell=n-d}^{n} r_{\ell,i} r_{\ell-N,i} > \frac{1-\gamma}{2} \left(\sum_{\ell=n-d}^{n} r_{\ell,i}^2 + \sum_{\ell=n-d}^{n} r_{\ell-N,i}^2 \right) \qquad (17.53)$$

In this last form, the search is for correlation peaks rather than periodicity metric nulls. The computation could be simplified by replacing the right-hand side by a fixed threshold. However, the sum on the right-hand side is an estimate of the average power and provides an automatic threshold normalization.

During data transmission, the special sync symbol is sent after every 68 regular data symbols. A correlation peak search for the sync symbol could also be used to aid symbol synchronization.

17.4.4 Remaining Blocks

The remaining blocks in the receiver do the reverse of the corresponding blocks in the transmitter. The constellation points at the output of the FEQ are quantized and mapped back into a bit sequence. This sequence is deinterleaved, processed by the Reed-Solomon decoder, descrambled, and CRC checked. The resulting output bits are sent to the data sink. Any CRC errors are reported to the maintenance functions.

The descrambler is described by (17.5). The Reed-Solomon decoding algorithm is beyond the scope of this book. Interested readers can find the details in books on error correcting codes like Wicker [F.19]. The CRC check is performed by shifting a codeword including the CRC bits into the CRC generator shown in Figure 17.3 starting with all the delay elements cleared. The register delay elements should all be 0 when the last bit is shifted in for the check to pass.

17.5   Making a Simplified ADSL Transmitter and Receiver

In this experiment you will make a simplified version of an ANSI T1.413-1998 [E.1] ADSL ATU-R transmitter and ATU-C receiver. Set the TMS320C6713 DSK sampling rate to $f_1 = 16$ kHz. The transmitter should use an $N = 64$ point IFFT and the receiver a 64-point FFT. The subcarriers are at frequencies that are multiples of $16000/64 = 250$ Hz. The subcarriers for $k = 0$ and 32 should be set to zero and not used. The subcarriers for $k = 1, \ldots, 31$ will be used to send QPSK (4-QAM) or 16-QAM constellations corresponding to 2 and 4 bits per constellation point, respectively. The subcarrier points for $k > 32$ are determined by the constellation points for $k < 32$ by the complex conjugate symmetry rule

$$Z_{64-k} = \overline{Z}_k \text{ for } k = 1, \ldots, 31 \qquad (17.54)$$

which will force $z_n = \text{IFFT}\{Z_k\}$ to be real.

Use a cyclic prefix of $\nu = 4$ samples. You may have to lengthen this after measuring the channel impulse response. Then the length of regular data symbols will be $N' = 68$ samples. The number of symbols transmitted per second will be $16000/68 = 235.294$. The minimum number of bits transmitted per symbol is $2 \times 31 = 62$ when QPSK is used for all the used subcarriers giving a data rate of $62 \times 16000/68 = 14588.235$ bits per second if no sync symbols are used. The maximum number of bits per symbol is $4 \times 31 = 124$ when 16-QAM is used for all the used subcarriers giving a data rate of $124 \times 16000/68 = 29176.47$ bits per second if no sync symbols are used. This rates are scaled by $68/69$ when a sync symbol is inserted after every 68 data symbols.

Carry out the steps in the following subsections to make and test your transmitter and receiver.

17.5.1   Making a 64-Point IFFT and a 64-Point FFT

You will need a 64-point IFFT for your transmitter and a 64-point FFT for your receiver. Modify the decimation-in-time FFT function from Chapter 4 to be a 64-point IFFT. This function should accept the 64-point constellation array in natural order, shuffle it into bit-reversed order, and then perform the equivalent of the "decimation-in-time" FFT algorithm to get the output time sequence in natural order. Read required cosine and sine values for the "twiddle factors" from a pre-computed table rather than compute them recursively like in the FFT function from Chapter 4. Only a single sine table is needed, not both a sine and cosine table, because $\cos \theta = \sin(\theta + \pi/2)$. You can achieve some computational savings in the last stage by using the fact that the output must be real and not computing the imaginary part. Test your IFFT by putting in some sequences that have known outputs. For example, $Z_k = \cos(2\pi mk/64)$ has the IFFT, $z_n = 0.5\delta[n - m] + 0.5\delta[n - (64 - m)]$.

Modify the decimation-in-time FFT function in Chapter 4 to be a 64-point FFT. Again, read the sine/cosine twiddle factors from a table. Some computational savings can be achieved by using the fact that the input is real in the first stage. Test your function by using some inputs with known FFT's.

17.5.2 Making a Scrambler, Constellation Point Mapper, and Their Inverses

Implement the 23-stage self synchronizing scrambler presented in Section 17.3.2 with input $x(n) = 1$ for all n to generate simulated customer random data. Bypass the CRC generator, Reed-Solomon encoder, and convolutional interleaver for the time being. You can implement these later as optional exercises. Initially you will test your transmitter with QPSK constellations for all of the 31 used subcarriers. In this mode, you should generate blocks of 62 bits with the scrambler. Later you will use 16-QAM for all 31 used subcarriers which will require blocks of 124 bits.

Make a function that maps an array of 62 input bits into a 64 element complex array Z with $Z[0] = Z[32] = 0$, $Z[1], \ldots, Z[31]$ QPSK constellation points $X_k + jY_k$ corresponding to successive pairs of input bits using the rule stated on page 287 and illustrated in Figure 17.7, and $Z_{64-k} = \overline{Z}_k$ for $k = 1, \ldots, 31$.

Make another function for 16-QAM that maps an array of 124 input bits into a 64 element complex array Z with $Z[0] = Z[32] = 0$, $Z[1], \ldots, Z[31]$ 16-QAM constellation points $X_k + jY_k$ corresponding to successive groups of four input bits using the rule stated on page 287 and illustrated in Figure 17.7, and $Z_{64-k} = \overline{Z}_k$ for $k = 1, \ldots, 31$.

Make the inverse functions. That is, make functions to convert a 64 element complex array back into an array of 62 bits for QPSK and 124 bits for 16-QAM. These functions should quantize a complex point $R[k]$ corresponding to an element of the output of the FFT in the receiver to the ideal constellation point shown in Figure 17.7 that is nearest to it in Euclidean distance, and put the two or four data bits corresponding to the quantized point into a 62 or 124 element array for $k = 1, \ldots, 31$. Implement a descrambler using (17.5). Test your inverse mappers and descrambler separately. Then test the cascade of the scrambler, mapper, inverse mapper, and descrambler for a string of transmitted DMT symbols. Do this internally in a program for one DSP without worrying about getting data to and from the codec. The output of the descrambler should be all 1's except, possibly, for a short initial transient if the scrambler and descrambler initial states differ.

Once the cascade is working, insert the transmitter IFFT and receiver FFT. First leave the IFFT output as a floating-point array. Use this array as the FFT input. Do this internally in the single DSP program without worrying about getting data to and from the codec. Next convert the IFFT output array to an integer format scaled appropriately for the right codec channel and use it internally in the program as the receiver FFT input and check that the cascade works. You will have to structure the inverse mapper to account for the scaling.

17.5.3 Measuring the Channel Impulse Response Duration

You will send the samples generated in the DSP for the transmitter through the DAC in the AIC23 codec to the Line Out, through a cable from the Line Out to the Line In of the receiver, and through the ADC of the codec. Initially, you will loop the Line Out of the DSK back to the Line In of the same DSK. Later, you will connect the Line Out of one station to the Line In of another station. The channels for both of these cases should be very similar. The duration of the channel impulse response must be known to select the length of the cyclic prefix. Write a program to first send 128 zeros to the right channel ADC to clear any initial startup transients. Then send periodic pulses to the right channel ADC. For example, send

$$s[n] = \sum_{k=2}^{\infty} 15000 \, \delta[n - 64k] \qquad (17.55)$$

The scale factor, 15000, was chosen to use a significant portion of the range of the ADC. This signal is a periodic train of unit pulses occurring every 64 samples with zeros in between. Loop the Line Out back to the Line In of the same DSK. Collect 128 samples from the Line In in an array starting when you send the first pulse. Examine the array using a watch window or, perhaps, more informatively by using the graphing capabilities of Code Composer and determine the duration of the channel impulse response. You can also connect the Line Out to the oscilloscope and observer the impulse response through the ADC to the Line Out. Another approach would be to internally within the DSP program, loop the samples received from right channel Line In back out to the left channel Line output and observe the repetitions of the impulse response on the oscilloscope. Choose a value, ν, for the cyclic prefix length that is somewhat larger than the duration of the channel impulse response.

17.5.4   Completing the Transmitter

Now it is time to put together the transmitter blocks along with an initialization sequence and operate in real-time. Use ping-pong buffers to send output samples to the left channel DAC when transmitting regular data symbols. Define two integer arrays, `ping[]` and `pong[]`, each of which can hold the output samples for one DMT symbol. Samples should be sent to the DAC from one of the arrays using an interrupt service routine triggered by interrupts from the McBSP1 XRDY1 flag. You could also use the EDMA to send out the samples. While the samples are going out of one array, your program should generate the samples for the next DMT symbol and put them into the other array and then wait for all the samples to be transmitted from the first array. Then the functions of the two arrays should be swapped with samples transmitted from the array just filled and samples for a new DMT symbol generated and put into the other array.

As an initial test of your ping-pong buffer scheme, select a subcarrier with index k_0 in the set $\{1, 2, \ldots, 31\}$ and a constant C_{k_0} to give a significant DAC output, and form the complex 64-dimensional array of subcarrier constellation points

$$Z_k = \begin{cases} 0 & \text{for } k = 0 \text{ and } 32 \\ C_{k_0} & \text{for } k = k_0 \\ \overline{C}_{k_0} & \text{for } k = 64 - k_0 \\ 0 & \text{otherwise} \end{cases} \qquad (17.56)$$

Send the 64-point time sequence z_n repeatedly and without a cyclic prefix. Perform the IFFT for each symbol, even though the results will be the same each time, to make sure it is integrated into your ping-pong scheme correctly. The transmitted signal should be a cosine wave at the subcarrier frequency. Check that this is what you see on the oscilloscope. Try this for several subcarrier frequencies.

Once your ping-pong buffer scheme is working, create a program to make the transmitter send the following sequence of signals:

1. When your transmitter program first starts, send about one second of silence. That is, send samples that are all 0 volts for about one second. This will allow any initial transients that might occur when your program is loaded and started to disappear. It will also allow the receiver to recognize that the transmitter is not yet sending a signal and get into a waiting state.

2. Next, send the R-REVERB symbol described on page 289 without a cyclic prefix 1024 times. This will allow the receiver to detect the presence of a signal, estimate symbol clock timing frequency offset, and initially train its equalizer. You should compute the real 64-point R-REVERB time sequence just once, scale it to use a significant range of the DAC, convert it to the integer format for the left DAC channel, and store the resulting 64-point sequence in an integer array for repeated transmission. You can use interrupts or the EDMA with linking to repeatedly transmit the R-REVERB 64-point time sequence. Since you have already implemented ping-pong buffers, a good approach would be to store the 64-point sequence in both ping[] and pong[] and let your ping-pong scheme alternate between sending the two buffers.

3. After all the R-REVERB symbols have been sent, send the R-SEGUE symbol ten times. This symbol is defined on page 290. The R-SEGUE symbol is a 180 degree phase reversal of the R-REVERB signal and provides a timing mark to signal the receiver indicating that actual data signal will be transmitted next.

4. Next, go into an endless loop and transmit data symbols including a cyclic prefix. The data blocks for the symbols should be generated by the 23-stage scrambler with all 1's as its input. Use QPSK for all the subcarriers. Send the sync symbol described on page 290 after every 68 data symbols.

5. Once you get your receiver to work with QPSK, you can use 16-QAM for the subcarriers. You might want to send QPSK for some know time period initially, and then switch to 16-QAM. Send a sync symbol after every 68 data symbols.

17.5.5 Making the Receiver

Receivers are usually much more difficult to make than transmitters. They require functions to detect signal presence, do automatic gain control, acquire and track sample and symbol clocks, perform adaptive equalization, implement algorithms for optimum detection of signals in noise, and implement sophisticated decoders for error correction. For systems where the signal is modulated on a high frequency carrier, the receiver often must also include components to acquire and track the carrier frequency and phase, and demodulate the signal to baseband. In this experiment, you will implement the receiver in a different DSK than the transmitter so sample clock acquisition and tracking, and symbol location issues must be addressed. You might find it useful to simulate some of the algorithms you will have to implement for the receiver in MATLAB before trying to get them to work in real-time on the DSK. Connect the Line Out of your transmitter DSK to the Line In of the receiver DSK. Perform the tasks listed below to make your ADSL receiver.

1. Make a function to detect when an ADSL signal is present at the receiver input. One approach is to detect signal energy by squaring the input samples and averaging the squares with a first-order exponential averager. Let $d(n)$ be the averager output and $r(n)$ the received signal. Then the power detection system should compute

$$d(n) = \beta d(n - 1) + (1 - \beta)r^2(n) \qquad (17.57)$$

where β is a positive constant slightly less than 1 that determines the averaging time constant. The closer β is to 1, the slower the averager will respond but the smoother its output will be. When $d(n)$ rises above a threshold for several samples, decide an ADSL signal is present. The threshold determines the false detection probability when the channel is noisy. When the threshold is low, an ADSL signal will be detected quickly but false alarms will also be frequent. When the threshold is high, false alarms will be rare but it will take longer to detect the signal. Of course, if the threshold is too high there will be no false alarms but the signal will never be detected. Experimentally select values for β and the threshold so the start of a received ADSL signal is reliably detected by the end of one or two 64-sample symbols.

   On channels with very little noise, quick detection can be achieved by detecting when the absolute value of the input exceeds a threshold for the first time and, possibly, several nearby times. This quick detection approach could be used for an initial estimate of the DMT symbol timing.

   Before a signal is detected, keep the receiver functions reset and waiting. When a signal is detected, start a counter to keep track of how many R-REVERB symbols have been received.

2. The receiver must track sample clock frequency offset and DMT symbol alignment. Also, the FFT you implemented does an in-place computation and overwrites the input array with the output array. These considerations make it inconvenient to use the ping-pong buffer approach in the receiver. An approach you can use is to set up a circular buffer of floats for received samples whose length is two symbols long, that is, $2N' = 2(64 + \nu)$ elements. The received samples should be converted to floats when stored in the circular buffer. A variable pointing to the start of a symbol can be defined. Sample clock adjustments and symbol alignment can be made by modifying this pointer. The 64 samples for a symbol, excluding the cyclic prefix, can be copied to a separate input array for the FFT.

3. Begin collecting samples in the circular buffer once a signal is detected. Declare a 64-element complex array for the FFT input. Copy successive blocks of 64 samples to the real part of this array and zero the imaginary part. Form the FFT's of the blocks. During the first 200 R-REVERB symbols, estimate the sample clock fractional period error, ϵ, using the "open loop" method described on page 294 using (17.43).

4. For the next several symbols, compute an estimate of the reciprocal of the channel frequency response, $1/H_k$, for each subcarrier. These will be used as the initial frequency domain equalizer (FEQ) coefficients. You already know the FFT values for the R-REVERB symbol from the definition of this signal. Let $R_{k,i}$ be the receiver's FFT output at index k for symbol i, and Z_k be the transmitted FFT value. The estimate for one received symbol is

$$W_k(i) = \frac{Z_k}{R_{k,i}} = \frac{Z_k \overline{R}_{k,i}}{|R_{k,i}|^2} \tag{17.58}$$

The arithmetic average of these estimates over several symbols can be computed to reduce noise. You can do the divisions using the standard C floating-point division operator. Compute these estimates for $k = 1, \ldots, 31$.

5. Initialize the FEQ coefficients to the values computed in the previous task. Then begin adaptively updating the equalizer coefficients using the LMS algorithm given by (17.29) in Section 17.4.1. Begin running the FEQ to compute the equalized values $\tilde{Z}_{k,i} = W_k(i)R_{k,i}$. Use an update constant μ large enough to track phase rotations caused by sample clock frequency offset but small enough to have relatively small coefficient jitter. You will have to turn on the slicer that quantizes FEQ outputs to the nearest ideal constellation points. Do not start the sample clock tracking algorithm yet. Run in this mode for 256 symbols. The constellations you observe at the equalizer output should be the QPSK input constellations. You might ask why this works with no cyclic prefix? The answer is that the R-REVERB symbol is being repeated. The end of one symbol acts as the cyclic prefix for the next.

6. A good diagnostic tool and informative way of observing the receiver operation is a *constellation display*. This display shows the real versus imaginary parts of the FEQ outputs for the equalized symbol stream for a particular subcarrier in an X-Y display on an oscilloscope. When the channel noise is small and the equalization is good, well defined constellation points will be observed. If additive noise with approximately Gaussian characteristics is present, the ideal constellation points will smear into circular clusters. Poor equalization will also cause clusters. When sample clock frequency offset is present and the equalizer updates are very slow, the observed constellation will slowly rotate.

To make a constellation display, send the real part of the equalizer output for a subcarrier to the left channel of the DAC and the imaginary part to the right channel for each of the $64 + \nu$ output samples during a symbol. Display the two channels as an X-Y plot on the oscilloscope. Freeze the equalizer coefficients by making μ zero and repeat R-REVERB symbols continuously. Look at a constellation display to see if there is constellation rotation caused by sample clock frequency offset.

7. After the 256 symbols of LMS equalizer updating (456 symbols after the start of signal detection) set the equalizer update scale factor to a small value and turn on the sample clock algorithm specified by (17.49) and (17.50) and perform the constellation counter-rotations. Initialize δ_i according to the open loop value computed during the first 200 received symbols. The clock PLL should track significantly faster than the equalizer. Adjust the timing phase when $|\tau_i|$ exceeds its threshold by shifting the 64-sample record

copied from the circular buffer forward or backward by one sample and correcting τ_i as discussed on page 296.

8. Continue running the FFT, FEQ, clock tracker, and symbol counter. When the counter gets close to 1024, begin looking for the 180 degree phase shift that occurs when R-SEGUE begins. Use your ingenuity to come up with an algorithm for detecting the phase shift. Your algorithm should operate on the received sample sequence and detect the phase shift within 4 samples. Count for 10 symbols from this timing mark to determine when regular data transmission begins with symbols including a cyclic prefix. The timing mark should allow you to determine with reasonable accuracy when the cyclic prefix of the first data symbol starts.

9. Turn on the cyclic prefix correlation algorithm presented in Section 17.4.3 for determining symbol alignment when the reception of regular data symbols begins. You should already have a ball park estimate of when a symbol begins. Search around this point in the circular buffer for the phasing that gives a null in the periodicity metric or a peak in the cyclic prefix correlation. For each received symbol, skip over the cyclic prefix and copy 64 good symbol samples to the FFT input array. Perform the FFT, FEQ, sample clock tracking, slicing, inverse map to a bit sequence, and descrambling. If everything is working properly, the descrambler output should be all 1's.

10. Additional optional tasks you can perform are:

    (a) Add Gaussian noise to the transmitted samples and make a plot of the bit-error rate as a function of signal-to-noise ratio.

    (b) Implement the CRC generator in the transmitter and CRC checker in the receiver.

    (c) Implement the interleaver in the transmitter and deinterleaver in the receiver.

    (d) Pass the transmitter output samples through a filter simulating a worse channel before sending them to the DAC in the transmitter and observe system performance.

    (e) Come up with a scheme for estimating the signal-to-noise ratio at the receiver for each subcarrier when there is noise on the channel and the amplitude response varies significantly. Do some research and implement an algorithm for optimally assigning the number of bits to each subcarrier to achieve a desired error probability.

    (f) Implement the Reed-Solomon encoder in the transmitter and decoder in the receiver. This is a major task and is worthy of being counted as an entire experiment.

    (g) Browse through the ADSL standards and implement additional features they contain.

17.6   Additional References

There are many books on wireless systems and DSL that have sections on multi-carrier modulation. See Golden, et al., [E.3] for a very up-to-date and technically in-depth treatment of DSL. Other books on DSL include Bingham [E.2] and Starr, et al., [E.16]. Haykin and

Moher [E.5] is a well written book on wireless communications that discusses multi-carrier modulation. Goldsmith [E.4], Schwartz [E.15], and Stüber [E.17] have technical sections on multi-carrier systems. The ITU-T Recommendations G.992.3, G.992.4, and G.992.5 for ADSL can be downloaded free of charge as PDF files from the ITU-T web site:

`http://www.itu.int/rec/T-REC-G/e`

The IEEE 802.11 standards for Wi-Fi can be downloaded free of charge from:

`http://standards.ieee.org/getieee802/802.11.html`

and the IEEE 802.16 standards for WiMax from:

`http://standards.ieee.org/getieee802/802.16.html`

Chapter 18

Suggestions for Additional Experiments

It would take a long time (maybe a couple of years?) to complete all the experiments in this book. However, in light of the recent ABET requirement in engineering education for capstone design projects, several additional topics for experiments related to current communication techniques are very briefly described below. References to get you started on each project are included.

18.1 Elementary Modem Handshake Sequence

As an introduction to modem handshake procedures, implement the V.22bis transmitter and receiver handshake sequences. Build the transmitter in one station and the receiver in another. See the ITU-T V.22bis recommendation [II.D.4] for a detailed specification of these sequences. You will have to figure out methods for detecting critical points in the sequences.

After you have the V22.bis transmitter and receiver working in separate DSP's, integrate the two into a single DSP. The challenge will be to figure out how to time-share the DSP between the transmitter and receiver programs. Remember that the transmitter and receiver in a single modem must operate independently in a full duplex fashion, so assume that the transmit and receive symbol clocks will be different in phase and slightly different in frequency. Implement a second transmitter and receiver on another station. Connect the two together in a full-duplex mode and make them talk to each other without errors. You can connect the modems through the TAS telephone line simulator and check that your modems work with a variety of lines and impairments. After completing this project, you will have built almost a complete V.22bis modem.

18.2 Make an ITU-T V.21 Frequency Shift Keyed (FSK) Modem

The V.21 modem transmits data at 300 bps using binary, continuous phase frequency modulation. This is one of the oldest and simplest types of modems. Each input data bit selects one of two frequencies to be transmitted. The originate and answer modems use different frequency pairs to achieve full-duplex transmission. The transmitted signal is generated by using an FM modulator like the one described in Chapter 8 with a two level input and has continuous phase between bit transitions. Data bytes are usually transmitted in a asynchronous mode using start and stop bits. Tasks you could perform for this project include:

1. Make a transmitter on one DSK taking the binary data input from the BERT test set.

2. Make a receiver including timing recovery on another DSK. No elaborate timing recovery system is needed for asynchronous byte transmissions. The start of a byte is signaled by a transition from a stop bit to a start bit. The data bits for the byte are spaced by $1/300$ seconds so the center of each bit can be found by simply counting samples. The timing start reference can be reset with the stop to start bit transition for each byte.

3. You can use the TAS telephone line emulator to make bit error rate vs. SNR tests along with a second BERT tester.

4. Make a full duplex modem by implementing the transmitter and receiver on one DSK and have two modems talk to each other in full-duplex mode.

5. Compute the theoretical spectrum for the FSK signal and compare it with the spectrum for your FSK modulator output.

Lucky, Salz, and Weldon [II.D.29] has a detailed chapter on FSK including a derivation of the power spectral density. ITU-T Recommendation V.21 specifies all the requirements for this modem. The class web page www.ee.umd.edu/courses, Fall 1999, ENEE 429W, Lecture Notes and Handouts, Continuous Phase Frequency Shift Keying should be useful.

18.3 Fast Equalizer Training Using Periodic Sequences

Equalizer training using the LMS algorithm with an ideal reference sequence takes a significant amount of time. The training time can be reduced very significantly by transmitting a sequence with a period equal to the equalizer duration in symbols and using FFT methods to compute the optimum equalizer coefficients. The training signal probes the channel with a bank of sinusoids whose frequencies are multiples of the repetition rate of the sequence. The equalizer is switched to the LMS decision directed mode after the initial fast training. This fast training method was included in a modem standard for the first time in the V.34 recommendation. A clear and detailed explanation of the method is presented by Chevillat, Maiwald, and Ungerboeck in [II.D.5]. Also see Tretter [II.D.39, Chapter 11]. As a project, you could incorporate this fast train technique into the modem transmitter and receiver you built in Chapters 13, 14, and 15.

18.4 Trellis Coded Modulation

Ungerboeck [II.F.14] made a major breakthrough in coding for narrow-band channels in the late 1970's now called *trellis coded modulation*. It has been included in every V series modem recommendation since its discovery and some ADSL standards. The method involves doubling the normal number of points in the constellation, using a convolutional code in the transmitter to select constellation points, and performing soft-decision Viterbi decoding in the receiver. Performance gains of 4 to 5 dB are readily achieved. See Ungerboeck [II.F.15] for a more recent and tutorial version of his original paper. The book by Biglieri, *et al.*, [II.F.3] gives a detailed presentation of trellis coding. Other references to trellis coding and soft decision Viterbi decoding can be found in Lee and Messerschmitt [II.D.26, Section 7.4 and Chapter 12], Gitlin, Hayes, and Weinstein [II.D.11, Sections 3.5 and 5.7], Proakis [II.D.32, Sections 5.3.2 and 5.4], and Tretter [II.D.39, Chapter 3].

As a project, you could implement the V.32 trellis code and a soft decision Viterbi decoder. See the ITU-T V.32 recommendation [II.D.4], Lee and Messerschmitt [II.D.26, p. 517], or Proakis [II.D.32, p. 504] for complete details on the V.32 encoder. This encoder was invented by Lee-Fang Wei [II.F.17] at Bell Telephone Laboratories. To avoid processing time problems in the receiver, you could implement the system using the V.22 symbol rate of 600 baud. This would also allow you to use most of the programs you have previously generated.

If you are interested in pursuing trellis coding further, implement one or more of the V.34 4-dimensional 16, 32, or 64 state codes. See Wei [II.F.18] for a discussion of multi-dimensional trellis codes. See Tretter [II.D.39] for the V.34 codes.

18.5 Reed-Solomon Encoder and Decoder

Reed-Solomon codes are cyclic block codes that are used in many applications. Applications include error correction for compact disks, digital subscriber line modems, many types of wireless systems, and satellite communication systems. They can be viewed as a type of BCH code with code symbols that are blocks of bits. These codes are good for correct isolated random errors as well as error bursts. Encoding is relatively easy. Efficient algorithms exist for doing the much more difficult decoding task. See books on error correcting codes like Gallager [F.7] and Wicker [F.19] for details.

As a project, you could make a Reed-Solomon encoder and decoder for an ADSL transceiver. This code has symbols consisting of eight bits (a byte) and a natural block length of 255 bytes. The codes have a wide range of error correcting capabilities. They require $2t$ check symbols to correct t symbol errors and have $255 - 2t$ information symbols.

18.6 Turbo Codes

Turbo codes were first disclosed at a public conference in 1993 by Berrou, Glavieux, and Thitimajshima [F.1]. They presented simulation results showing performance of their encoder and decoding algorithm remarkably close to the Shannon capacity limit with a reasonable computational complexity. Their results were initially met with skepticism, but others quickly duplicated them and turbo codes were rapidly used in applications and included in standards. Turbo codes use parallel concatenation of recursive systematic convolutional codes with interleaved input streams and soft-decision iterative decoding. A very good reference for these codes is Vucetic [F.16]. A substantial project would be to learn about turbo codes and implement an encoder and decoder for the code specified in the WiMax IEEE recommendation 802.16 [E.8].

18.7   Low Density Parity Check Codes

Robert G. Gallager invented and studied low-density parity-check codes (LDPC codes)during his doctoral research in the Department of Electrical Engineering at the Massachusetts Institute of Technology and published his work in his doctoral dissertation in 1960. An expanded version of his dissertation was published as an M.I.T. Press monograph in 1963 [F.6]. These codes were ignored for many years because the decoding algorithm was too computationally complex to be economically justifiable given the hardware technology of the times. They were rediscovered in the late 1990's and shown to have performance near to the Shannon capacity limit with reasonable complexity and cost for the current digital processor technology. The decoding algorithm is an iterative one and can be viewed as message passing on a graph using a sum-product algorithm. The complexity is comparable to or less than that of turbo codes and the performance is at least as good. LDPC codes are at the forefront of current research on error correcting codes. A good reference for LDPC codes is Fan [F.4]. The web site, www.inference.phy.cam.ac.uk/mackay/CodesGallager.html, of David MacKay is a good source of articles on LDPC codes and software for designing and simulating them. LDPC codes have been used in some commercial products and are included in the mobile WiMax 802.16e standard [E.9] as optional codes. A good project would be to learn about the theory of LDPC codes, implement encoders and decoders for the IEEE 802.16e LDPC codes, and perform simulations to get error probability versus signal-to-noise ratio curves.

18.8   V.34 Constellation Shaping by Shell Mapping

Additional performance can be gained by shaping the constellation to minimize the average signal power while maintaining the minimum distance between points. For example, in two dimensions a circle with the same area as a square has less average power. In N dimensions, a sphere has less average power than any other figure with the same volume. The V.34 recommendation specifies a 16-dimensional shaping technique called *shell mapping* that closely approximates the gains of ideal spherical shaping. In previous V series modems, the two-dimensional constellation points are used with equal likelihood. With shell mapping, the points in the two-dimensional constellation nearest the origin are more likely than the outer points. The two-dimensional points approximately have a circular Gaussian distribution. See the ITU-T V.34 recommendation [II.D.21] for a cookbook formula for shell mapping. An explanation of the theory can be found in Laroia, Farvardin, and Tretter [II.F.9][II.F.10] and in Tretter [II.D.39, Chapter 8].

As a project, you could implement the V.34 shell mapping algorithm and its inverse. You should weigh using pre-computed tables against on-the-fly computation to achieve the required processing speed. This trade-off between memory and computation often arises in DSP applications.

18.9 Nonlinear Precoding for V.34

The V.34 modem can use a noise whitening filter in the receiver to improve the Viterbi decoder performance. The intersymbol interference introduced by the noise whitening filter is compensated for at the transmitter by a nonlinear precoding technique. Nonlinear precoding was first suggested by Tomlinson [II.D.36] and Harashima [II.D.16]. The method was generalized to QAM systems and improved by Laroia [II.D.25] and quickly further improved by Laroia and Betts of AT&T and, independently, by Cole of General DataComm during development of the V.34 recommendation. The result was presented at a TIA meeting in July 1993 [II.D.15]. The final version of the nonlinear precoder can be found in the V.34 recommendation [II.D.21] and in Tretter [II.D.39, Chapter 5].

As a project, implement the V.34 shell mapper, nonlinear precoder, and 4D 16-state trellis encoder in the transmitter. See Tretter [II.D.39, Chapters 9 and 10] to learn how these components are connected. Implement the inverses in your receiver. Transmit the signal through the TAS channel simulator with additive noise and compare the bit-error rate performance of the system with and without precoding. Also compare the performance with and without shell mapping.

18.10 Speech Codecs

DSP's are used extensively for speech codecs in telephones for digital exchanges, codecs at the local office, in wireless telephones, and in secure military applications. The ITU has generated a G series of recommendations for speech codecs with various degrees of complexity. The simplest is the pulse code modulation (PCM) Recommendation G.711. The next in complexity is G.726 for adaptive differential pulse code modulation (ADPCM). Then there are several code excited linear prediction (CELP) recommendations like G.729 that are quite complex. There are a variety of cellular phone standards that each have their own unique speech codecs. There are C reference programs available from ITU and on the web for implementing many of these codecs. You could easily work for several semesters making an array of speech codecs to work on the DSK.

Appendix A

Generating Gaussian Random Numbers

This appendix describes how to convert a sequence of uniformly distributed random numbers into a pair of sequences that approximate a pair of white, uncorrelated, Gaussian random sequences. These sequences can be used to simulate the inphase and quadrature noise components in passband communication systems. This method is convenient because many programming languages have a function for generating uniformly distributed random numbers. Limitations of the 'C6713 C compiler random number generator are discussed.

A.1 The 'C6713 C Compiler Pseudo Random Number Generator

The 'C6713 optimizing C compiler library contains the functions int rand(void) and void srand(unsigned int seed) for generating a random sequence of integers uniformly distributed over the range $[0, \text{RAND\_MAX}]$ where RAND_MAX is defined to be $2^{15} - 1 = 32767$ in stdlib.h. These numbers occupy only the least significant 15 bits of the DSP's 32-bit word. The source code for these functions is shown in Program A.1 below. The function srand sets the value of the random number generator seed so that subsequent calls of rand produce a new sequence of pseudo-random numbers. The srand function does not return a value. If rand is called before srand is called, it generates the same sequence as if srand was called first with a seed value of 1. The function rand generates the same sequence whenever srand sets the same seed. This random number generation technique is often referred to as the multiplicative congruential method.

Program A.1 'C67x C Compiler Programs rand() and srand()

```
#include <stdlib.h>

 unsigned long next = 1;
```

```
 int rand(void)
{
    int r;
    next = next * 1103515245 + 12345;
    r = (int)((next/65536) % ((unsigned long)RAND_MAX + 1));
    return r;
}

 void srand(unsigned seed)
{
    next = seed;
}
```

A.2   A Better Uniform Random Number Generator

It is shown in Section A.4 that a finite word-length uniform random number generator causes a limit on the peak value of the approximately Gaussian random variables generated by the method presented in this appendix. This puts an upper limit on the signal-to-noise ratio (SNR) that will cause bit errors in a simulation of a digital communication system. At SNR's above this limit, no errors will occur. The random numbers generated by the 'C6713 C compiler function use only 15 bits of the 32-bit DSP word. The function rand() shown in Program A.2 generates 31-bit random numbers uniformly distributed over the integers in the range $[0, \text{RAND\_MAX}]$ where now $\text{RAND\_MAX} = 2^{31} - 2$ and increases the SNR limit. This function was used in the old TMS320C3x/4x C compiler library.

Program A.2   32-bit Uniform Pseudo Random Number Generator Function

```
/*****************************************************************************/
/* rand.c for TMS320C3x/4x                                                 */
/*                                                                         */
/* NOTE:  This file should be compiled with the -mm (short multiply)       */
/*        switch for best results.                                         */
/*****************************************************************************/
#include <stdlib.h>

static unsigned next = 1;

/*****************************************************************************/
/* rand() - COMPUTE THE NEXT VALUE IN THE RANDOM NUMBER SEQUENCE.           */
/*                                                                         */
/*     The sequence used is x' = (A*x) mod M,  (A = 16807, M = 2^31 - 1).  */
/*     This is the "minimal standard" generator from CACM Oct 1988, p. 1192.*/
/*     The implementation is based on an algorithm using 2 31-bit registers */
/*     to represent the product (A*x), from CACM Jan 1990, p. 87.          */
```

```
/*                                                                      */
/**************************************************************************/
#define A 16807u                        /* MULTIPLIER VALUE    */

int rand()
{
    unsigned x0 = (next << 16) >> 16;  /* 16 LSBs OF SEED       */
    unsigned x1 = next >> 16;          /* 16 MSBs OF SEED       */
    unsigned p, q;   /* MSW (31 bits), LSW OF PRODUCT (31 bits) */

    /*---------------------------------------------------------------*/
    /* COMPUTE THE PRODUCT (A * next) USING CROSS MULTIPLICATION OF   */
    /* 16-BIT HALVES OF THE INPUT VALUES. THE RESULT IS REPRESENTED AS 2 */
    /* 31-BIT VALUES.  SINCE 'A' FITS IN 15 BITS, ITS UPPER HALF CAN BE  */
    /* DISREGARDED.  USING THE NOTATION val[m::n] TO MEAN "BITS n THROUGH */
    /* m OF val", THE PRODUCT IS COMPUTED AS:                        */
    /*   q = (A * x)[0::30]  = ((A * x1)[0::14] << 16) + (A * x0)[0::30] */
    /*   p = (A * x)[31::60] =  (A * x1)[15::30]       + (A * x0)[31]  + C */
    /* WHERE C = q[31] (CARRY BIT FROM q).  NOTE THAT BECAUSE A < 2^15, */
    /* (A * x0)[31] IS ALWAYS 0.                                     */
    /*---------------------------------------------------------------*/
    q = ((A * x1) << 17 >> 1) + (A * x0); /* q[31] is the carry    */
    p = ((A * x1) >> 15) + (q >> 31); /* q>>31 moves the carry to the lsb */
    q = q << 1 >> 1;   /* CLEAR CARRY which is q[31]*/

    /*---------------------------------------------------------------*/
    /* IF (p + q) < 2^31, RESULT IS (p + q).  OTHERWISE, RESULT IS   */
    /* (p + q) - 2^31 + 1.  This can be proved as follows:           */
    /* A * x = [p, q] = p * 2^31 + q  = p * [(2^31 - 1) + 1] + q     */
    /*       = p * (2^31 - 1) + (p + q)                              */
    /* Thus mod(a*x, 2^31 - 1) = mod( p+q, 2^31 - 1)                 */
    /*                         = p + q for  p + q < 2^31 -1          */
    /*                         = p + q - 2^31 + 1 otherwise      */
    /* The last line follows by observing that p+q is always less than */
    /* 2^32.                                                         */
    /*---------------------------------------------------------------*/
    p += q;  /* form p + q and now p is p + q */

    /* Now p has the value of p + q.  If p[31] = 1, p+q > 2^31 -1 and if */
    /* p[31] = 0, p+q <= 2^31 - 1.  Thus p >> 31 adds 0 or 1 as needed. */
    /* ( ( ... ) << 1 ) >> 1 clears bit 31 which does nothing if it is  */
    /* already 0 but subtracts 2^31 if it is 1.                     */

    return next = ((p + (p >> 31)) << 1) >> 1; /* ADD CARRY, THEN CLEAR IT */
```

```
}

/*************************************************************************/
/* srand() - SET THE INITIAL SEED FOR rand().                          */
/*************************************************************************/
void srand(unsigned seed)
{
    next = seed;
}
```

A.3   Turning Uniformly Distributed Random Variables into a Pair of Gaussian Random Variables

The first step is to convert a random variable V which is uniformly distributed over $[0, 1)$ into a random variable R that has the Rayleigh probability density function

$$f_R(r) = \frac{r}{\sigma^2} e^{-\frac{r^2}{2\sigma^2}} u(r) \tag{A.1}$$

and cumulative distribution function

$$F_R(r) = \left[1 - e^{-\frac{r^2}{2\sigma^2}}\right] u(r) \tag{A.2}$$

V can be generated by calling **rand**, converting the returned integer to a floating-point number, and dividing the result by RAND_MAX + 1 = 32768 = 2^{15} for the 'C6713 compiler. The random variable V can be converted into R by the transformation

$$R = \sqrt{-2\sigma^2 \log_e(1 - V)} \tag{A.3}$$

Notice that this function is the inverse of the cumulative distribution function $F_R(r)$. A proof that this gives the desired result can be found in most introductory probability textbooks. Then number returned by **rand** is divided by RAND_MAX + 1 rather than RAND_MAX so $1 - V$ can never become zero and cause a numerical computation problem for the $\log_e$ function.

The next step is to convert V into a pair of uncorrelated Gaussian random variables. Let Θ be a random variable uniformly distributed over $[0, 2\pi)$ and independent of V. Θ can be generated by calling **rand** again, dividing the result by RAND_MAX + 1 to get a random number uniformly distributed over $[0, 1)$, and then multiplying the result by 2π. Finally, it can be shown that

$$X = R\cos\Theta \tag{A.4}$$

and

$$Y = R\sin\Theta \tag{A.5}$$

are two uncorrelated Gaussian random variables, each with zero mean and variance σ^2. That is, they each have a probability density function of the form

$$f(x) = \frac{1}{\sigma\sqrt{2\pi}} e^{-\frac{x^2}{2\sigma^2}} \tag{A.6}$$

This method is good for simulating noise in communication systems because it approximates the tails of the Gaussian distribution well. Notice that as V approaches 1, R, X, and Y approach infinity in magnitude. Other methods that are based on the Central Limit Theorem and involve taking the sum of a fixed number of uniform independent random variables approximate the peak of the pdf well but approximate the tails poorly. Problems in communication systems are usually caused by infrequently occurring large values of the noise, that is, the tails of the pdf.

A.4 Limit on the Peak Magnitude of the Approximately Gaussian Random Variables Generated by This Method

To add a little generality, consider a random number generator that outputs N-bit words in the range $[0, \text{RAND\_MAX}]$ with $\text{RAND\_MAX} = 2^N - 1$. The largest value of V that can be reached in (A.3) is

$$V_{\max} = \frac{\text{RAND\_MAX}}{\text{RAND\_MAX} + 1} = 1 - 2^{-N} \tag{A.7}$$

Then, the largest value that R, X, or Y can reach is

$$R_{\max}^{[N]} = X_{\max}^{[N]} = Y_{\max}^{[N]} = \sqrt{-2\sigma^2 \log_e(1 - V_{\max})} = \sigma\sqrt{2N \log_e 2} \tag{A.8}$$

For the 'C6713 compiler the result is

$$R_{\max}^{[15]} = X_{\max}^{[15]} = Y_{\max}^{[15]} = \sigma\sqrt{30 \log_e 2} = 4.56009\sigma \tag{A.9}$$

and for the 'C30 old compiler it is

$$R_{\max}^{[31]} = X_{\max}^{[31]} = Y_{\max}^{[31]} = \sigma\sqrt{62 \log_e 2} = 6.55554\sigma \tag{A.10}$$

The ratio of these two numbers is

$$\frac{R_{\max}^{[31]}}{R_{\max}^{[15]}} = \sqrt{\frac{31}{15}} = 1.4376 \tag{A.11}$$

Appendix B

A TTL/RS-232C Interface for McBSP0

The pins for the DSP peripherals including the McBSP serial ports and timers are available on the peripheral expansion connector on the TMS320C6713 DSK. The signal levels have the TTL values of 0 and 5 volts. The experiments for Chapter 10 require the use of McBSP0 and Timer0 to exchange serial data with the Navtel bit-error rate tester which uses RS-232C levels. In the RS-232C protocol, a logical 0 is called a space and is represented by +12v while a logical 1 is called a mark and is represented by −12v. The RS-232C interface protocol is described in detail in Chapter 10. The circuit diagram for a TTL to RS-232C converter for use with the Navtel tester is shown in Figure B.1. McBSP0 was selected for the RS-232C channel because McBSP1 is normally connected to the AIC23 codec on the TMS320C6713 DSK. This allows a communication system to be made where an external device like the Navtel bit-error rate tester supplies and receives binary data simulating a terminal, and the codec can be used to transmit and receive the analog modulated signals transmitted over a channel.

The converter uses a MAXIM MAX238 RS-232 Driver/Receiver chip. The MAX238 just requires a single +5v power source and three external capacitors. It contains a voltage doubler and inverter to generate the ±12v levels. The package contains four TTL to RS-232C drivers and four RS-232C to TTL receivers.

A DB25 connector for the RS-232C cable is attached to a bracket for a PC card slot on the back of the PC. The Received Data, Transmitted Data, Receiver Clock, and Transmitter Clock leads are connected to the MAX238 chip. Request to Send (RTS) is looped back to Clear to Send (CD) and Data Carrier Detect (DCD) on the connector. Also, Data Terminal Ready (DTR) is looped back to Data Set Ready (DSR). In this way, when the Navtel tester is turned on and a bit-error rate test is selected, it raises RTS and DTR which automatically raise CTS, DCD, and DSR.

Transmitted Data on the RS-232C side is sent to the Receive Data pin of McBSP0. Similarly, Transmit Data of McBSP0 is routed to Received Data of the RS-232C connector. A similar name swap occurs with the clocks. This is because the RS-232C connector on the converter is configured to look like a modem or DCE. In this case, pin 3 is data received by the modem and pin 2 is data transmitted by the modem.

The Timer0 output pin TOUT0 is looped back to the sample rate generator 1 external clock input pin CLKS0. This signal is divided down in the sample rate generator to form the bit clocks and frame syncs. The transmit and receive bit clocks should be output on the CLKX and CLKR pins to drive the Navtel tester.

Details on how to set the McBSP0 and Timer0 control registers are provided in Section 10.4.

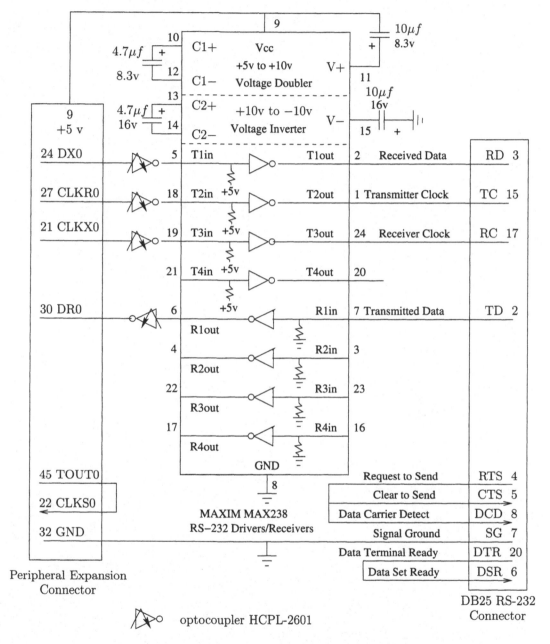

Figure B.1: TTL to RS-232C Converter

Appendix C

Equipment List for Each Station

This equipment list was included to provide guidance to those setting up a similar laboratory and is not intended to be an endorsement of any particular manufacturer. Clearly, many companies make equipment with equivalent capabilities that can be substituted for the specific items listed below.

- A modern PC running Windows XP

- Texas Instruments TMS320C6713 Digital Starter Kit (TMS320C6713 DSK)

  32-bit floating point DSP module which includes stereo ADC and DAC for each PC

- Network cards installed in each PC

- Tektronix 2205 Oscilloscope

  Dual channel oscilloscope

- GW INSTEK Function Generator, Model GFG-3015

  Sine, triangular, and square wave outputs; frequency counter; AM and FM modulated signals

- Navtel DATATEST3 – bit error rate tester

  U of MD homemade RS232/TTL converter to interface Navtel with TMS320C6713 McBSP0 and Timer0 on the DSK

- Penril Alliance V.32 modem

  A V.32 and V.22bis modem with built in diagnostic constellation X-Y outputs. These are not made anymore and it is unlikely you will find them. Substitute a recent V.90 or 92 modem.

- Microphone and speaker (These can be part of the PC system.)

- 3 foot cables with 3.5-mm stereo audio plugs on one end and left and right RCA connectors on the other end.

- RCA to BNC adapters to hook the left and right channels to the oscilloscopes and signal generators.

- Two BNC to alligator clip cables

- Three BNC T connectors

- Three BNC female to banana/screw connectors

Additional Shared Equipment

- Two TAS 111 Telephone Network Emulators

  Emulate amplitude and phase distortion, nonlinear distortion, additive noise, frequency translation and carrier phase jitter

- Two Tektronix 495P Spectrum Analyzers

- HP laser printer

References

I. List of Manuals

1. Brian W. Kernighan and Dennis M. Ritchie, *The C Programming Language*, second edition, Prentice Hall (1988).

2. Spectrum Digital, *TMS320C6713 DSK Technical Reference*, 2004.

3. Texas Instruments, *Code Composer Studio User's Guide*, SPRU328B.

4. Texas Instruments, *TLV320AIC23 Stereo Audio CODEC Data Manual*, SLWS106C, July 2001.

5. Texas Instruments, *TMS320C6000 Assembly Language Tools User's Guide*, SPRU186.

6. Texas Instruments, *TMS320C6000 Chip Support Library API Reference Guide*, SPRU401.

7. Texas Instruments, *TMS320C6000 CPU and Instruction Set Reference Guide*, SPRU189F.

8. Texas Instruments, *TMS320C6000 DSP/BIOS*, SPRU303.

9. Texas Instruments, *TMS320C6000 Optimizing Compiler User's Guide*, SPRU187k.

10. Texas Instruments, *TMS320C6000 Peripherals Reference Guide*, SPRU190D.

11. Texas Instruments, *TMS320C6000 Programmer's Guide*, SPRU198F.

12. Texas Instruments, *TMS320C6713B Floating-Point Digital Signal Processor*, SPRS294B.

II. Selected Reference Books and Papers

A. DSP Laboratory Books Using DSP Hardware

1. Rulph Chassaing, *Digital Signal Processing with C and the TMS320C30*, Wiley-Interscience (1992).

2. Rulph Chassaing, *Digital Signal Processing Laboratory Experiments Using C and The TMS320C31 DSK*, Wiley-Interscience (1999).

3. Rulph Chassaing, *DSP Applications Using C and the TMS320C6x DSK*, Wiley-Interscience (2002).

4. Rulph Chassaing, *Digital Signal Processing and Applications with the C6713 and C6416 DSK*, Wiley Inter-Science (2005).

5. Rulph Chassaing and Darrell W. Horning, *Digital Signal Processing with the TMS320C25*, Wiley-Interscience (1990).

6. Vinay K. Ingle and John G. Proakis, *Digital Signal Processing Laboratory Using the ADSP-2101 Microcomputer*, Prentice Hall (1991).

7. Douglas L. Jones and Thomas W. Parks, *A Digital Signal Processing Laboratory*, Prentice Hall (1988).

8. Nasser Kehtarnavaz and Mansour Keramat, *DSP System Design Using the TMS320C6000*, Prentice Hall (2001).

9. Nasser Kehtarnavaz and Burc Simsek, *C6x-Based Digital Signal Processing*, Prentice-Hall (2000).

10. Sen M. Kuo, Bob H. Lee, and Wenshun Tian, *Real-Time Digital Signal Processing, Implementations and Applications*, John Wiley & Sons (2006).

11. Henrik V. Sorensen and Jianping Chen, *A Digital Signal Processing Laboratory Using the TMS320C30*, Prentice Hall (1997).

B. DSP Laboratory Books Using Software Simulation

1. Oktay Alkin, *PC-DSP*, Prentice Hall (1990).

2. Oktay Alkin, *Digital Signal Processing, A Laboratory Approach Using PC-DSP*, Prentice Hall (1993).

3. John R. Buck, Michael M. Daniel, and Andrew C. Singer, *Computer Explorations in Signals and Systems Using MATLAB*, Prentice Hall (2002).

4. C. Sidney Burrus, James H. McClellan, Alan V. Oppenheim, Thomas W. Parks, Ronald W. Schafer, and Hans W. Schuessler, *Computer-Based Exercises for Signal Processing Using MATLAB*, Prentice Hall (1994).

5. Alan Kamas and Edward A. Lee, *Digital Signal Processing Experiments Using a Personal Computer with Software Provided*, Prentice Hall (1989).

6. Russell M. Mersereau and Mark J.T. Smith, *Digital Filtering: A Computer Laboratory Textbook*, Wiley (1994).

7. Sanjit K. Mitra, *Digital Signal Processing Laboratory Using MATLAB*, McGraw-Hill (1999).

8. John G. Proakis and Masoud Salehi, *Contemporary Communication Systems Using MATLAB*, PWS Publishing Company (1998).

9. John G. Proakis, Masoud Salehi, nd Gerhard Bauch, *Contemporary Communication Systems Using MATLAB and Simulink*, 2nd Ed., Thomson-Brooks/Cole (2004).

10. Mark J.T. Smith and Russell M. Mersereau, *Digital Signal Processing, A Computer Laboratory Textbook*, Wiley (1992).

C. Books and Papers on Digital Signal Processing

1. S. Banerjee, H.R. Sheikh, L.K. John, B.L. Evans, and A.C. Bovik, "VLIW DSP vs. Superscalar Implementations of a Baseline H.263 Video Encoder," *Proc. IEEE Asilomar Conference on Signals, Systems, and Computers*, Oct. 29–Nov. 1, 2000, Vol. 2, pp. 1665–1669.

2. Maurice Bellanger, *Digital Processing of Signals*, 2nd ed., Wiley (1989).

3. Charles Sidney Burrus and Thomas W. Parks, *DFT/FFT and Convolution Algorithms*, Wiley-Interscience (1985).

4. Ronald E. Crochiere and Lawrence R. Rabiner, *Multirate Digital Signal Processing*, Prentice-Hall (1983).

5. Paul M. Embree and Bruce Kimble, *C Language Algorithms for Digital Signal Processing*, Prentice Hall (1991).

6. C.W. Farrow, "A Continuously Variable Digital Delay Element," *Proceedings of the 1988 IEEE International Symposium on Circuits and Systems*, Helsinki, Finland, June 1988, pp. 2641–2645.

7. IEEE Press, *Programs for Digital Signal Processing* (1979).

8. Leland B. Jackson, *Digital Filters and Signal Processing*, Kluwer Academic Publishers (1986).

9. G.M. Jenkins and D.G. Watts, *Spectral Analysis and Its Applications*, Holden-Day (1968).

10. Steven M. Kay, *Modern Spectral Estimation: Theory and Application*, Prentice Hall (1988).

11. Dimitris G. Manolakis, Vinay K. Ingle, and Stephen M. Kogon, *Statistical and Adaptive Signal Processing*, McGraw-Hill (2000).

12. Alan V. Oppenheim and Ronald W. Schafer, *Digital Signal Processing*, Prentice Hall (1975).

13. Alan V. Oppenheim and Ronald W. Schafer, *Discrete-Time Signal Processing*, Prentice Hall (1989).

14. T.W. Parks and C.S. Burrus, *Digital Filter Design*, Wiley-Interscience (1987).

15. John G. Proakis and Dimitris G. Manolakis, *Digital Signal Processing*, Macmillan (1992).

16. Lawrence R. Rabiner and Bernard Gold, *Theory and Application of Digital Signal Processing*, Prentice Hall (1975).

17. Steven A. Tretter, *Introduction to Discrete-Time Signal Processing*, Wiley (1976).

D. Books and Papers on Communications

1. John B. Anderson and Seshadri Mohan, *Source and Channel Coding*, Kluwer Academic Publishers (1991).

2. Albert Benveniste and Maurice Goursat, "Blind Equalizers," *IEEE Transactions on Communications*, Vol. COM-32, No. 8, August 1984, pp. 871–883.

3. John A.C. Bingham, *The Theory and Practice of Modem Design*, Wiley-Interscience (1988).

4. CCITT Blue Book, Volume VIII–Fascicle VIII.1, *Data Communication Over the Telephone Network*, Series V Recommendations, IX-th Plenary Assembly, November 1988.

5. Pierre R. Chevillat, Dietrich Maiwald, and Gottfried Ungerboeck, "Rapid Training of a Voiceband Data-Modem Receiver Employing an Equalizer with Fractional-T Spaced Coefficients," *IEEE Transactions on Communications*, Vol. COM-35, No. 9, September 1987, pp. 869–876.

6. G.J. Foschini, "Equalization Without Altering or Detecting Data," *Bell System Technical Journal*, Vol. 64, October 1985, pp. 1885–1911.

7. L.E. Franks and J.P. Burbrouski, "Statistical Properties of Timing Jitter in a PAM Timing Recovery Scheme," *IEEE Transactions on Communications*, Vol. COM-22, No. 7, July 1974, pp. 913–920.

8. Floyd M. Gardner, *Phaselock Techniques*, 2nd ed., Wiley-Interscience (1979).

9. Jerry D. Gibson, *Principles of Digital and Analog Communications*, 2nd ed., Macmillan (1993).

10. R.D. Gitlin and J.F. Hayes, "Timing Recovery and Scramblers in Data Transmission," *The Bell System Technical Journal*, Vol. 54, No. 3, March 1975, pp. 569–593.

11. Richard D. Gitlin, Jeremiah F. Hayes, Stephen B. Weinstein, *Data Communications Principles*, Plenum (1992).

12. R.D. Gitlin, H.C. Meaders, Jr., and S.B. Weinstein, "The Tap-Leakage Algorithm: An Algorithm for the Stable Operation of a Digitally Implemented, Fractionally Spaced Adaptive Equalizer," *The Bell System Technical Journal*, Vol. 61, No. 8, October 1982, pp. 1817–1839.

13. Dominique N. Godard, "Passband Timing Recovery in an All-Digital Modem Receiver," *IEEE Transactions on Communications*, Vol. COM-26, No. 5, May 1978, pp. 517–523.

14. Dominique N. Godard, "Self-Recovering Equalization and Carrier Tracking in Two-Dimensional Data Communication Systems," *IEEE Transactions on Communications*, Vol.COM-28, No. 11, November 1980, pp. 1867–1875.

15. Yuri Goldstein, Vedat Eyuboglu, and S. Olafsson, "Precoding for V.fast," Subcommittee Contribution, Technical Subcommittee TR30.1, TIA meeting, Newton, MA, July 1993.

16. Hiroshi Harashima and H. Miyakawa, "Matched-Transmission Techniques for Channels with Intersymbol Interference," *IEEE Transactions on Communications*, Vol. COM-20, No. 4, August 1972, pp. 774–780.

17. Simon Haykin, *Communication Systems*, 3rd ed., Wiley (1994).

18. Simon Haykin, *Adaptive Filter Theory*, 2nd ed., Prentice-Hall (1991).

19. Simon Haykin, *Digital Communications*, Wiley (1988).

20. ITU-T Recommendation V.32, "A Family of 2-Wire Duplex Modems Operating at Data Signalling Rates up to 9600 bit/s for use on the General Switched Telephone Network and on Leased Telephone-Type Circuits," www.itu.ch, March 1993.

21. ITU-T Recommendation V.34, "A Modem Operating at Data Signalling Rates of Up to 33600 bit/s for use on the General Switched Telephone Network and on Leased Point-to-Point 2-Wire Telephone-Type Circuits," www.itu.ch, February 1998.

22. ITU-T Recommendation V.90, "A Digital Modem and Analogue Modem Pair for use on the Public Switched Telephone Network (PSTN) at Data Signalling Rates of up to 56 000 bit/s Downstream and up to 33 600 bit/s Upstream," www.itu.ch, September 1998.

23. ITU-T Recommendation V.92, "Enhancements to Recommendation V.90," www.itu.ch, November 2000.

24. Neil K. Jablon, "Joint Blind Equalization, Carrier Recovery, and Timing Recovery for High-Order QAM Signal Constellations," *IEEE Transactions on Signal Processing*, Vol. 40, No. 6, June 1992, pp. 1383–1398.

25. Rajiv Laroia, "ISI Coder – Combined Coding & Precoding," Subcommittee Contribution, Technical Subcommittee TR30.1, TIA meeting, Baltimore, MD, June 1993.

26. Edward A. Lee and David G. Messerschmitt, *Digital Communication*, Kluwer Academic Publishers (1988).

27. F. Ling and S.U.H. Qureshi, "Convergence and Steady-State Behavior of a Phase-Splitting Fractionally Spaced Equalizer," *IEEE Transactions on Communications*, Vol. 38, No. 4, April 1990, pp. 418–425.

28. R.W. Lucky, "Automatic Equalization for Digital Communication," *The Bell System Technical Journal*, Vol. XLIV, No. 4, April 1965, pp. 547–588.

29. Robert W. Lucky, J. Salz, and E.J. Weldon, Jr., *Principles of Data Communication*, McGraw-Hill (1968).

30. K.H. Mueller, and J.J. Werner, "A Hardware Efficient Passband Equalizer Structure for Data Transmission," *IEEE Transactions on Communications*, Vol. COM-30, No. 3, March 1982, pp. 538–541.

31. G. Picchi and G. Prati, "Blind Equalization and Carrier Recovery Using a Stop-and-Go Decision Directed Algorithm," *IEEE Transactions on Communications*, Vol. COM-35, September 1987, pp. 877–887.

32. John G. Proakis, *Digital Communications*, 2nd ed., McGraw-Hill (1989).

33. Y. Sato, "A Method of Self-Recovering Equalization for Multilevel Amplitude-Modulation Systems," *IEEE Transactions on Communications*, Vol. COM-23, June 1975, pp. 679–682.

34. J.J. Stiffler, *Theory of Synchronous Communications*, Prentice-Hall (1971).

35. Gary L. Sugar and Steven A. Tretter, "Convergence Properties of Optimal Adaptive Carrier Phase Jitter Predictors of Sinusoidal Jitter," *IEEE Transaction on Communications*, Vol. 43, Nos. 2/3/4, February 1995, pp. 225–228.

36. M. Tomlinson, "New Automatic Equaliser Employing Modulo Arithmetic," *Electronics Letters*, Vol. 7, Nos. 5/6, March 25, 1971, pp. 138–139.

37. J.R. Treichler and B.G. Agee, "A New Approach to Multipath Correction of Constant Modulus Signals," *IEEE Transactions on Acoustics, Speech, and Signal Processing*, Vol. ASSP-31, April 1983, pp. 459–472.

38. John R. Treichler, C. Richard Johnson, Jr., and Michael G. Larimore, *Theory and Design of Adaptive Filters*, Wiley-Interscience (1987).

39. Steven A. Tretter, *Constellation Shaping, Nonlinear Precoding, and Trellis Coding for Voiceband Telephone Channel Modems with Emphasis on ITU-T Recommendation V.34*, Kluwer Academic Publishers (2002).

40. J.J. Werner, "An Echo-Cancellation-Based 4800 Bit/s Full-Duplex DDD Modem," *IEEE Journal on Selected Areas in Communications*, Vol. SAC-2, No. 5, September 1984, pp. 722–730.

41. B. Widrow and S.D. Stearns, *Adaptive Signal Processing*, Prentice-Hall (1985).

E. References for Wireline and Wireless Multi-Carrier Modulation

1. American National Standards Institute (ANSI), T1.413-1998, "Network and Customer Installation Interfaces – Asymmetric Digital Subscriber Line (ADSL) Metallic Interface," November 1998.

2. John A.C. Bingham, *ADSL, VDSL, and Multicarrier Modulation*, John Wiley & Sons, (2000).

3. Philip Golden, Hervé Dedieu, and Krista S. Jacobsen, Ed.'s, *Fundamentals of DSL Technology*, Auerbach Publications (2006).

4. Andrea Goldsmith, *Wireless Communications*, Cambridge University Press (2005).

5. Simon Haykin and Michael Moher, *Modern Wireless Communications*, Pearson Prentice Hall (2003).

6. IEEE Std 802.lla, "Wireless LAN Medium Access Control (MAC) and Physical Layer (PHY) specifications, High-speed Physical Layer in the 5 GHz Band," (1999).

7. IEEE Std 802.llg, "Wireless LAN Medium Access Control (MAC) and Physical Layer (PHY) specifications, Amendment4: Further Higher Data Rate Extensions in the 2.4 GHz Band," (2003).

8. IEEE Std 802.16, "Air Interface for Fixed Broadband Wireless Access Systems," (2004).

9. IEEE Std 802.16e, "Air Interface for Fixed and Mobile Broadband Wireless Access Systems, Amendment for Physical and Medium Access Control Layers for Combined Fixed and Mobile Operation in Licensed Bands," (June 2005).

10. ITU-T Recommendation G.992.3, "Asymmetric digital subscriber line transceivers 2 (ADSL2)," (01/2005).

11. ITU-T Recommendation G.992.4, "Splitterless asymmetric digital subscriber line transceivers 2 (splitterless ADSL2)," (07/2002).

12. ITU-T Recommendation G.992.5, "Asymmetric Digital Subscriber Line (ADSL) transceivers – Extended bandwidth ADSL2 (ADSL2+)," (01/2005).

13. Hyung G. Myung, Junsung Lim, and David J. Goodman, "Single Carrier FDMA for Uplink Wireless Transmission," *IEEE Vehicular Technology Magazine*, Sept. 2006, pp. 30–38.

14. John G. Proakis and Masoud Salehi, *Fundamentals of Communication Systems*, Pearson Prentice Hall (2005).

15. Mischa Schwartz, *Mobile Wireless Communications*, Cambridge University Press (2005).

16. Thomas Starr, John M. Cioffi, and Peter J. Silverman, *Understanding Digital Subscriber Line Technology*, Prentice-Hall (1999).

17. Gordon L. Stüber, *Principles of Mobile Communication*, Kluwer Academic Publishers (1996).

18. Wikipedia, "3GPP Long Term Evolution," `http://en.wikipedia.org/wiki/3GPP_Long_Term_Evolution`.

F. Books and Papers on Error Correcting Codes

1. Claude Berrou, Alain Glavieux, and Punya Thitimajshima, "Near Shannon Limit Error-Correcting Coding and Decoding: Turbo-Codes (1)," IEEE International Conference on Communications '93, Geneva, Switzerland, *Technical Program, Conference Record*, Volume 2/3, pp. 1064–1070.

2. Ezio Biglieri, *Coding for Wireless Channels*, Springer Science+Business Media (2005).

3. Ezio Biglieri, Pariush Divsalar, Peter J. McLane, and Marvin K. Simon, *Introduction to Trellis-Coded Modulation with Applications*, Macmillan (1991).

4. John L. Fan, *Constrained Coding and Soft Iterative Decoding*, Kluwer Academic Publishers (2001).

5. G. David Forney, Jr., "Burst-Correcting Codes for the Classic Bursty Channel," *IEEE Transactions on Communications Technology*, Vol. COM-19, No. 5, October 1971, pp. 772–781.

6. Robert G. Gallager, *Low-Density Parity-Check Codes*, MIT Press (1963).

7. Robert G. Gallager, *Information Theory and Reliable Communication*, J. Wiley (1968), Chapter 6.

8. S.W. Golomb, *Shift Register Sequences*, Holden-Day (1967).

9. Rajiv Laroia, Nariman Farvardin, and Steven Tretter, "On SVQ Shaping of Multidimensional Constellations – High-Rate Large-Dimensional Constellations," *Proceedings of the 26th Annual Conference on Information Sciences and Systems*, Princeton University, March 1992, pp. 527–531.

10. Rajiv Laroia, Nariman Farvardin, and Steven A. Tretter, "On Optimal Shaping of Multidimensional Constellations," *IEEE Transactions on Information Theory*, Vol. 40, No. 4, July 1994, pp. 1044–1056.

11. Todd K. Moon, *Error Correction Coding*, Wiley-Interscience (2005).

12. W.W. Peterson and E.J. Weldon, Jr., *Error Correcting Codes*, MIT Press (1972).

13. John L. Ramsey, "Realization of Optimum Interleavers," *IEEE Transactions on Communications Theory*, Vol. IT-16, No. 3, May 1970, pp. 338–345.

14. Gottfried Ungerboeck, "Channel Coding with Multilevel/Phase Signals," *IEEE Transactions on Information Theory*, Vol. IT-28, No. 1, January 1982, pp. 55–67.

15. Gottfried Ungerboeck, "Trellis-Coded Modulation with Redundant Signal Sets," Pts. I and II, *IEEE Communications Magazine*, Vol. 25, February 1987, pp. 5–21.

16. Branca Vucetic and Jinhong Yuan, *Turbo Codes, Principles and Applications*, Kluwer Academic Publishers (2000).

17. L.-F. Wei, "Rotationally Invariant Convolutional Channel Coding with Expanded Signal Space – Part II: Nonlinear Codes," *IEEE Journal on Selected Areas in Communications*, Vol. SAC-2, No. 5, September 1984, pp. 672–686.

18. L.-F. Wei, "Trellis-Coded Modulation with Multidimensional Constellations," *IEEE Transactions on Information Theory*, Vol. IT-33, No. 4, July 1987, pp. 483–501.

19. Stephen B. Wicker, *Error Control Systems for Digital Communication and Storage*, Prentice Hall (1995).

III. Interesting Web Sites

1. `www.ee.umd.edu/~tretter`, Steven A. Tretter, University of Maryland web site for ENEE 428 Communications Design Laboratory.

2. `www.ti.com`, Texas Instruments web site.

3. `www.ece.utexas.edu/~bevans`, Papers, class lecture notes, and project reports and suggestions by Dr. Brian Evans at the University of Texas at Austin on applications using TMS320C6000 DSP's for wireline and wireless systems.

Index

Printed in the United States
By Bookmasters